Capsicum

Capsicum
The genus *Capsicum*

Edited by Amit Krishna De

CRC PRESS

Boca Raton   London   New York   Washington, D.C.

Printed and bound in India by
Replika Press Pvt. Ltd.

ISBN 10 : 0-415-29991-8
ISBN 13 : 978-0-415-29991-6

FOR SALE IN SOUTH ASIA ONLY.

Contents

Figures

Tables

About the editor

Dr Amit Krishna De completed MSc in biochemistry and PhD on the topic "Biochemical and Pharmacological Investigations on Capsaicin from Indian Chillies" from Calcutta University. His area of specialisation is Biochemical Pharmacology and Nutritional Toxicology and during the last 15 years he has carried out extensive work with Indian spices. He has won several awards for his work on chilli which includes Kamal Satbir Young Scientist Award from Indian Council of Medical Research, S R Dasgupta Award from Indian Pharmacological Society, S S Rastogi Award and R B Singh Felicitation Award from International College of Nutrition, Indu Vasudevan Award from Indian Association of Biomedical Chemists, APSI Gold Medal from Academy of Plant Sciences, India, etc.

Dr De is involved in teaching as Honorary lecturer in several undergraduate colleges and universities and is at present Assistant Executive Secretary of the Indian Science Congress Association at Calcutta. He is also Fellow of National Environmental Science Academy (FNESA), Academy of Environmental Biology (FAEB), International College of Nutrition (FICN) Academy of Plant Sciences (FAPS) and Indian Association of Biological Scientists (FIABS).

Dr De has several research publications in journals of International repute and is author of several books including *Chilli, Medicinal Values of Indian Spices, Trace Elements in Health and Diseases, Tobacco and Smoking, Selected Questions and Answers in Biochemistry*, etc. He has also edited some books like *Sustainable Development and Environment, Recent Trends in Spices and Medicinal Plant Research, Environment and Man, Spices: Traditional Uses and Medicinal Properties*, etc. He has also delivered lectures on different aspects of "Chilli" in different Institutions/Organisations/Universities all over the world.

Contributors

Andrés Alvarruiz-Bermejo, Escuela Tecnica Superior de Ingenieros Agronomos, Universidad de Castilla-La Mancha, Campus Universitario, s/n, E-02071 Albacete, Spain. Tel.: +34-(9)67-599200; E-mail: Andres.Alvarruiz@uclm.es

Moór Andrea, Vegetable Crops Research Institute PC, Budapest 1775 POB 95 Hungary

Saikat Kumar Basu, Guest Lecturer, Environmental Studies, Rabindra Bharati University, 162/B/378 P A Shah Road, Calcutta 700 045, India. E-mail: aryan_173@rediffmail.com

H. J. Buckenhüskes, Gewürzmüller GmbH, Klagenfurter Str. 1-3, 70469 Stuttgart, Germany. E-mail: buckenhueskes@gewuerzmueller.de

J. Chakrabarti, Director, Central Food Laboratory, 3 Kyd Street, Calcutta 700 001, Fax: 91-33-22498897; E-mail: cflcal@cal.vsnl.net.in

A. Covaci, Toxicological Centre, University of Antwerp, Building S, 6th floor, Universiteitsplein 1, 2610 Wilrijk, Belgium. Tel.: +32 3 8202704, +32 3 8202704; E-mail: covaci@uia.ua.ac.be

Amit Krishna De, Indian Science Congress Association, 14 Biresh Guha Street, Calcutta 700 017, India. Tel.: 91-33-23503887; E-mail: amitkde@satyam.net.in, amitkde_2000@yahoo.com

Ricardo Gómez-Ladrón de Guevara, Escuela Tecnica Superior de Ingenieros Agronomos, Universidad de Castilla-La Mancha, Campus Universitario, s/n, E-02071 Albacete, Spain. Tel.: +34-(9)67-599200; E-mail: Ricardo.Gomez@uclm.es

W. E. Eipeson, Ex-Scientist, Fruit and Vegetable Technology Department, Central Food Technological Research Institute, Mysore 570 013, India. E-mail: eipeson@sancharnet.in

P. Giridhar, Scientist, Plant Cell Biotechnology Department, Central Food Technological Research Institute, Mysore

Manuel González-Ramos, Escuela Tecnica Superior de Ingenieros Agronomos, Universidad de Castilla-La Mancha, Campus Universitario, s/n, E-02071 Albacete, Spain. Tel.: +34-(9)67-599200

Rame Gowda, Associate Professor, Department of Seed Technology, University of Agricultural Sciences, GKVK, Bangalore

P. Indira, Associate Professor, Department of Olericulture, College of Horticulture, Vellanikkara, Kerala Agricultural University, KAU Post Office, Thrichur District, 680656, Kerala, India

Keiko Ishikawa, Japan Horticultural Production and Research Institute, 207, kamishiki, Mastudo-shi, Chiba Japan, 270-2221. E-mail: pimento@smail.plala.or.jp or enken@green.ocn.ne.jp

T. Sudhakar Johnson, Research Scientist, Dabur Research Foundation, Ghaziabad, Uttar Pradesh

P. Manirakiza, Toxicological Centre, University of Antwerp, Building S, 6th floor, Universiteitsplein 1, 2610 Wilrijk, Belgium

Pék Miklós, Red Paprika Development-Research PBC. 6300 Kalocsa Obermayer square 9 Hungary

C. Mini, Assistant Professor, Department of Olericulture, College of Horticulture, Vellanikkara, Kerala Agricultural University, KAU Post Office, Thrichur District, 680656, Kerala, India. E-mail: miniviswan@yahoo.com

L. B. Naik, Sr Scientist and In charge Head section of Seed Science and Technology, Indian Institute of Horticultural Research, Hessaraghatta Lake, Bangalore 560 089, Karnataka, India. Fax: 0091-080-8466291; E-mail: lbnaik@iihr.kar.nic.in

José E. Pardo-González, Escuela Tecnica Superior de Ingenieros Agronomos, Universidad de Castilla-La Mancha, Campus Universitario, s/n E-02071 Albacete, Spain. E-mail: Jose.PGonzalez@uclm.es

K. V. Peter, Vice Chancellor, Kerala Agricultural University, KAU Post Office, Thrichur District, 680656, Kerala, India. Fax: 91-487 370019; E-mail: kvptr@yahoo.com

V. Prakash, Director, Central Food Technological Research Institute, Mysore 570 013, India. Fax: 91-821-516308; E-mail-prakash@cftri.com

J. S. Pruthi, Director, Agmark Laboratories, 344 Defence Officers Enclave, Dhaula Khan, Part II, New Delhi 110 010, India

S. Ramachandra Rao, Research Associate, School of Materials Science, Japan Advanced Institute of Science and Technology, 1-1, Asahidai, Tatsunokuchi, Ishikawa 923-1292, Japan

G. A. Ravishankar, Scientist and Head, Plant Cell Biotechnology Department, Central Food Technological Research Institute, Mysore 570 013, India. Fax: 0091-0821-517233; E-mail: pcbt@cscftri.ren.nic.in

B. R. Roy, Ex-director, Central Forensic Laboratory and Ex-Professor, Department of Food Technology, EC-184, Salt Lake, Calcutta 700 064, India. E-mail: cflcal@cal.vsnl.net.in

P. Schepens, Head of Toxicological Centre, University of Antwerp, Building S, 6th floor, Universiteitsplein 1, 2610 Wilrijk, Belgium. E-mail: paul.schepens@ua.ac.be

Norbert Somogyi, Ambassade de Hongrie, 7-9 Square Vergennes, 75015 Paris, France. Fax: 331 56360268; E-mail: somogy@freemail.hu, agriculture@ambhongrie org

B. Suresh, Senior Research Fellow, Plant Cell Biotechnology Department, Central Food Technological Research Institute, Mysore

J. Szolcsányi, Department of Pharmacology and Pharmacotherapy, University Medical School, Pecs, H-7643, PO Box 99, Hungary. Fax: 36-72-211-761; E-mail: szolcs@apacs.pote.hu

P. S. S. Thampi, Chief Editor, Indian Spices/Spices India, Spices Board, P.B. No. 2277, Cochin 682 025, India. Fax: 0091-0484-331429/334429; E-mail: spicesboard@vsnl.com/ pub@indianspices.com

Ramón Varón-Castellanos, Escuela Politecnica, Universidad de Castilla-La Mancha, Campus Universitario, s/n, E-02071 Albacete, Spain. Tel.: +34-(9)67-599200; E-mail: Ramon.Varon@uclm.es

H. S. Yogeesha, Senior Scientist, Section of Seed Science and Technology, IIHR, Hessaraghatta Lake Post Bangalore 560 089, India. E-mail: hsy@iihr.kar.nic.in

Preface to the series

There is increasing interest in industry, academia and the health sciences in medicinal and aromatic plants. In passing from plant production to the eventual product used by the public, many sciences are involved. This series brings together information which is currently scattered through an ever increasing number of journals. Each volume gives an in-depth look at one plant genus, about which an area specialist has assembled information ranging from the production of the plant to market trends and quality control.

Many industries are involved, such as forestry, agriculture, chemical, food, flavour, beverage, pharmaceutical, cosmetic and fragrance. The plant raw materials are roots, rhizomes, bulbs, leaves, stems, barks, wood, flowers, fruits and seeds. These yield gums, resins, essential (volatile) oils, fixed oils, waxes, juices, extracts and spices for medicinal and aromatic purposes. All these commodities are traded worldwide. A dealer's market report for an item may say 'Drought in the country of origin has forced up prices'.

Natural products do not mean safe products and account of this has to be taken by the above industries, which are subject to regulation, For example, a number of plants which are approved for use in medicine must not be used in cosmetic products.

The assessment of safe to use starts with the harvested plant material which has to comply with an official monograph. This may require absence of, or prescribed limits of, radioactive material, heavy metals, aflatoxin, pesticide residue, as well as the required level of active principle. This analytical control is costly and tends to exclude small batches of plant material. Large scale contracted mechanized cultivation with designated seed or plantlets is now preferable.

Today, plant selection is not only for the yield of active principle, but for the plant's ability to overcome disease, climatic stress and the hazards caused by mankind. Such methods as *in vitro* fertilization, meristem cultures and somatic embryogenesis are used. The transfer of sections of DNA is giving rise to controversy in the case of some end-uses of the plant material.

Some suppliers of plant raw material are now able to certify that they are supplying organically-farmed medicinal plants, herbs and spices. The Economic Union directive (CVO/EU No 2092/91) details the specifications for the *obligatory* quality controls to be carried out at all stages of production and processing of organic products.

Fascinating plant folklore and ethnopharmacology leads to medicinal potential. Examples are the muscle relaxants based on the arrow poison, curare, from species of *Chondrodendron*, and the anti-malarials derived from species of *Cinchona* and *Artemisia*. The methods of detection of pharmacological activity have become increasingly reliable and specific, frequently involving enzymes in bioassays and avoiding the use of laboratory animals. By using bioassay linked fractionation of crude plant juices or extracts, compounds can be specifically targeted which, for example, inhibit blood platelet aggregation, or have antitumour, or antiviral, or any other required activity. With the assistance of robotic devices, all the members of a genus may be readily screened. However, the plant material must be *fully* authenticated by a specialist.

The medicinal traditions of ancient civilisations such as those of China and India have a large armamentaria of plants in their pharmacopoeias which are used throughout south-east Asia. A similar situation exists in Africa and South America. Thus, a very high percentage of the World's population relies on medicinal and aromatic plants for their medicine. Western medicine is also responding. Already in Germany all medical practitioners have to pass an examination in phytotherapy before being allowed to practise. It is noticeable that throughout Europe and the USA, medical, pharmacy and health related schools are increasingly offering training in phytotherapy.

Multinational pharmaceutical companies have become less enamoured of the single compound magic bullet cure. The high costs of such ventures and the endless competition from me too compounds from rival companies often discourage the attempt. Independent phytomedicine companies have been very strong in Germany. However, by the end of 1995, eleven (almost all) had been acquired by the multinational pharmaceutical firms, acknowledging the lay public's growing demand for phytomedicines in the Western World.

The business of dietary supplements in the Western World has expanded from the Health Store to the pharmacy. Alternative medicine includes plant-based products. Appropriate measures to ensure the quality, safety and efficacy of these either already exist or are being answered by greater legislative control by such bodies as the Food and Drug Administration of the USA and the recently created European Agency for the Evaluation of Medicinal Products, based in London.

In the USA, the Dietary Supplement and Health Education Act of 1994 recognized the class of phytotherapeutic agents derived from medicinal and aromatic plants. Furthermore, under public pressure, the US Congress set up an Office of Alternative Medicine and this office in 1994 assisted the filing of several Investigational New Drug (IND) applications, required for clinical trials of some Chinese herbal preparations. The significance of these applications was that each Chinese preparation involved several plants and yet was handled as a *single* IND. A demonstration of the contribution to efficacy, of *each* ingredient of *each* plant, was not required. This was a major step forward towards more sensible regulations in regard to phytomedicines.

My thanks are due to the staffs of Harwood Academic Publishers and Taylor & Francis who have made this series possible and especially to the volume editors and their chapter contributors for the authoritative information.

Roland Hardman, 1997

Preface

The land and sea of the world carry a variety of medicinal plants that have been used traditionally for the treatment of various diseases. Most of these plants are known to the local tribes, yet many remain scientifically unexplored. Medicinal uses of plants have been reported in a wide variety of ancient literature. From time immemorial, human beings have depended on plants/herbs for curative treatment of diseases. Some of these plants have been grouped together as spices and they are regularly consumed not only for their nutritive properties but also because of their aroma and flavour. Spices have been of general interest in order to increase the quality of food and to decrease body ailments.

In medieval Asia, in addition to fine textiles and ivory, spices were considered to be highly important. In this respect, India has been regarded as the 'Kingdom of spices'. In ancient economics, spices ranked in precious inventories of royal possessions. Even today, spices are playing a significant role in the development of national economies. In general, spices are used in perfumes, offerings, medicine, preservatives, cultural marks, cosmetics, insecticides and as dyes, condiments, food flavouring extracts and colouring agents.

Among the spices, hot pepper or chilli (*Capsicum*) has been popular from ancient time. Many members of the genus *Capsicum* are of commercial interest, not only for their taste and colour, but also because of their essential oil and active principles. The *Capsicum* genus represents a diverse plant group, from the well-known fruits of the sweet green bell pepper and bright red paprika to the fiery hot small red chilli fruit. When Columbus tasted chilli, which he found on his voyage, he believed he had reached India and called it red pepper. *Capsicum* has been domesticated for 7,000 years. How then, can the most consumed spice in the world be considered a new crop? One reason is that *Capsicum* is an essential ingredient in the fastest growing food sector in the United States, 'Mexican or Southwestern food.' In addition, many of the new uses of chilli are hidden within manufactured products. Chilli is being used as a food flavouring, a colouring agent, a pharmaceutical ingredient, and in other innovative ways. At present, *Capsicum* is an indispensable and important ingredient in the food of many countries, specially the tropical countries. The use of the essential oil and oleoresins from *Capsicum* species in therapeutic applications is increasing day by day. This is leading to a high demand of oleoresins and active principles in the pharmaceutical industry. Moreover, some species of *Capsicum* are highly valued for their colours and are used by dye industries.

Species of *Capsicum* which are native to South and Central America, Mexico and the West Indies are now widely cultivated throughout temperate, subtropical Europe, the Southern US, tropical Africa, India, East Africa and China. Indian chillies are medium to highly pungent while African chillies are the hottest ones. It is believed that about 50 species are distributed throughout the world, and range from the hot variety to the sweet variety. The essential oil is obtained by steam distillation. Oleoresins are produced by solvent extraction. These are used by spice industries and have a very high future potential. Oleoresin of paprika is still the single largest selling oleoresin in the world market. Currently, this is manufactured in countries like Spain, Morocco, Ethiopia, Mexico and the United States.

Capsaicin, the active hot principle of the *Capsicum* species, has many medicinal properties. With the discovery of a capsaicin-like receptor in the human body in 1997, the demand for capsaicin in the therapeutic field for the treatment of pain and inflammatory related diseases is of growing importance. Recently, capsaicin has also been used as an antiobesity agent because of its cholesterol lowering properties. The antimicrobial and antiseptic properties of some *Capsicum* species are also well documented. The future prospects of *Capsicum* are very promising, especially for the new hybrids developed by tissue culture and biotechnology methods. Their high content of capsaicin with bright red colour will provide a cheap available source of valuable ingredients for industrial uses.

This book deals comprehensively with the genus *Capsicum* by using contributions from eminent scientists working on different aspects of the subject: the history, taxonomy, trade, chemistry, cultivation, micropropagation and tissue culture, harvesting, processing, contamination, quality standardisation, pharmacology and toxicity, therapeutic implications, future prospects, etc. The book covers topics on the History, Trade, Chemistry, Biosynthesis, Irrigation, Cultivation, Post-Harvest Technologies, Food Industry and Future Perspectives of *Capsicum*. There are 16 Chapters in total. As often happens in a multi-authored volume, some overlap has occurred in several chapters. Such matter has not been removed as it would have changed the spirit of the contributions.

This book will be particularly helpful to the students, teachers, researchers and practitioners of modern complementary and traditional herbal medicine. Companies manufacturing products from natural ingredients will find this publication very useful. The data in this book will be especially useful to those engaged in these disciplines whether as students, teachers or professionals: botany, taxonomy, agronomy, pharmacognosy, pharmacy, pharmacology, toxicology, health sciences, chemistry and biochemistry.

I am grateful to Dr Roland Hardman, the Editor of the series, for his invitation to edit this volume and for giving continuous advice. I also warmly acknowledge the cooperation extended by the different contributors of this book, from all over the world, in submitting the manuscript well in time.

Amit Krishna De

1 *Capsicum*: historical and botanical perspectives

Saikat Kumar Basu and Amit Krishna De

Chilli is known by different names in different parts of the world. The genus *Capsicum*, which is commonly known as "red chile", "chilli pepper", "hot red pepper", "tabasco", "paprika", "cayenne", etc., belongs to the family Solanaceae. The original distribution of this species appears to have been from the south of Mexico, extending into Columbia. The taxonomy of the genus *Capsicum* is confounded within certain species complexes. It is generally believed that about 20 *Capsicum* species are distributed worldwide. The five major species of *Capsicum* cultivated are: *Capsicum annuum*, *Capsicum frutescens*, *Capsicum chinense*, *Capsicum pendulum* and *Capsicum pubescens*. However, the question of exactly how many species are involved is still controversial. The authors have reorganized and reconstructed the available literature in a new and original form to give a brief introduction to the readers regarding the historical and botanical perspectives of different *Capsicum* species.

Introduction

The term "chilli" is a rather confusing terminology; "chile", "aji", "paprika", "chili", "chilli" and "*Capsicum*" are all used frequently and interchangeably for "chilli pepper" plants under the genus *Capsicum*, which belongs to a dicotyledonous group of flowering plants. A particular species of *Capsicum* is called "chile pepper" in parts of Mexico, southwestern United States and parts of Central America. To make matters still more confusing a "sweet bell pepper" is often referred to as "*Capsicum* pepper", whereas the term "chilli pepper" is used for a "hot pepper". The term "bell pepper" is used to refer to a non-pungent, chunky, sweet chilli pepper type, whereas "chilli pepper" generally refers to a pungent chilli variety.

Red peppers have been familiar to all Spanish South Americans by the Arawakan name "aji" and by the Nahuatlan name "chilli" in Mexico and Central America. The genus *Capsicum*, which is commonly known as "red chile", "chilli pepper", "hot red pepper", "tabasco", "paprika", "cayenne", etc., belongs to the family Solanaceae (Night shade family), that includes tomato, potato, petunias and tobacco (Hawkes *et al.*, 1979; Macrae, 1993). According to some references, the popular name "chile" or "chilli" orginates from the hot pepper species cultivated in the South American country of Chile. However, the name "chilli" seems to have nothing to do with the country name and, on the contrary, it is believed to have originated from a district of Central America (De, 1994, 2000).

One of the very first sources of life amongst the Inca, the *Commentarios Reales*, written by Garcilaso de la Vega, "El Inca", in 1609, mentions the common or even daily use of chillies. According to de la Vega, there are three different kinds of chilli, two of which can be identified as "ají" and "rocoto" while the third is only insufficiently described.

Even though the chilli is regarded by different names within the same country, and even in different states or provinces, the botanical name of chilli is the Latin name *Capsicum*. The word comes from a Greek based derivative of Latin "Kapto" meaning "to bite", a certain reference to heat or pungency. The word "chile" is a variation of "chil", derived from the Nahuatl (Aztec) dialect, which referred to plants now known as *Capsicum*, whereas "aji" is a variation of "axi" from the extinct Arawak dialect of the Caribbean. This brings up the point of the correct way to spell "chile" (Domenici, 1983). The "*e*" ending in chile is the authentic Hispanic spelling of the word, whereas English linguists have changed the *e* to an *i*. From the Nahuatl dialect of Aztec language, the name "Chiltepin" has been derived. This was the name given to the earliest known variety of *Capsicum*. The name has been traced back by linguists to be a concatenation of the word chile and tecpintl, the combination resulting in "Flea Chile", which is believed to allude to the sharp biting taste of the chilli pepper. Down the ages the original name has been slowly reformulated as "chile" + "tecpintl" to "chiltecping" to "chiltepin" to "chilepiquin", the latter two names being frequently interchangeable. The version used depends upon the source of information. However, *C. annuum* var. *aviculare* is the modern scientific name of this earliest known variety (Macrae, 1993).

However, a multilinguistic nation like India represents a unique case, where the same specimen *C. annuum* L. is referred to by different names in different parts of the nation. In original Sanskrit the plant is known as "Mairichi phalam", "Katuvira" and "Bruchi". However, in modern Indian languages it is known by different names, and even by more than one name within the same language as given in Table 1.1.

The chilli pepper is one of the very first domesticated plants of Middle America. In the Valle de Tehuacan (Puebla), which is one of the best documented examples of early settlement in Mesoamerica, archaeological evidence for the consumption of chilli peppers dates back to the seventh millennium BCE, long before the cultivation of maize and beans. The early findings of peppers (in coproliths and charred remains) were probably harvested in the wild. However, domesticated chillies similar to modern varieties in both size and shape can be found from the fifth millennium BCE onwards. In pre-Columbian Mexico, chilli was one of the preferred tributes which dependent city-states had to deliver to their hegemonial powers. The paying of tribute, including in the form of chilli peppers, was later continued by the new Castilian rulers.

Capsicum has been known since the beginning of civilization in the Western Hemisphere. It has been a part of the human diet since about 7500 BCE (MacNeish, 1964). It was the ancient ancestors of the native peoples who took the wild chilli piquin and selected the various types known today. Heiser (1976) states that between 5200 and 3400 BCE, the Native Americans were growing chilli plants. This places chilli among the oldest cultivated crops of the Americas. Seeds found in cave dwellings indicate that the natives were enjoying peppers in 7000 BCE, along with potatoes in the Andes. In Mexico dry pepper fruits and seeds were recovered from 9000-year-old burials in Tamaulipas and Tehuacan. Domestication might have taken place 10,000–12,000 years ago. Christopher Columbus is believed to be the first European to discover chilli in one of his legendary travels to America around 1493. He was believed to have been looking for an alternative source of black pepper which at that time was the favourite spice of Europe. What he discovered was a small fiery pod that had for centuries provided seasoning for native Americans, the hot chilli pepper. But to mention it specifically, chilli or *Capsicum* is not related to *Piper* genus, which contains *Piper nigrum* L. of the family Piperaceae, the source of black and white pepper. Within a century after its discovery hot chilli peppers attained a worldwide distribution. According to some other sources, the American origin of *Capsicum* was first reported in 1494 by Chanca, a physician who accompanied Columbus on his second voyage to the West Indies (Macrae, 1993).

Chilli peppers grow as a perennial shrub in suitable climatic conditions. A genus that usually represents glabrous, perennial, woody subshrubs or shrubs, some tending to be vines, rarely

Table 1.1 Capsicum in different languages

Language	Common name
Pharm	*Fructus Capsici acer*
Arabic	Felfel, Bisbas
Amharic	Mit'mita, Berbere
Assami	Jolokia
Bengali	Lanka, Morich
Burmese	Nga yut thee, Nil thee
Chinese	Lup-Chew
Danish	Chili
Dutch	Spaanse peper, Cayennepeper
English	Cayenne pepper, Red pepper, Chilli, Chili
Estonian	Kibe paprika
Finnish	Chilipippuri
French	Poivre rouge, Piment enragé, Piment fort, Piment-aiseau, Poivre de Cayenne
German	Roter Pfeffer, Cayenne-Pfeffer, Chili-Pfeffer, Beißbeere
Gujarati	Lal marcha (red), Lila marcha (green)
Hebrew	Pilpel adom
Hindi	Lal mirch (red), Hari mirch (green)
Hungarian	Csilipaprika, Igen erôs apró, Cayenne bors, Cayenni bors, Macskakpöcs paprika, Aranybors, Ördögbors, Chilipaprika
Icelandic	Chilipipar, Cayennepipar
Indonesian	Lombok, Cabé, Cabai
Italian	Peperone, Diavoletto, Peperoncino, Pepe di Caienne, Pepe rosso picante
Japanese	Togarashi
Kannada	Menashinakayi
Laotian	Mak phet kunsi
Malay	Lada merah
Malayalam	Mulagu
Marathi	Lal mirchya (red), Hirvya mirchya (green)
Oriya	Lankamaricha
Pashto	Murgh
Portuguese	Pimentão, Piri-piri, Pimenta de caiena
Punjabi	Lal-mircha
Sanskrit	Marichiphala, Ujjvala
Singhalese	Rathu miris, Gasmiris
Spanish	Chile, Guindilla, Cayena inglesa, Pimienta de Cayena, Pimienta picante, Ají
Swahili	Pilipili hoho
Swedish	Chilipeppar
Tagalog	Siling labyo, Sili
Tamil	Mulagu
Telegu	Mirapakaya
Thai	Pisi hui, Prik khee, Prik
Tibetan	Sipen marpo, Si pan dmar po
Turkish	Aci kirmizi biber
Urdu	Lalmarach
Vietnamese	Ot

herbs, and are native to Central and South America live for a decade or more in tropical conditions of their natural habitat, but it is mostly cultivated as an annual elsewhere. Chilli is native to the Western Hemisphere and probably evolved from an ancestral form in the Bolivia–Peru area. The first chillies consumed were probably collected from wild plants. Prehistoric

Americans took the wild chilli "Piquin" and selected it for the various pod-types known today. However, domesticated chillies were apparently not grown prehistorically in New Mexico (Macrae, 1993).

In fact, it is not known exactly when chillies were introduced into New Mexico. Chillies may have been used by the indigenous peoples as a medicine, a practice common among the Mayans. By the time the Spanish arrived in Mexico, Aztec plant breeders had already developed dozens of varieties. Undoubtedly, these chillies were the precursors to the large number of varieties found in Mexico today. Whether chillies were traded and used in New Mexico pueblos is still not clear (Macrae, 1993).

Capsicum species have been thought to be of Central American origin, but one species has been reported to be introduced in Europe in the fifteenth century. By the middle of the seventeenth century, the *Capsicum* was cultivated throughout southern and middle Europe as a spice and/or medicinal drug. One species was introduced to Japan and about five species were introduced into India, of which *C. annuum* L. and *C. frutescens* L. were cultivated on a large scale (The Wealth of India, 1992).

The British Pharmacopoeia (BPC, 1988) consists of drugs (chillies, red peppers) characterized by the presence of dried, ripe fruits of *C. annuum* var. *minimum* (Mill.) Heiser, and small fruited varieties of *C. frutescens* L. In commerce the description given applies to various African commercial varieties and these and the Japanese variety are sold in the UK as chillies, while the larger but less pungent Bombay and Natal fruits are sold as *Capsicums*. Very large *Capsicum* fruits that resemble tomatoes in texture and are practically non-pungent are widely grown in southern Europe as vegetables (Evans, 1996).

Records of prehistoric *Capsicum* species around burial sites in Peru indicate that the original home of the chillies may be tropical South America. There seems to have been a diffusion from there to Mexico, or an independent origin in the latter country, where a great diversity of the genus is found. While *C. annum* has not been recorded in the wild state and *C. frutescens* doubtfully so, they have now naturalized in the tropics of many countries and are easily disseminated by birds (The Wealth of India, 1992).

The plant was introduced into Spain by Columbus, from where it spread widely. Subsequently, the prolonged viability, easy germination and easy transportation assisted its spread all across the globe. The original distributions of this species appear to have been from the South of Mexico extending into Columbia (The Wealth of India, 1992). "Ginnie Pepper" was well known in England in 1597 and was grown by Gerarde (Evans, 1996).

The Portuguese introduced chilli into India. Chilli is used as a condiment in large quantities in India, Africa and tropical America, where the fruit develops greater pungency than in colder regions. It has now, however, become a popular condiment all over the world. The long, thin fruits constitute the source of dry chillies used for commerce. The wide popularity of chilli and its extensive cultivation are due to its being a short duration crop and its ease of cultivation under a wide range of climatic and edaphic conditions, particularly in comparison with black pepper (*P. nigrum* L.). The cultivation and utility of both *Capsicum* spp. are similar, except for local peculiarities (The Wealth of India, 1992).

Capsicums are mentioned in a classic text of the Tibetan medical tradition, the "Blue Beryll": "*Capsicum* (tsi-tra-ka) increases digestive warmth of the stomach, and is the supreme medication for the alleviation of oedemata, haemorrhoids, animalcules, leprosy and wind". Another passage tells us "bad-kan-nad-sel tshe-'phel tsi-tra-ka mar sbrang sbyar – to alleviate diseases of phlegm and prolong the lifespan: [use] *Capsicum* mixed with butter and honey". Interestingly, the spice translated as cayenne pepper has another name, gYer-ma, and is generally used to "treat wind disorders".

In pre-hispanic medicine, chilli peppers were used to treat a host of conditions, often in combination with other plant and mineral substances. For example, chilli was used to treat diseases of the gastro-intestinal tract (infections, diarrhoea), in addition to tooth pain, cough and lack of appetite. It was popular as an aphrodisiac, as well. For that reason, the Spanish Padre Acosta, travelling through New Spain in the sixteenth century, warned that high consumption of chilli peppers would be detrimental to the "soul's health" because it "promoted sensuality" (Acosta, 1985). Chilli peppers had other uses as draconic punishment in childrens' education, and even in warfare – the enemy was driven out of his fortification by employing the acrid smoke of smoldering chillies. This tactic was employed not only in pre-Columbian times, but also during the Mexican revolution at the beginning of the twentieth century.

Chilli is essentially a crop of the tropics and grows better in hotter regions. It is cultivated over large areas in all Asian countries, Africa, South and Central America, parts of USA and southern Europe, both under tropical and subtropical conditions. The major chilli growing countries are India, Nigeria, Mexico, China, Indonesia and the Korean Republic. Japan has shown the highest yield of green chillies, followed by India (Prod. Yearb, FAO, 1988).

Although chilli is popular as food and spice, it was first used as a medicinal plant to treat respiratory ailments and for pain. Today they are used in ointments to relieve muscular pain. Other application of chilli is the use of capsaicin as the main ingredient for spray as a personal defence instrument or for crowd control.

Taxonomy

Various authors ascribe 25 species to the genus, with new species to be discovered and named as exploration of the New World tropics expands. Exploration and plant collection throughout the New World have given us a general but false impression of speciation in the genus. Humans selected several taxa and in moving them toward domestication selected the same morphological shapes, size and colours in at least three distinct species. Without the advantage of genetic insight these early collectors and taxonomists named these many size, shape and colour forms as distinct taxa, giving us a plethora of plant names that have only recently been sorted out, reducing a long list to four domesticated species.

The early explorations in Latin America were designed to sample the flora of a particular region. Thus, any collection of *Capsicum* was a matter of chance and usually yielded a very limited sample of pepper from that area. Only with the advent of collecting trips designed to investigate a particular taxon did the range of variation within a species begin to be understood. One needs only to borrow specimens from the international network of herbaria to appreciate what a limited sample exists for most taxa, particularly for collections made prior to 1950. The domesticated *C. pubescens*, e.g. that is widespread in the mid-elevation Andes from Colombia to Bolivia, is barely represented in the herbarium collections of the world.

Most herbarium collections of *Capsicum*, with the exception of *C. annuum* holdings, are woefully inadequate. Furthermore, besides *C. annuum*, very little attention has been paid to the many cultivars of each of the domesticated species. Often material is unusable because it was collected only in fruit, neglecting the most important and critical characters associated with floral anatomy and morphology. The advent of germplasm collecting programmes during the past three decades, and concomitant improvement in herbarium collections have helped in a better understanding of the nature of variation in the genus *Capsicum*. The increasing number of *Capsicum* herbarium specimens permits renewed interest and debate on the proper species classification (IBPGR, 1983).

One of the more perplexing questions regarding the taxonomy of *Capsicum* is defining the genus (Eshbaugh, 1977, 1980b; Hunziker, 1979). The taxonomy of the genus *Capsicum* is

confounded within certain species complexes, e.g. *C. baccatum sensu lato*. Major taxonomic difficulties below the species level in other taxa, e.g. *C. annuum*, also exist. What taxa are ultimately
included in *Capsicum* may indeed change if the concept of the genus is broadened to include taxa
with non-pungent fruits but which have other common morphological and anatomical traits,
such as the nature of the anther, the structure of nectaries and the presence of giant cells on the
inner surface of the fruit (Pickersgill, 1984). The data from plant breeding and cytogenetics confirm that the domesticated species belong to three distinct and separate genetic lineages. Earlier
studies have suggested two distinct lineages based upon white and purple flowered groupings
(Ballard *et al.*, 1970) but an evaluation of more recent data argues for the recognition of three
distinct genetic lineages.

According to Pickersgill (1988), the status of *C. annuum*, *C. chinense* and *C. frutescens* as separate
species is extremely doubtful. Several other workers, including Eshbaugh *et al.* (1983), have
illustrated that at a primitive level the species are not distinguishably separable, although we
treat the three as separate domesticated taxa. However, the corresponding wild forms integrate
in such a manner that it often becomes a problem to distinguish them with distinct botanical
identities and taxonomic names. Eshbaugh *et al.* (1983), Mcleod *et al.* (1979, 1983), Loaiza-
Figuerosa *et al.* (1989) suggested that "these species form an allozymically indistinguishable
association of a single polytypic species" based on isoenzyme data analysis of the three species.
There is a very close relationship of all these three taxa based on data from several studies (Smith
and Heiser, 1957; Pickersgill, 1980). Stuessy (1990) has observed that "the ability to cross does
not just deal with a primitive genetic background, it deals with the degree of genetic compati-
bility developed in a particular evolutionary line".

Despite the different views of different workers, it can be suggested that the *C. annuum–
C. chinense–C. frutescens* complex is one of the biggest taxonomic problems. Preliminary studies
by Gounaris *et al.* (1986) and Mitchell *et al.* (1989) suggest that the molecular data may provide
valuable links in solving this complicated problem. According to others, like D'Arcy and
Hunziker, the problem would be solved by merging the three taxon into one.

There is a difference of opinion regarding the global number of species of *Capsicum*. Robert
Morrison, as far back as 1680, published the names of 33 species of *Capsicum* in Plantarum
Historiae Universalis Oxoniensis. Joseph Pitton de Tournefort mentioned 27 species in 1700.
Only two species were mentioned by Carlous Linneaus (1753), while Fingerhurth (1832)
reported 25 species and Felix Dunal (1852) a maximum of 50 species. H. C. Irish, in his famous
work "A Revision of the Genus *Capsicum*", in 1898, mentioned only two species like his fore-
runner Linnaeus (1753). However Bailey (1923), who has been regarded as the "greatest horti-
culturist who ever lived", mentioned and named only one species. In 1957, P. Smith and
C. Heiser recognized five species only, which gives the "Big Five" species of *Capsicum*. Further,
recent works by W. Hardy Eshbaugh, Barbara Pickersgill and Armado. T. Hunziker identified
22 other wild species (Macrae, 1993).

After much work by taxonomists concerning the classification of the presently domesticated
species of *Capsicum*, they have been considered to belong to one of five species (in total, there are
some 26 species of pepper known at present, but these are mostly found only in the wild and are
considered undomesticated) (Macrae, 1993). The primary characteristics are based upon flower
and seed colour, shape of the calyx, the number of flowers per node and their orientation
(Hawkes *et al.*, 1979). The big five species are *C. annuum* (containing the NuMex, Jalapeno and
Bell varieties), *C. frutescens* (containing the Tabasco variety), *C. chinense* (containing the Habanero
and Scotch Bonnet varieties), *C. baccatum* (containing the aji varieties) and *C. pubescens* (containing the Rocoto and Manzano varieties) (Macrae, 1993).

It is generally believed that about 20 *Capsicum* species are distributed worldwide. However, the question of exactly how many species are involved is still controversial. The five major species of *Capsicum* cultivated are believed to have originated in Mexico, *C. frutescens* in swamps around the higher regions of Amazon river in Colombia and Peru, *C. baccatum* var. *baccatum* in southern Peru and northern Bolivia (De, 1994). According to Lawrence, Solanaceae is a large family consisting of about 85 genera and in excess of 2,200 species, distributed primarily in tropical America and South America, where about 38 endemic genera have been reported.

A short key, comprising of few prominent genera of Solanaceae including *Capsicum*, as prepared by Sir David Prain (in *Bengal Plants*, 1905, Vol. 2, pp. 742–743) is given below:

Key to the genera

* Fruit indehiscent, a berry with many compressed, subdiscoid seeds; embryo curved or subspiral, corolla-lobes plaited or valvate:
  - ⚥ Corolla rotate or wide campanulate:
    Anthers convenient in a cone, longer than the filaments, not dehiscing throughout their length:
    Anthers dehiscing introrsely by longitudinal slits, the tips empty; leaves pinnatisect
    ... ... *Lycopersicum*
    Anthers dehiscing by apical pores or by short apical slits; leaves entire, lobed or pinnatifid
    ... ... *Solanum*
    Anthers not convenient in a cone and not longer than the filaments, dehiscing throughout their length by lateral slits; flowers pedicelled:
    Calyx in fruit small; flowers solitary or in pairs ... ... *Capsicum*
    Calyx in fruit enlarged; overtopping the berry; flowers solitary ... ... *Physalis*
  - ⚥ Corolla urceolate; anthers not convenient in a cone; calyx in fruit enlarged, overtopping the berry; flowers clustered, subsessile:
* Fruit capsular, valves completely or partially separating:
  Five stamens, all perfect:
  Flowers axillary, solitary; corolla-lobes plaited, seeds somewhat compressed; embryo curved ... ... *Datura*
  Flowers in terminal panicles; corolla-lobes induplicate-valvate; seeds hardly compressed; embryo straight ... ... *Nicotiana*
  Four Stamens only, perfect, didynamous; flowers solitary or in unilateral stamens ... ... *Browallia*

Wettestein (1895) regarded the family to orginate in several lines. It is closely related to Scrophulariaceae and can be differentiated from them by actinomorphic plicate corolla, the presence of four–five stamens and the bicollateral vascular bundle. Bentham and Hooker (1883), Bessey (1915) and Engler and Prantl (1925) considered this to be a primitive member of the Tubiflorae (and derived from Linaceae). Hutchinson (1959, 1960) included it in Solanales, together with Convolvulaceae, and, thought it to be ancestral to the Personales containing families like Scrophulariaceae and Acanthaceae.

In the Pflanzenfamilien, Wettestein arranged the genera belonging to Solanaceae into series, and subdivided these into tribes. He placed the genus *Capsicum* within Tribe 2 Solaneae under Series A. Engler and Diels placed Solanaceae under Order Tubiflorae consisting of eight suborders and 23 families. Solanaceae has been conserved over other names for the family (Lawrence, 1951). Simply, the taxonomic position of *Capsicum* can be represented as follows:

Kingdom: Plantae
Division: Magnoliophyta
Class: Magnoliopsida
Order: Solanales
Genus: *Capsicum*
Species: *chinense/annuum/pubescens/*etc.

The significant diagnostic characteristics of *Capsicum are*: Annual or perennial glabrous herbs or undershrubs; *leaves* that alternate, are entire or repand; *Flowers* pedicelled, axillary, solitary or two–three together; *Sepals* connate in a subentire or minutely five-toothed calyx, much shorter than the fruit; *Petals* five, connate in a rotate corolla; tube short; lobes valvate in bud; *Stamens* five, adnate nearly to base of corolla-tube, filaments short; *Anthers* not exceeding filaments; dhiscence longitudinal; *Carpels* connate in a two-celled, rarely a three-celled ovary; style linear; stigma subcapitate; *Fruit* globose or elongated or irregularly shaped, many seeded berry; *Seeds* discoid, smooth or subscabrous; *Embryo* peripheric (Prain, 1905).

Some taxonomists divide *Capsicum* L., into *Capsicum* L. and *Tubocapsicum* Makino. *Capsicum* comprises a very wide range of forms, varying in shape, colour, pungency and position of fruits. The floral characteristics, however, show considerable consistency and are useful for taxonomic identification. Based on the morphological characteristics, the genus has been classified by various taxonomists. More than 100 binomials have been proposed for the cultivars, many of which should be referred to one or the other of the two cultivated species, namely, *C. annuum* and *C. frutescens*. Irish reduced all the cultivated forms to two species, *C. annuum* and *C. frutescens*, whereas Bailey included all of them under *C. frutescens*, on the basis that all the peppers are perennial in their native habitat. Recent treatment of the genus recognizes both the species based upon the lack of crossability between the two, and their slight but distinct morphological differences. The majority of the commercial chillies belong either to *C. annuum* or *C. frutescens*, but mostly to the former. All the domesticated forms commonly grown in the old world are within this group. Five or six species are under cultivation, and about 20 wild species have now been recognized in this genus. In addition to *C. annuum, C. frutescens, C. baccatum* L. var. *pendulum* (Willd.), Eshbough (syn. *C. pendulum* Willd.), *C. chinense* Jacq. (syn. *C. angulosum* Mill.) and *C. pubescens*, Ruiz & Pav. have been introduced from South America (The Wealth of India, 1992).

The species has been classified earlier and based on the characteristics of fruits is now divided into the following varieties: (i) var. *annuum* syn. *C. frutescens* C. B. Clarke (Fl Br Ind) in part, non Linn.; *C. purpureum* Roxb., an erect and much-branched, 45–100 inches tall herb or subshrub, with a single, terminal white flower (ii) var. *glabriusculum* (Dunal) Heiser and Pickers. syn. *C. minimum* Roxb.; C. B. Clarke (Fl Br Ind) in part; auct. including the wild or spontaneous forms; and (iii) var. *grossum* (Willd.) Sendt. syn. *C. grossum* Willd. including var. *cerasiformis* C. B. Clarke (TheWealth of India, 1992).

Most of the cultivars grown in India are attributed to var. *annuum* while a few belong to var. *grossum*, the latter grown for their big, lobed fruits with their mild flavour and pungency, principally used as a vegetable (The Wealth of India, 1992). The cultivars of *C. annuum* var. *annuum* of American origin have been classified into five groups: (i) *Cerasiform* – Cherry-pepper, fruit erect or declined, globose, up to 2.5 cm across, yellow or purplish green, very pungent;

(ii) *Conoides* – Cone-pepper, fruit usually erect, conical or oblong cylindric, up to 5 cm long; (iii) *Fasciculatum* – Cluster-pepper, fruit erect, clustered, very slender, up to 7.5 cm long, red, very pungent; (iv) *Grossum* – Bell-pepper, *Capsicum*, Green-pepper, Pimento, Sweet-pepper, plants tall, stout, fruit large, thick-fleshed, inflated, with depression at the base, sides usually furrowed, broadly oblong bell or apple-shaped yellow or red when mature, non-pungent, mild in flavour and used as vegetable and salad and the source of pimento; (v) *Longum* – *Capsicum*-pepper, Cayenne-pepper, Chilli-pepper, Long-pepper, Red-pepper, fruit mostly drooped and elongated, up to 30 cm long, tapering to apex, often 5 cm across at base, very pungent, the principal condiment of pepper, source of chilli powder, paprika and medicinal *Capsicum* (The Wealth of India, 1992).

Although, the general perception is that the bigger and larger varieties are regarded as "*Capsicum*", the other forms of chillies are not "*Capsicum*" is wrong. All forms of chillies belong to the same taxonomic genus of *Capsicum* within the family Solanaceae. Among a few other varieties is one which is bright red in colour and has ovoid to globose shaped *Capsicum* fruits called "Ruby King". Since most of them are found in California (USA) they are also called the "Californian wonder" (De, 1994).

Among other less known commercially important varieties are *C. annuum* L. cv. "*Jalapena*", *C. frutescens* L. var. *mulagueta*, *C. annuum* L. var. *Kuloi*, *C. annuum* L. var. *annuum*, *C. annum* L. var. *Scotch-Bonnet*, *C. frutescens* Mill cv. *annuum*, etc.

Distribution

Distribution of the important species of *Capsicum* are given in Table 1.2 (Hunziker, 1956; Eshbaugh, 1977; Pickersgill, 1984).

The detailed taxonomic and other related biological, agricultural and geographical aspects of the major species of genus *Capsicum* are described as follows:

Capsicum annuum Linn. (purple, red, yellow chilli, pepper)

A suffrutescent or herbaceous, short-lived perennial (cultivated as annual) up to 1 m in height, cultivated throughout India from sea level up to an altitude of 2,100 m. Leaves oblong, glabrous; flowers solitary, rarely in pairs, pure white to bluish white, very rarely violet; berries green, maturing into yellow, orange to red shading into brown or purple, pendent, rarely erect, extremely variable in size (up to 20 cm long and 10 cm in diameter), shape and pungency, sometimes lobed, seeds white or cream to yellow, thin, almost circular, having long placental connections (The Wealth of India, 1992).

Red pepper is native to tropical America, most probably of Brazil. The crop is cultivated for its fruit throughout the plains of India and in the lower hills of Kashmir and the Chenub valley. When grown on the hills, it is said to be very pungent. There are several varieties differing chiefly in length, shape and colour of the fruit. The fruits, according to varieties, may be round or oblong, lipid or acute, smooth or rugose, red, white yellow or variegated (De, 1994).

A light, well-manured soil is best for all kinds. Crops which are raised during the rainy season should preferably be sown in well drained heavy soil. The crop prefers a limy soil. The seedlings may be planted at least 10 cms apart, when they attain a length of 7.5 cms. A subsequent replenishing of the soil, light earth and a good supply of water are the necessary requirements for the crop. Ammonium sulphate as a fertilizer is recommended for a heavy yield of the crop (De, 1994).

This type of fruit is used for the preparation of arrow poisons of the tribal people Dyaks of Borneo and Youri Tabocas of Brazil. It is useful in seasickness, typhoid fever and also in chronic fever (De, 1994).

Table 1.2 Synopsis of the genus *Capsicum* (Solanaceae)[a]

Capsicum	New world distribution
annuum L.	Colombia north to southern United States
baccatum L.	Argentina, Bolivia, Brazil, Paraguay, Peru
buforum Hunz.	Brazil
campylopodium Sendt.	South Brazil
cardenasii Heiser & Smith	Bolivia
chacoense Hunz.	Argentina, Bolivia, Paraguay
chinense Jacq.	Latin and South America
coccineum (Rusby) Hunz.	Bolivia, Peru
cornutum (Hiern) Hunz.	South Brazil
dimorphum (Miers) O.K.	Colombia
dusenii Bitter	Southeast Brazil
eximium Hunz.	Argentina, Bolivia
glapagoensis Hunz.	Ecuador
geminifolium (Dammer) Hunz.	Colombia, Ecuador
hookerianum (Miers) O.K.	Ecuador
lanceolatum (Greenm.) Morton & Standley	Mexico, Guatemala
leptopodum (Dunal) O.K.	Brazil
minutiflorum (Rusby) Hunz.	Argentina, Bolivia, Paraguay
mirabile Mart ex. Sendt.	South Brazil
parvifolium Sendt.	Colombia, Northeast Brazil, Venezuela
praetermissum Heiser & Smith	South Brazil
pubescens Ruiz & Pav.	Latin and South America
scolnikianum Hunz.	Peru
schottianum Sendt.	Argentina, South Brazil, Southeast Paraguay
tovarii Eshbaugh, Smith & Nickrent	Peru
villosum Sendt.	South Brazil

Note

a The following *Capsicum* species have been omitted: *C. anomalum, C. breviflorum* and *C. ciliatum* following the earlier suggestion of Eshbaugh (1983). Also, *C. flexuosum* Sendt. has been treated as a variety of *C. schottianum* by Hunziker. The treatment of *C. frutescens* L. remains to be resolved and some may choose to retain it as a distinct species of *Capsicum* while others include it in *C. chinense*, as suggested earlier. Finally, *C. eximium* var. *tomentosum* Eshbaugh & Smith is so distinctive that it may deserve species status.

Capsicum frutescens Linn. (spur pepper, goat pepper and chillies, bird chilli, red pepper)

A shrubby, hardy, finely pubescent perennial up to 2 m height, occurring wild or semi-wild in the tropics. Stems angular, leaves broadly ovate, usually wrinkled, or less pubescent; flowers greenish white, two or more at node, rarely erect, constriction between the base of the calyx and pedicel absent; fruits ovoid, obtuse or oblong, acuminate, immature and green, sometimes with dark pigmentation, nature red, rarely orange, erect, soft-fleshed, calyx embracing the base of fruit; seeds cream to yellow (The Wealth of India, 1992).

This herb originally introduced from South America is now cultivated throughout India. Of the cultivated species in India this is perhaps the most common and the largest. It is grown in cold weather in light sandy soil in most parts of the country, especially in Bengal, Orissa and Madras. The fruit when dry is bright red. These fruits are collected in large quantities, dried in the sun, and made ready for marketing.

Chillies are often used for flavouring pickles. By pouring vinegar upon the fresh fruits, all the essential qualities are preserved owing to their oleaginous properties. Hence, chilli-vineger is famous as a flavouring substance. In India a sort of chilli extract is produced by country folk which in colour and consistency is almost like treacle. The dried powder of this chilli is often sold in the market for ready use. When used as a spice, the preparation is generally mixed with water. Dried chilli powder is sometimes mixed with food by certain people in India. Mundas of Chota Nagpur use mustard oil in which roots of the chilli are mixed to shampoo the extremities to promote circulation of the blood. In Madagascar this is actually given to those suffering from delirium tremens.

Capsicum minimum Roxb. (Bird's eye chilli)

Some workers consider C. *frutescens* and C. *minimum* to be synonymous. The original home of this spice is America. The bird's eye chillies are cultivated throughout India but not so extensively as other chillies. The fruits are very small, suberect and almost oblong. These are found in many parts of India, principally in the Southern district, growing in waste places and gardens in an apparently wild state. It is also found abundantly in Java and other parts of the Eastern Archipelago under similar conditions. It is now cultivated to a large extent in the tropical regions of the world.

In India, it is not much used as a spice but in Europe it is used for stews, chops and other food preparations when mixed with vinegar and salt. In West Indies, the fruit is used to treat scarlet fever. It is used medicinally in Madagascar, where it is regarded as a stimulant, a promotive of salivation, a digestive, a laxative and an antiseptic. In Cambodia, it is much praised as a drug that brings about profuse perspiration. In West Indies, a stomachic preparation called "mandram" is still used today by the American Indians. It is prepared by adding cucumbers, shallots, lime juice and wine to mashed pods of bird's eye chillies. This concoction cures stomach aches and aids digestion. Externally it is also used as a counter-irritant to cure inflammation, boils and rheumatism (De, 1992–1994, 2000).

Capsicum pubescens Ruiz & Pav.

Capsicum pubescens forms a distinct genetic lineage. This pepper, first described by Ruiz and Pavon (1794) has not received wide attention from taxonomists until recently (Eshbaugh 1979, 1982). Morphologically, it is unlike any other domesticated pepper, having large purple or white flowers infused with purple and fruits with brown/black seeds. Genetically, it belongs to a tightly knit group of wild taxa including C. *eximium* (Bolivia and northern Argentina), C. *cardenasii* (Bolivia) and C. *tovarii* (Peru). C. *pubescens* is unique among the domesticates as a mid-elevation Andean species. C. *pubescens* is still primarily cultivated in South America, although small amounts are grown in Guatemala and southern Mexico, especially Chiapas. This species remains virtually unknown to the rest of the world. A small export market seems to have reached southern California. Two of the major difficulties in transferring this species to other regions include (1) its growth requirements for a cool, freeze-free environment and long-growing season and (2) the fleshy nature of the fruit that leads to rapid deterioration and spoilage.

Capsicum pubescens ranges throughout mid-Andean South America. An analysis of the fruit size of this domesticate indicates that fruits of a smaller size occur in Bolivia, while fruits from accessions outside Bolivia are somewhat larger, suggesting that Bolivian material approaches a more primitive size (Eshbaugh, 1979). Eshbaugh (1979, 1982) has argued that the origin of this domesticate can be found in the "ulupicas", C. *eximium* and C. *cardenasii*. Clearly, these two taxa are closely related to each other and to *C. pubescens* genetically. Natural hybrids between these taxa have been reported and evaluated (Eshbaugh, 1979, 1982). Furthermore, the two species that show the highest isoenzyme correlation with *C. pubescens, C. eximium* and *C. cardenasii*, occur primarily in Bolivia (Jensen *et al.*, 1979; Eshbaugh, 1982; McLeod *et al.*, 1983). All three of these taxa form a closely knit breeding unit with the two wild taxa, hybridizing to give fertile, progeny with viable pollen above the 90% level. Crosses between the wild *taxa C. eximium* and *C. cardenasii* and the domesticate *C. pubescens* most often show hybrid pollen viability greater than 55%. These factors lend to the conclusion that domesticated *C. pubescens* originated in Bolivia and that *C. eximium–C. cardenasii* is the probable ancestral gene pool. This does not prove that these two taxa are the ancestors of *C. pubescens*, but of the extant pepper taxa they represent the most logical choice. One perplexing question remains to be investigated and that is the origin of the brown/black seed coat in domesticated *C. pubescens*, a colour unknown in any of the other pepper species.

Capsicum chinense Jacq.

Examples of this pepper include Habanero and the Scotch Bonnet. It is found mostly in the area from the Amazonian Basin to Bolivia. Probably domesticated in the Bolivian Andes. The popular pepper Habanero in the Yucatan peninsula belongs to this species. The flowers are two or more at each node, corolla is greenish-white, filament purple; best criterion is the fruits' distinct aroma.

Capsicum baccatum var. *pendulum*

Capsicum baccatum var. *pendulum* represents another discrete domesticated genetic line. Eshbaugh (1968, 1970) notes that this distinct South American species is characterized by cream coloured flowers with gold/green corolla markings. Typically, the fruits are elongated with cream coloured seeds. The wild gene pool, tightly linked to the domesticate, is designated *C. baccatum* var. *baccatum* and is most common in Bolivia, with populations in Peru (rare) and Paraguay, northern Argentina and southern Brazil. This lowland to mid-elevation species is widespread throughout South America, and is particularly adjacent to the Andes. Known as "aji", it is popular not only as a hot spice but also for the subtle bouquet and distinct flavours of its many cultivars. This pepper is little known outside South America, although it has reached Latin America (Mexico), the Old World (India) and the United States (Hawaii). It is a mystery as to why it has not become much more wide spread, although the dominance of the *C. annuum* lineage throughout the world at an early date may be responsible.

    Capsicum baccatum var. *pendulum* is widespread throughout lowland tropical regions in South America. The wild form, recognized as *C. baccatum* var. *baccatum*, has a much more localized distribution but still ranges from Peru to Brazil. These two taxa have identical flavonoid (Ballard *et al.*, 1970; Eshbaugh, 1975) and isoenzyme profiles (Jensen *et al.*, 1979; McLeod *et al.*, 1979, 1983) and are morphologically indistinguishable except for the overall associated size differences found in the various organ systems of the domesticated taxon (Eshbaugh, 1970). The wild form of *C. baccatum* exhibits a high crossability index with domesticated *C. baccatum* var. *pendulum*, with the progeny typically exhibiting pollen viability in excess of 55% (Eshbaugh, 1970). The greatest centre of diversity of wild *C. baccatum* var. *baccatum* is in Bolivia, leading to the conclusion that this is the centre of origin for this domesticate.

    The wild gene pool, tightly linked to the domesticate, is designated *C. baccatum* var. *baccatum* and is most common in Bolivia, Brazil, Chile and Argentina. Pronounced as "bah-KAY-tum".

Flowers are solitary at each node. Pedicels erect or declining at anthesis. Corolla white or greenish-white, with diffuse yellow spots at the base of corolla lobes on either side of mid-vein (flower is white with yellowish spots, anthers are white but turn brownish-yellow with age); corolla lobes usually slightly revolute. Calyx of mature fruit without annular constriction at junction with pedicel (though sometimes irregularly wrinkled), veins prolonged into prominent teeth. Fruit flesh firm. Seeds straw-coloured. Chromosome number $2n = 24$, with one pair of acrocentric chromosomes, e.g. Escabeche (Peru).

Subdivisions:
- *Capsicum frutescens* var. *baccatum* Synonym for *Capsicum baccatum* var. *baccatum*.
- *Capsicum baccatum* var. *baccatum* The wild subspecies (common name: Locoto).
- *Capsicum baccatum* var. *microcarpum* (common name: Aji or Peruvian Pepper).
- *Capsicum baccatum* var. *pendulum* The cultivated subspecies (common name: Aji).
- *Capsicum baccatum* var. *praetermissum* (common name: Ulupica). Pods are spherical, 6 mm in diameter, red in colour and grow on bushy plants. Local name from Itaberai, Brazil is "Pimenta Cumari". Also known as *Capsicum praetermissum*.

The five major species are morphologically identifiable by the following traits:

Species	Flower colour	Number flw/node	Seed colour	Calyx constriction
C. annuum	White	1	tan	absent
C. frutescens	Greenish	2–5	tan	absent
C. chinense	White/ greenish	2–5	tan	present
C. baccatum	White with yellow spot	1–2	tan	absent
C. pubescens	purple	1–2	black	absent

The haploid chromosomal count of the cultivated and wild species is 12. There are wide variations among and within the species, whether wild or cultivated, as no karyotype is characterized by any single species and certain characteristics are studied among the majority of the members. Natural polyploidy is reported in the case of *Capsicum*, although a spontaneous tetraploid has also been reported in an intravarietal cross. Induced polyploidy with colchicine has also been reported where the induced polyploid exhibits a high vitamin C content profile. Diploids showing mitotic abnormalities and irregularities have also been reported (The Wealth of India, 1992).

In general, there appears to be a well-developed sterility barrier between cultivated species. It is impossible to cross *C. pubescens* with other species. Several crosses between *C. annuum, C. frutescens* and *C. baccatum* var. *pendulum* have produced a few F_1 hybrids but are mostly highly sterile. In the case of favourable crosses, the success or failure depends upon the direction of the cross. Reciprocal differences were observed in the case of fertility. The result of hybridization also differed according to the parental cultivars. Viable seeds have easily been produced from *C. annuum* $\times$ *C. chinensis* and *C. frutescens* $\times$ *C. pendulum*. The crosses *C. annuum* $\times$ *C. frutescens* and *C. frutescens* $\times$ *C. chinensis* have yielded a few viable F_1, F_2 and bud cross seeds (The Wealth of India, 1992).

Thus, on the basis of the above study, we may come to the conclusion that the genus *Capsicum* represents a very wide and divergent taxonomic group consisting of both wild and cultivated species. Some workers consider *Capsicum* to consist of three principal species, *C. annuum, C. frutescens* and *C. chinense*, but others have divided the genus into a divergent spectrum of species.

Sometimes superficial morpho-anatomical and conventional biochemical and cytological techniques are not sufficient to differentiate between closely associated species. But with enhancement and advancement in the realms of Molecular Biology, Molecular Genetics, Phytochemistry and Cell Biology, the genus stands the possibility of being further split. Only further scientific investigation in new directions can broadly highlight the taxonomic status of the genus *Capsicum*.

The authors do not claim any originality in the presentation of this chapter. It is only an effort on the part of the authors to reorganize and reconstruct the available literature to give a brief introduction to the readers.

References

Acosta, José de (1985) Historia natural y moral de las Indias. México.

Bailley, L. H. (1923) *Cultivated Evergreens*, Macmillan, New York, pp. 177–178.

Ballard, R. E., McClure, J. W., Eshbaugh, W. H. and Wilson, K. G. (1970) A chemosystematic study of selected taxa of *Capsicum, Amer. J. Bot.*, 57, 225–233.

Bentham, G. and Hooker, J. D. (1883) *Genera Plantarum*, Vol. I–III.

Bessey, C. E. (1915) The phylogenetic taxonomy of flowering plants' Ann Missour, *Botanical Garden*, II, 109.

British Pharmacopoeia (BPC) (1988) International edition, London, HMSO. ISBN: 0-11-321456-1.

De, A. K. (1992) Capsaicin: the wonder drug from chilli, *Everyman's Science*, 27, 47.

De, A. K. (1993) The wonders of chilli, *Indian Spices*, 29, 15.

De, A. K. (1994) *Chilli*, Books for All, Delhi.

De, A. K. (2000) *Recent Trends in Spices and Medicinal Plant Research*, edited by A. K. De, Associated Publishing Company, Delhi.

Domenici, P. (1983) The correct way to spell chile, *Congressional Record*, 129 (149) (Nov. 3).

Dunal, M. F. (1852) Solanaceae. In: *de Candolle, A. Prodromus* 13(1b), 4–690.

Engler, A. and Prantl, K. (1925) Band XXI., Die naturlichen planzenfamilien, Leipzig.

Eshbaugh, W. H. (1968) A nomenclatural note on the genus *Capsicum, Taxon*, 17, 51–52.

Eshbaugh, W. H. (1970) A biosystematic and evolutionary study of *Capsicum baccatum* (Solanaceae), *Brittonia*, 22, 31–43.

Eshbaugh, W. H. (1975) Genetic and biochemical systematic studies of chili peppers (*Capsicum* – Solanaceae), *Bull. Torrey Bot. Club*, 102, 396–403.

Eshbaugh, W. H. (1977) The taxonomy of the genus *Capsicum* – Solanaceae, pp. 13–26. In: E. Pochard (ed.): *Capsicum* 77. Comptes Rendus 3me Congres EUCARPIA Piment, Avignon-Montfavet, France.

Eshbaugh, W. H. (1979) Biosystematic and evolutionary study of the *Capsicum pubescens* complex, pp. 143–162. In: *National Geographic Society Research Reports*, 1970 Projects. National Geographic Society, Washington, DC.

Eshbaugh, W. H. (1980a) Chili peppers in Bolivia, *Plant Genet. Resources Newslett.*, 43, 17–19.

Eshbaugh, W. H. (1980b) The taxonomy of the genus *Capsicum* (Solanaceae) – 1980, *Phytologia*, 47, 153–166.

Eshbaugh, W. H. (1981) Search for chili peppers in Bolivia, *Explorers J.*, 58, 126–129.

Eshbaugh, W. H. (1982) Variation and evolution in *Capsicum eximium* Hunz, *Baileya*, 21, 193–198.

Eshbaugh, W. H. (1983) The genus *Capsicum* in Africa, *Bothalia*, 14, 845–848.

Eshbaugh, W. H., Guttman, S. I. and McLeod, M. J. (1983) The Origin and evolution of domesticated *Capsicum* species, *J. Ethnobiol*, 3, 49–54.

Evans, W. E. (1996) Trease and Evan's Pharmacognosy, Harcourt Brace & Co. Asia pte. Ltd, 14th edn.

Fingerhurth, A. (1832) Monographic genesis Capsici eldorpii, Sumptibus, Arnz & Comp.

Gounaris, L., Michalowshi, C. B., Bohnert, H. J. and Price, C. A. (1986) Restriction and gene maps of plastid DNA from *Capsicum annuum, Current Science*, 11, 7–16.

Hawkes, J. G., Lester, R. N. and Skelding (1979) *The Biology and Taxonomy of the Solanaceae*, Academic Press, London.

Heiser, C. B. (1976) Peppers *Capsicum* (Solanaceae), pp. 265–268. In: N.W. Simmonds (ed.), *The Evolution of Crops Plants*, Longman Press, London.

Hutchinson (1959, 1960) *Families of Flowering Plants*, Vols I and II, London.

Hunziker, A. T. (1956) Synopsis of the genus Capsicum. Huit. Congr. Int. deBot., Paris, 1954 Compt. Rend. Des Seanc et Communic. Desposes lors du Congresdans Sec, 3,4,5,6 Sec 4, 73–74.

Hunziker, A. T. (1979) South American Solanaceae: A Synoptic Survey, pp. 49–85. In: J. G. Hawkes, R. N. Lester and A. D. Skelding (eds), *The biology and taxonomy of the Solanaceae*. Academic Press, London.

IBPGR (1983) Genetic resources of *Capsicum*. Int. Board for Plant Genetic Resources, Rome.

Jensen, R. J., McLeod, M. J., Eshboagh, W. H. and Guttman, S. I. (1979) Numerical taxonomic analysis of allozyme variation of the genus *Capsicum* (Solaneacae), *Taxon*, 28, 315–327.

Lawrence, G. H. M. (1951, 1960, 1964) *Taxonomy of Vascular Plants*, Oxford & IBH Publishing Co.

Linneaus, C. (1753) *Species Plantarum*,

Loaiza-Figuerosa, Ritland, F. K., Cancino, J. A. L. and Tansley, S. D. (1989) Patterns of genetic variation of the genus Capsicum (Solanaceae) in Mexico, *Plant Syst. Evol.*, 165, 159–188

McLeod, M. J., Eshbaugh, W. H. and Guhman, S. I. (1979). A preliminary biochemical systematic study of the genus *Capsicum* – Solanaceae, pp. 701–714. In: J. G. Hawkes, R. N. Lester and A. D. Skelding (eds), *The Biology and Taxonomy of the Solanaceae*, Academic Press, London.

McLeod, M. J., Eshbaugh, W. H., Guhman, S. I. and Rayle, R. E. (1983) An electrophoretic study of the evolution in Capsicum (Solanaceae), *Evolution*, 37, 562–574.

Macrae, R. (ed.) (1993) *Encyclopedia of Food Science, Food Technology and Nutrition (Peppers and Chillies)*, Academic Press, 3496–3505.

MacNeish, R. S. (1964) Ancient Mesoamerican civilization, *Science*, 143, 531–537.

Mitchell, C. D., Eshbaugh, W. H., Wilson, K. G. and Pittman (1989) Patterns of chloroplast DNA variation in Capsicum (Solanaceae): a preliminary study, *Int. Org Plant Biosystematics Nesl.*, 12, 3–11.

Morrison, R. (1680) Plantarum Historiae Universalis Oxoriensis.

Pickersgill, B. (1980) Some aspects of interspecific hybridisation. In: *Capsicum*, unpublished and preliminary report at the IV Eucerpia capsicum working group meetings in Wageningen, The Netherlands.

Pickersgill, B. (1984) Migrations of chili peppers, *Capsicum* spp., in the Americas, pp. 105–123. In: D. Stone (ed.), Pre-Columbian plant migration. Papers of the Peabody Museum of Archeology and Ethnology, Vol. 76. Harvard University Press, Cambridge, MA.

Pickersgill, B. (1988) The genus *Capsicum*: a multidisciplinary approach to the taxonomy of the cultivated and wild plants, *Biol. Zent.*, 107, 381–389.

Prain, D. (1905) The Vegetation of the Districts of Hughle-Howrah and the 24 Pergunahs, *Bengal Plant*, Vols I and II, Rep. 1977, 208 pp.

Prod. Yearb, (1988) FAO 42, 188.

Smith P. G. and Heiser, C. B. Jr. (1957) Breeding behaviour of cultivated peppers, *Proc. Amer. Soc. Hort. Sci.*, 70, 286–290.

Stuessy, T. (1990) *Plant Taxonomy*, Columbia University Press, New York.

The Wealth of India: A Dictionary of Indian Raw Materials and Industrial Products (1992) Capsicum, Vol. 3. Ca-Ci, PID CSIR, New Delhi, pp. 218–264.

Wettstein, R. (1895) Scrophulariaceae. In: A. Engler and K. Prantl (eds), *Die Nat. Pfl. Fam.*, 39–107.

2 A glimpse of the world trade in *Capsicum*

P. S. S. Thampi

Capsicum is cultivated in different parts of the world. Of the total world consumption chillies proper account for one third of the total, while paprika comprises two thirds. The present scenario in *Capsicum* trade is presented in this paper.

Introduction

The voyage of Columbus that led to the discovery of red peppers made this very fiery variety of spice known to the world. As a result, black pepper, which was a highly prized commodity in Europe, started facing stiff challenges from the new species of red peppers.

Chillies were historically known to be used to impart flavour and hotness to food. Many civilizations were known to use this species, especially the Mayans and the Aztecs. The very foundation of Mexican food is based on the essence of chillies.

Columbus' discovery of the New World resulted in the discovery of the chillies which was found to be an excellent substitute to black pepper. Called the chilli pepper, this chilli, which was once native to the warm temperature and tropical regions of America, found its way to the Indian coast. It had its roots in the West Asian coast, and also in the islands of East Africa. This augmented the use of chillies in the food of various nationalities and societies and has become an essential part of indigenous diets. Chillies are cultivated in different parts of the world, but many non-commercial varieties are still cultivated and consumed in the American regions of Mexico.

Over 25 varieties of chillies, including the domesticated varieties, are widespread in the mid-elevation Andes from Colombia to Bolivia and in the Amazon regions of America. The result of voyages and colonization has helped to spread the cultivation of chillies to the tropical and subtropical countries of Africa, India, Japan, Turkey, Hungary, Morocco and China. The power of chillies has been tested even politically. In 1978–79 Korea faced a shortage of red chilli. It was feared that the Korean government would fall if the supply of chilli was not adequate. That year, however, Korea imported upto 50,000 tonnes from all over the world.

Belonging to the genus *Capsicum*, chillies have various species of long, hot, mild, fiery and sweet types of peppers. The dried fruits of the species *Capsicum* are used as a condiment or culinary supplement. The consumption of chillies per head is very high in India.

Though the origin of chillies is traced to certain American countries, chillies are now grown worldwide. The prominent producers are India, China, Pakistan, Korea, Mexico and Bangladesh. Even though production figures are available, the total estimate of international trade in *Capsicum* is very difficult to estimate. Besides this, the genus *Capsicum* has many varieties and strains. The domesticated species is *Capsicum annuum*, which is the best known species and very hot. The following chillies are predominant in world trade:

Capsicum pubescens   This variety is hairy and is very popular, though there is very limited consumption.

Capsicum baccatum   Meaning berry-like. This species is common in Bolivia, Peru, Paraguay, North Argentina and Brazil. Known as Ajis, this species is very hot and has distinct flavours in its various cultivars. Its reach has extended to the borders of Mexico, Hawaii and India.

Capsicum chinense   Although the name implies a Chinese origin, this species originated and was widely cultivated in the Amazon basin of America.

Capsicum frutescens   This means shrubby or bushy and includes the South American Ricottas and Mexican Manzanos.

Chillies are known worldwide by different names. Popular names are red pepper, cayenne pepper, sweet pepper, bell pepper, pimento, sweet banana, etc.

With regard to the international trade in chillies, ITC Geneva has made its estimate for world consumption and production. Of the total world consumption, chillies proper account for one-third of the total, while paprika comprises two-thirds. During the late 1990s it was estimated that the total world production of chillies was in the region of 25,00,000 tonnes. The figures relating to the consumption has been estimated to be over 50,000 tonnes, of which paprika varieties accounts for two-thirds.

World production of chillies is going up year after year as more consumers develop a taste for chillies. One of the reasons is the growing popularity of 'ethnic food'.

Though there are no official figures for the production of chillies worldwide, it is estimated to be 25,00,000 tonnes. India tops the list with 8,50,000 tonnes, followed by China with 4,00,000 tonnes, Pakistan 3,00,000 tonnes, South Korea 1,50,000 tonnes, Mexico 3,00,000 tonnes, Bangladesh 1,00,000 tonnes and other countries combining to produce 4,00,000 tonnes.

Now let us examine the other trade figures in *Capsicum* chillies. According to ITC estimates, the total imports of *Capsicum* in dried, crushed and ground form has totalled 10,95,200 tonnes. The major source of imports by countries like Australia, Austria, Belgium, Luxembourg, Brazil, Canada, Egypt, France, Germany, China, Italy, Japan, Korea, Malaysia, Mexico, The Netherlands, Pakistan, Poland, Russian Federation, Saudi Arabia, Singapore, South Africa, Spain, Sweden, Switzerland, the UK and the USA were from the producing and the non-producing centres for *Capsicum*. The main producing centres listed in the source of imports were India, Mexico, Chile, China, Morocco, Pakistan, South Africa, Vietnam, Zimbabwe, Brazil, Hungary and Spain. Countries like Egypt, Israel, Indonesia, Jamaica, Turkey, Myanmar, Malawi and Malaysia also figured quite prominently. The countries like Austria, Australia, Cuba, Croatia, Italy, Kenya, Korea, Yugoslavia, Uzbekistan, Ukraine, Zambia, Thailand, Peru, Poland, Nigeria and Sri Lanka are also new known sources for exports. European countries like the UK, Switzerland, The Netherlands, France and Belgium are also found to be exporting chillies. Huge quantities were found to be re-exported by countries like Germany, The Netherlands and Singapore.

Though Singapore is not a producer country for *Capsicum* it has exported a quantity of 16,000 tonnes between 1994–98. The ITC figures compiled for the import by Singapore during 1994–98 are 86,426 tonnes. This reveals that over 65% of the total imports are re-exported to different countries. Singapore has mainly supplied to Amsterdam, Hong Kong, Japan and The Netherlands.

Yet another prominent importer and re-exporter is The Netherlands. During 1994–98, the total import was estimated to be 27,293 tonnes, of which nearly 11,578 tonnes were re-exported to countries like Russia, Belgium, Sweden, Switzerland, the UK, the USA and Italy.

Germany is also a major importer and a re-exporter. Over 69,000 tonnes of chillies were imported and an approximate quantity of 13,000 tonnes were re-exported during the period 1994–98.

The share of *Capsicum* in the total imports of different spices has been consistent around 15% to 17%. The latest figures available from the ITC Geneva for the year 1998 confirm these figures. The total imports of spices for 1998 have been estimated at US$2,338.54 million. Pepper tops the list with 40%, followed by *Capsicum* at 16%, spice mixtures at 10%, seed spices at 8%, ginger at 6%, cinnamon at 6%, nutmeg, mace, cardamom at 6%, thyme, saffron, bay leaf at 3% and vanilla at 3%. However, in 1994 the share of *Capsicum* was 17% of the total imports at US$158.15 million. In that year *Capsicum* also came second with 17% next to pepper at 25%. The share of other spices were, spice mixtures at 13%, seed spices at 10%, cinnamon at 9%, vanilla at 8%, ginger at 7%, thyme, saffron and bay leaf at 4%, nutmeg, mace and cardamom at 4% and cloves at 2%. The unit price of world imports of *Capsicum* was 1.45 dollars per kg in 1994 and this has shown a rising trend of 1.97 dollars per kg in 1995 and up to 2.01 dollars per kg in 1998.

India has emerged as a major producer and supplier. The main varieties grown in the country are:

Sannam: This is a common variety and dominates over 75% of production. These are long and narrrow, ranging from 4 to 8 VMS with a capsaicin level ranging from 0.10% to 0.45%, depending on the area where they are grown.

Mundu: The Mundu is a roundish cherry-shaped chilli and the approximate production is 10,000 tonnes.

The wrinkled variety: The wrinkled variety is mainly used for the manufacturing of paprika-type oleoresin, and the annual production is 40,000 tonnes.

Birds eye chillies: This variety has high capsaicin levels ranging from 0.45% to 0.90% and is small in size with length ranging from 1 cm to 4 cm. The Indian exports of *Capsicum*, which was 50,051 tonnes during 1996–97, went up to 68,019 tonnes in 1998–99 and slightly declined to 61,000 tonnes during 2000–01. Chillies contributed to a 12% share in the total spice export from India during 2000–01. It was 14% during the previous year. The important markets for Indian chillies are the USA, the South East Asian countries, Sri Lanka, Bangladesh and the Middle East countries.

There are also many other varieties like Ellachipur Sannam which has an ASTA colour value of 70.40%, Hindpur S 7 having a colour value of 33%, Kanthari white with a capsaicin value of 0.504%, Kashmir chilli with a capsaicin value of 0.325%, Madras pari having a capsaicin value of 0.206%, the Nalchetti variety having an ASTA colour value of 77.03%, Ramnadu Mundu having a capsaicin value of 0.166%, the Sangli sannam variety having a capsaicin value of 0.215%, Sattur S 4 having an ASTA colour value of 59.1%, Scotch Bonnet with a capsaicin value of 0.878%, etc. The Tomato chillies have a colour value of 125.26%.

Good quantities of chillies are exported in the form of oils and oleoresins. Of all the countries, India is the major producer and only 10% of the total production is exported. The total estimated production in India is 9,50,000 tonnes of which the exports account for only 65,000 tonnes. Other than exports in bulk, exports are in the form of oil, oleoresins, powders, pastes and mixtures. The country-wise exports from India for the last five years for chillies/powders, chilli oils and chilli oleoresin are shown in Tables 2.1, 2.2 and 2.3, respectively.

Challenges in international trade

The international trade in *Capsicum* is now concerned with issues like aflatoxins, pesticides, residues, microbial contaminations, and capsaicin and colour values. Major importing countries in Europe and America have fixed quality requirement standards for chillies. The American Spice Trade Association's cleanliness specifications and the US Food and Drug Administration's defect action levels for chillies prescribed levels for dead insects, mammalian excreta,

Table 2.1 Country-wise export of chilli from India during 1996–97 to 2000–01 (Quantity (Qty) in tonnes; value in Rs. lakhs)

Country	1996–97		1997–98		1998–99		1999–2000		2000–01[a]	
	Qty	*Value*	*Qty*	*Value*	*Qty*	*Value*	*Qty*	*Value*	*Qty*	*Value*
Australia	222.51	117.95	176.54	84.43	252.56	153.13	215.02	135.56	229.18	119.07
Argentina		—	14.85	6.31	2.25	1.19	7.50	4.15		—
Austria		—	1.20	0.24	67.00	10.01	0.11	0.54		—
Baharain	128.86	55.95	102.79	45.51	83.25	42.05	80.21	39.07	73.10	38.75
Belgium	64.42	38.69	117.27	40.33	104.33	40.56	99.25	33.27	77.66	39.93
Bangladesh	184.55	68.22		—	10,892.62	4,454.37	4,722.12	2,367.03	2,289.57	662.70
Burma (Myanmar)	627.05	190.23		—	0.21	0.11		—		—
Brazil	35.50	23.56	67.37	32.57	139.17	100.65	119.45	80.19	114.98	63.87
Barbados		—	14.00	3.89	10.98	7.25	12.00	5.04		—
Bahamas		—		—	1.90	0.26		—		—
Brunei	25.00			—	6.00	3.93	14.65	7.87		—
Belarus		13.86								
Canada	477.58	236.55	770.45	335.92	627.11	311.74	883.77	520.79	1,091.22	313.82
China	146.75	82.28	44.50	10.78	24.80	8.97	147.59	59.59	48.39	18.45
Cyprus	17.00	6.93		—		—		—		—
Canary Islands		—		—	2.50	1.38		—		—
Chile	24.52	8.32	69.43	24.81	0.04	0.04		—		—
Congo		—	0.22	0.16				—		—
Cuba		—		—	50.00	18.65		—		—
Croatia		—		—	0.09	0.07		—		—
Czech Republic	1.20	0.93	10.25	5.65	3.70	1.69		—		—
Denmark		—	38.56	8.03	0.80	0.14	2.04	0.31		—
Djibouti		—	0.20	0.13				—		—
Egypt (A.R.E.)	483.40	188.08	480.00	152.83	72.00	25.47	601.50	296.94	629.00	350.31
Ethiopia		—		—	79.00	24.97	10.28	6.37	18.00	3.47
Ecuador		—		—	2.00	1.07	11.38	4.58		—
Finland		—	3.50	0.98	29.09	13.12	32.00	11.15	16.50	7.75
France	624.19	240.53	437.39	150.70	203.74	85.05	190.22	83.19	229.64	89.40
Fiji		—		—	2.60	1.06	13.78	7.77	2.78	2.30
Greece	181.66	53.19	212.75	68.63	80.00	37.09	100.65	49.38	46.64	13.59

(*Continued*)

Table 2.1 (Continued)

Country	1996–97		1997–98		1998–99		1999–2000		2000–01[a]	
	Qty	Value	Qty	Value	Qty	Value	Qty	Value	Qty	Value
Germany	387.30	176.69	450.39	161.92	89.29	35.54	307.10	127.54	166.06	56.17
Guatemala	—	—	0.15	0.03	11.38	4.58	—	—	—	—
Guinea	—	—	—	—	—	—	—	—	—	—
Guadeloupe	—	—	—	—	12.50	6.11	—	—	—	—
Hong Kong	1,453.12	589.10	157.38	76.49	117.16	73.87	141.68	88.33	90.27	50.60
Iraq	—	—	—	—	18.00	3.55	—	—	—	—
Italy	587.11	306.05	626.51	293.53	385.49	215.68	518.60	281.48	460.36	261.15
Indonesia	433.96	160.58	1,174.01	230.12	550.44	184.14	1,621.27	499.32	1,613.63	411.38
Ireland	—	—	11.00	3.90	—	—	—	—	—	—
Iran	—	—	4.92	3.62	70.55	11.99	—	—	—	—
Israel	293.08	125.83	437.11	158.95	211.91	97.58	219.65	109.38	294.75	115.69
Ivory Coast	—	—	0.50	0.22	—	—	—	—	—	—
Japan	532.13	236.80	246.52	128.00	137.76	99.98	64.95	77.43	339.73	95.33
Jordan	51.00	13.64	160.60	31.09	42.50	8.87	44.00	11.19	33.80	5.52
Kuwait	174.82	79.99	208.00	77.27	109.85	61.16	313.94	169.61	196.45	83.59
Korea (South)	2.00	0.94	—	—	100.00	13.23	—	—	—	—
Korea (North)	9.00	7.84	—	—	2.00	1.20	—	—	—	—
Kenya	6.00	3.39	39.50	10.59	89.57	32.51	6.00	3.22	15.00	3.21
Kazakhstan	—	—	—	—	18.00	18.03	—	—	—	—
Libya	—	—	12.50	3.67	49.00	21.09	—	—	—	—
Lebanon	22.30	9.15	79.70	24.78	43.50	15.02	65.24	20.86	58.04	17.70
Liberia	—	—	—	—	1.15	0.21	—	—	—	—
Latvia	—	—	—	—	90.00	33.07	136.40	74.94	0.06	0.04
Lithuania	12.25	5.52	—	—	—	—	—	—	—	—
Morocco	—	—	—	—	19.00	6.01	19.75	6.84	—	—
Maldives	67.20	25.74	17.92	5.91	151.57	35.82	0.38	0.24	83.29	38.80
Malaysia	5,891.83	2,443.40	1,868.94	432.00	1,478.72	605.27	4,007.61	1,519.09	1,611.02	581.46
Mauritius	281.55	125.66	162.25	55.85	168.40	78.55	193.82	92.95	207.99	92.88
Mexico	1,183.15	513.04	3,382.28	884.68	742.20	222.08	379.94	161.91	2,640.15	1,014.75
Martinique	—	—	—	—	7.00	6.53	—	—	—	—
Macau	—	—	—	—	1.03	1.05	—	—	—	—

Mali	0.30	—	—	—	334.11	119.66	—	—	—	—
Malagasy Rep.	—	0.15	—	—	—	—	—	—	—	—
Norway	929.62	—	20.16	6.13	22.39	5.82	93.02	36.26	0.34	0.10
The Netherlands	68.85	417.25	1,041.81	362.79	300.55	158.31	710.41	183.53	617.80	234.64
Nepal	54.80	8.13	437.52	75.04	388.24	122.18	1,413.81	305.91	657.43	172.76
New Zealand	—	34.03	106.63	49.78	56.80	51.55	49.47	37.30	26.33	13.47
Niger	—	—	—	—	25.00	12.34	—	—	—	—
Netherland Antils.	—	—	—	—	13.00	6.67	—	—	—	—
Oman	143.35	53.66	251.56	106.97	155.64	90.79	151.71	88.29	124.66	60.77
Portugal	289.46	116.78	61.72	27.49	2.02	0.48	13.08	5.62	69.74	14.27
Poland	16.00	7.06	46.75	18.29	97.50	28.04	217.58	70.62	235.50	88.74
Pakistan	2,265.28	487.32	1,339.89	185.63	14,901.39	3,201.20	8,885.77	2,108.86	4,601.11	1,109.40
Paraguay	7.00	4.95	72.00	—	15.00	6.51	—	—	—	—
Peru	—	—	—	37.14	—	—	—	—	—	—
Phillipines	65.10	35.01	109.00	49.67	49.00	25.77	72.00	38.65	70.92	28.42
Qatar	6.75	1.96	72.23	17.68	18.65	6.86	9.04	0.93	102.29	30.56
Romania	—	—	—	—	4.05	0.82	—	—	37.50	12.68
Reunion	1.10	0.36	—	—	0.04	0.02	—	—	—	—
Russia	241.58	128.47	664.62	408.38	211.00	97.86	484.25	183.50	472.50	133.80
Singapore	7,279.79	2,808.76	2,344.31	536.80	1,016.52	412.76	897.50	381.49	666.98	228.29
Sweden	—	—	3.05	2.34	24.23	31.88	28.12	26.85	14.37	15.36
Spain	501.11	233.38	764.41	174.16	55.00	34.49	75.41	45.06	173.00	74.18
Switzerland	4.00	2.17	10.05	2.38	1.42	0.68	36.90	9.96	82.64	19.82
Senegal	—	—	—	—	26.85	8.46	—	—	—	—
Seychelles	6.55	2.52	6.90	2.15	1.13	0.87	—	—	—	—
South Africa	552.04	180.96	2,648.08	811.05	383.09	128.64	180.92	56.47	794.01	286.30
Sri Lanka	5,154.67	1,843.18	11,881.35	2,539.11	19,805.56	7,302.73	17,546.05	5,971.15	22,589.29	6,173.64
Sudan	—	—	—	—	0.34	0.19	—	—	—	—
Swaziland	—	—	—	—	0.48	0.13	—	—	—	—
Saudi Arabia	912.69	298.65	807.03	197.98	491.72	189.23	632.97	216.63	356.03	104.27
Slovenia	50.40	9.02	—	—	20.00	13.65	—	—	—	—
Thailand	793.55	314.93	52.98	13.00	10.61	5.35	12.00	4.98	51.60	33.27
Tanzania	—	—	4.28	2.17	0.07	0.06	4.70	3.76	4.91	4.28
Togo	—	—	—	—	—	—	—	—	—	—
Trinidad	3.00	1.79	—	—	15.50	8.22	—	—	22.17	8.02
Tunisia	120.25	39.19	93.73	31.98	36.00	18.87	—	—	—	—

(Continued)

Table 2.1 (Continued)

Country	1996–97		1997–98		1998–99		1999–2000		2000–01[a]	
	Qty	Value	Qty	Value	Qty	Value	Qty	Value	Qty	Value
Turkey	75.00	38.76	65.51	15.80	30.50	9.82	—	—	34.52	25.32
Taiwan	685.84	259.37	290.57	105.56	181.16	74.06	335.64	150.01	148.46	53.87
USA	9,205.68	4,206.55	12,056.91	4,919.53	7,441.40	4,408.22	11,799.28	5,987.20	10,998.79	4,334.57
UK	1,962.75	1,007.60	2,083.71	830.60	1,423.49	804.71	1,728.37	920.25	1,939.52	745.44
Uganda		—	0.30	0.33	1.20	0.93	0.60	0.19		—
Uruguay	14.00	9.24		—		—				
UAE	3,851.78	13,76.50	2,522.50	717.12	3,555.15	881.54	3,648.54	1,107.10	3,089.64	819.66
Ukraine	128.00	52.83	274.94	66.75	74.00	22.28	16.00	4.97	46.00	23.72
Uzbekistan		—		—	18.00	14.69		—		—
Venezuela			7.00	5.58	4.00	4.92		—		—
Vietnam	0.26	0.16	0.18	0.13	0.04	0.03		—		—
YAR	13.50	7.64	20.25	6.34	22.00	4.65	42.00	15.89	11.00	3.14
Zimbabwe	15.00	8.21	16.00	9.68		—		—		—
Total	50,051.04	20,145.17	51,779.38	15,890.05	68,018.95	25,287.29	64,775.90	25,065.87	61,000.00	19,523.51

Source: DGCI&S., Calcutta/S. Bill/Exporters returns.

Note
a Provisional.

Table 2.2 Country-wise export of chilli oil from India during 1996–97 to 2000–01 (Quantity (Qty) in tonnes; value in Rs. lakhs)

Country	1996–97		1997–98		1998–99		1999–2000		2000–01[a]	
	Qty	Value	Qty	Value	Qty	Value	Qty	Value	Qty	Value
Germany	0.01	0.08	—		0.20	3.77	—		—	
Japan	—		0.80	12.85	—		—		—	
Singapore	—		—		0.02	0.09	—		—	
USA	—		0.35	10.54	—		—		—	
Total	0.01	0.08	1.15	23.39	0.22	3.86	—		—	

Source: DGCI&S., Calcutta/S. Bill/Exporters returns.

Note
a Provisional.

Table 2.3 Country-wise export of chilli oleoresin from India during 1996–97 to 2000–01 (Quantity (Qty) in tonnes; values in Rs. lakhs)

Country	1996–97		1997–98		1998–99		1999–2000		2000–01[a]	
	Qty	Value	Qty	Value	Qty	Value	Qty	Value	Qty	Value
Australia	0.66	5.03	0.60	3.11	0.01	0.11	—		—	
Bangladesh	—		0.06	0.45	—		—		—	
Brazil	—		0.29	4.71	—		—		—	
Canada	0.66	8.80	0.40	4.42	—		—		—	
China	—		—		0.40	2.46	—		—	
France	0.72	9.75	0.30	4.20	0.25	3.48	0.65	5.35	2.55	26.07
Germany	1.35	11.33	1.09	13.74	—		—		—	
Italy	—		0.84	20.95	0.54	6.29	—		—	
Israel	0.58	7.49	—		—		—		—	
Japan	0.51	8.33	15.46	185.66	4.00	59.84	—		—	
Korea (South)	—		1.35	24.62	0.41	6.32	—		—	
Korea (North)	—		0.30		4.02	—		—		—	
Mauritius	—		1.03	16.38	—		—		—	
Mexico	0.20	6.33	11.19	162.31	26.98	192.03	0.04	2.15	26.07	184.39
The Netherlands	—		0.58	7.95	—		—		—	
Phillipines	—		0.60	6.14	—		—		—	
Russia	—		—		0.04	2.78	0.02	1.03	—	
Singapore	—		4.02	21.77	—		—		—	
Spain	2.10	23.90	3.01	42.92	—		—		—	
Turkey	—		2.16	19.47	—		—		—	
Taiwan	2.00	63.41	—		—		—		—	
USA	1.40	11.55	60.23	492.05	0.23	31.25	0.55	32.46	Neg.	0.04
UK	1.54	10.02	0.22	2.23	0.05	2.58	—		—	
Uruguay	—		0.15	1.51	—		—		—	
UAE	—		—		0.02	3.46	—		—	
YAR	—		1.44	28.68	—		—		—	
Total	11.72	165.94	105.02	1,063.27	27.52	229.45	1.70	48.66	33.89	288.00

Source: DGCI&S., Calcutta/S. Bill/Exporters returns.

Note
a Provisional.

molds, insect desified, extraneous foreign matters, rodent and insect filth in each and every consignment.

These standards have made trading in *Capsicum* more complex. With a view to produce clean chillies, organic farming practices are gaining momentum to produce pesticide-free chillies suitable for various international markets. In countries like India, quality awareness programmes and continued improvements in post-harvest operations are taken up on a consistent basis to export 100% clean chillies.

3 Chemistry and quality control of *Capsicums* and *Capsicum* products

J. S. Pruthi

"Quality control" is a "sine quonone" for the success of any industry, may it be Agro-food industry or Spice industry. This chapter covers the various quality control techniques whether physical, or physico-chemical etc. The quality evaluation of *Capsicums* whole, ground or processed, variations in chemical composition of *Capsicums*, major principles, methods of determination of pungent and colouring principles of all types of chillies, *Capsicums*, paprika, detection of adulteration, quality evaluation of oleoresins of different types and quality standards etc. have been discussed.

Both at National and International levels, there are numerous food bodies implementing standards for spices and spice products. For instance, there are ten international organizations like EC, FCC, FAO, WHO, ISO, Codex, ASTA, EOA, etc. and numerous national bodies (like, BIS (ISI) BS, Agmark) in each country, which guide the industry concerned in trade at both national and international levels. These standards are the result of conscientious efforts in standardization. Harmonization of national and international standards avoids costly litigation in international trade and hastens the marketing of products. However, due to a lack of space it is not possible to cover all such standards. Only ISO and BIS specifications have been listed while detailed Indian PFA and Agmark standards have been described to facilitate quality control.

Introduction

Quality control constitutes the very backbone of success in any industry, be it agriculture or spices.

The quality of spices is evaluated by the following three techniques: (i) Physical or sensory evaluation; (ii) Physico-chemical analysis; and (iii) Nutritional assessment.

Since chillies, like most of the spices, are used in small quantities both in vegetarian and non-vegetarian dishes as "Food additives" only, they contribute little to human nutrition. However, spices indirectly play an extremely valuable role in good nutrition by helping to increase the appeal and appreciation of foods that are nutritionally important to us. Except in most restrictive diets, spices may be used and they can make the difference that persuades a patient to stay faithful to an otherwise unappealing diet. These analyses should be particularly good news for waist watchers because they confirm that even the spice highest in calories (poppy seeds) does not add more than two or three calories per serving in normal usage. Considering the amounts used, other spices typically contribute not more than one calorie per serving and usually less. Having this definitive data at last, nutritionists and all those interested in good nutrition, may utilize spices to their full potential in making wholesome foods excitingly delicious as well as good for us.

Fortunately, quality has been a tradition in the spice trade of India and to maintain this tradition and to adhere to modern developments in the standardization of agricultural produce, the

Government of India has prescribed standards for almost all the important spices, including chillies. Chillies are graded compulsorily under law before export. The grades adopted for chillies (whole and ground) are those prescribed under the Agricultural Produce Grading and Marking Act and these grades are popularly known as "Agmark grades".

Quality evaluation of chillies and chilli products

Detailed quality evaluation techniques include, physical, chemical-microbiological, instrumental and sensory evaluation of spices, including chillies (whole and ground) and their processed products, like chilli oleoresin, etc. All modern methods have been described in detail in a recent book on "Quality Assurance", by Pruthi (1999). The chemical composition of *Capsicum*/chillies/paprika/ red pepper as a measuring quality is presented in Table 3.1, which is self-explanatory.

Chemical composition of Capsicums/*paprika*

Capsicum fruits contain colouring pigments, pungent principles, resin, protein, cellulose, pentosans, mineral elements and a very little volatile oil, while seeds contain fixed (non-volatile) oil. Chillies and paprika may be regarded as taking up positions at opposite ends of a spectrum of common properties. From chillies through *Capsicums* to paprika, there is a steady decrease in pungency level and an increase in the pigment content. The fruits of most *Capsicum* species contain significant amounts of vitamins B, C, E and provitamin A (carotene) when in a fresh state. The large type of *C. annuum* is among the richest known sources of vitamin C, which may be present up to 340 mg/100 g in some varieties (Purseglove *et al.*, 1981). The variation in the composition of 12 varieties of chillies is given under Table 3.2 composition and has been studied by Bajaj *et al.* (1978, 1980), Mauriya *et al.* (1983) and reviewed by Govindarajan (1986) and Pruthi (1999).

Table 3.1 Composition of pepper and paprika

Quality characteristics (Units within brackets)	Paprika (Imported)	Paprika (Domestic)	Pepper (Chilli)	Pepper (Red)
Chemical composition				
Moisture (g)	6.40	7.90	6.50	6.20
Food energy (cal)	385.00	390.00	415.00	420.00
Protein (g)	15.80	13.80	14.00	16.00
Fat (g)	10.10	10.40	14.10	15.50
Total carbohydrates (g)	58.00	60.30	58.20	54.30
Fibre carbohydrates (g)	19.40	19.00	15.60	26.00
Total ash (g)	9.70	7.60	7.20	8.00
Minerals				
Calcium (g)	0.20	0.20	0.10	0.10
Phosphorus (g)	0.27	0.30	0.32	0.32
Sodium (g)	0.02	0.02	0.01	0.01
Potassium (g)	2.40	2.40	2.10	2.10
Iron (mg)	23.10	23.10	9.90	9.90
Vitamins				
Thiamine (mg)	0.40	0.60	0.59	0.52
Riboflavin (mg)	1.17	1.36	1.66	0.93
Niacin (mg)	15.40	15.30	14.20	13.60
Ascorbic acid (mg)	48.60	58.80	63.70	29.40
Vitamin A (IU)	3,350	4,915	6,165	3,530

Source: Based on analyses performed by 900 American Laboratories, St Louis, MO, USA.

Table 3.2 Variations in the yield and biochemical constituents of 12 Indian chilli varieties

Varieties	Yield	Capsaicin ($kg\,ha^{-1}$)	Oleoresin (%)	Colour (%, ASTA units)	Ascorbic acid value (mg/100 g)
TC 1	1,542	0.156	13.0	85.0	135
TC 2	1,607	0.141	12.4	82.0	133
Sel 1	2,100	0.178	14.4	76.0	114
KCS 1	2,143	0.191	13.4	81.0	122
DPL (Cl)	1,984	0.313	11.6	88.0	127
Phule	1,627	0.344	12.9	85.0	153
Jawahar	1,548	0.244	12.2	82.0	122
Musalwadi	2,069	0.159	12.4	88.0	109
CA 586	1,894	0.490	13.2	85.0	164
LCA 206	2,358	0.465	12.0	92.0	154
LCA 235	1,521	0.525	13.5	95.0	166
LCA 248	2,400	0.250	13.2	85.0	173
Mean	1,983	0.288	12.9	85.0	139
S.Em	113.47	0.008	0.097	0.500	1.44
CD (0.05)	332.81	0.023	0.280	1.043	3.35

Source: Govinda Reddy *et al.* (1999).

The fixed oil (non-volatile) in chillies

The chilli fixed oil is comprised mainly of triglycerides (about 60%) of which linoleic and other unsaturated fatty acids predominate. The fat content and composition of paprika powder and its propensity to auto-oxidation are dependent upon whether the seeds are removed from the pod before grinding or not (Anu and Peter, 2000).

The volatile oil (essential oil) of Capsicums

Fruits of *Capsicum* species have a relatively low volatile oil content, ranging from about 0.1% to 2.6% in paprika. The characteristic aroma and flavour of the fresh fruit is imparted by the volatile oil. The composition of the volatile oil of fresh Californian green bell peppers has been examined by Buttery *et al.* (1969) using gas chromatography. Twenty-four components in this oil were positively identified. One of the major components, 2-methoxy-isobutyl pyrazine, was considered to posess an aroma characteristic of the fresh fruit and to dominate the organoleptic profile of paprika.

*Major principles in chillies/*Capsicums

*Pungent principles in chillies/*Capsicums

Several pungent compounds found in nature are derivatives of *o*-methoxyphenol. The major principles naturally present in *Capsicums* are capsaicin and dihydrocapsaicin (Kulka, 1967). The degree of pungency and the character of taste sensation vary markedly with different varieties of chillies. It was in 1846 that Thresh isolated for the first time the pungent principle from *Capsicums*, Nelson (1919) and Nelson and Dawson (1923) declared it to be an amide of vanillylamine and isodecanoid acid. Further work on the chemistry of capsaicin has been reviewed by

Newman (1953); Rogers (1966); Pruthi (1970a,b) and further reviewed by Pruthi (1980). The chemistry of pungent principles has been reviewed by Pruthi (1980, 1999), Govindarajan (1986) and more recently by Anu and Peter (2000) and by the Basic Synthite Chemicals, Cochin, India and is briefly discussed later.

Kosuge *et al.* (1965) established that the pungent principle of red pepper consists not of one chemical but actually of the unsaturated and saturated amides – capsaicin and dihydrocapsaicin. The mixture of these two amides was named capsaicinoid, which is odourless (cited by Pruthi, 1980). With regard to the pungency of this group of compounds, it can be stated that an aromatic ring having a phenolic hydroxyl group and an ether group such as methoxy in *ortho* position to each other is a basic pre-requisite. A side-chain is also necessary. The length and composition of this side-chain are important. The pungency is greatly enhanced by an acid amide group, in this instance vanillylamide, as found in the capsaicinoid molecule (Kulka, 1967).

On the basis of these observations, the synthetic compound nonenoylvanillylamide was prepared. This has considerable pungency and heat (Kulka, 1967). Friedrich and Rangoonwala (1965) studied the separation of capsaicin and nonanoic acid vanillylamide. Kosuge and Furuta (1970) further studied the pungent principles, which consist of capsaicin, dihydrocapsaicin, nordihydrocapsaicin, homodihydro capsaicin and two or more analogs of these materials. Thin layer chromatography and open tubular gas chromatography showed that the natural pungent mixture contains no *cis* isomer of capsaicin. The chemical structure of nordihydrocapsaicin was determined as N-(4-hydroxy-3-methoxybenzyl)-7-methyloctanamide by gas chromatography, infrared spectrometry, mass spectrometry and NMR spectroscopy. Homodihydro-capsaicin was identified as N-(hydroxy-4,3-methyloxybenzyl)-9-methyl decamide (Kulka, 1967).

The pungency is caused by a group of vanillyl amides named capsaicinoids located in the placenta of the fruit. The heat of *Capsicum* powder is measured by Scoville heat units (Scoville, 1912). One part per million concentration of capsaicinoids is measured as 15 Scoville units. The nature of pungency has been established as a mixture of seven homologous branched-chain alkyl vanillyl amides, named capsaicinoids (Table 3.3).

The distribution of the pungent principles in the fruit is uneven and is the greatest in the placenta. According to Govindarajan (1986), the group paprika contains less than 0.1% of capsaicinoids, the best grade of Spanish paprika having 0–0.0003% and for the pungent

Table 3.3 Seven capsaicinoids identified in *Capsicum* species

Structural formula	Names of capsaicinoids
$(CH_3)_2 \cdot CH \cdot CH{=}CH \cdot (CH_2)_4 - CO{-}R^a$	Capsaicin
$(CH_3)_2 \cdot CH \cdot (CH_2)_6 - CO{-}R$	Dihydrocapsaicin
$(CH_3)_2 \cdot CH \cdot (CH_2)_9 - CO{-}R$	Nordihydrocapsaicin
$(CH_3)_2 \cdot CH \cdot (CH_2)_9 - CO{-}R$	Homodihydrocapsaicin
$(CH_3)_2 \cdot CH \cdot CH{=}CH \cdot (CH_2)_5 - CO{-}R$	Homocapsaicin
$CH_3 \cdot (CH_2)_7 - CO{-}R$	Nonanoic acid vanillylamide
$CH_3 \cdot (CH_2)_8 - CO{-}R$	Decanoic acid vanillylamide

Source: Anu and Peter (2000).

Note
where R is

$-NH-CH_2-$ ⟨benzene ring⟩ $-OH$, OCH_3

grade a maximum of 5%. Pungency level in chillies varies from 0.1% to 1.4% (Anu and Peter, 2000). The chemistry of pungent principles has been reviewed by Pruthi (1980, 1999), Govindarajan (1986) and more recently by Anu and Peter (2000) and the Basic Synthite Chemicals, Cochin (India).

Analytical techniques for the determination of major pungent principles (capsaicins in Capsicums/*chillies)*

A number of analytical techniques have been reported from time to time since 1912 for assaying the pungency or capsaicin (Joint Committee of the Pharmaceutical Society, and the Society for Analytical Chemistry, 1959, 1964). The chronological developments in analytical techniques are summarized in Table 3.4, along with their working principle and the authors' critical remarks. The 35 methods suggested have been categorized as organoleptic, photometric, spectrophotometric polarographic titration, paper chromatographic, thin layer chromatographic and GLC techniques (Table 3.5) Barnyai and Szabolis (1976).

The pungency or capsaicin content of *Capsicums* varies widely among varieties, seasons, places of origin, etc. The available data on the capsaicin content of paprikas, peppers and chillies grown in different parts of the world (Africa, Ethiopia, Hungary, India, Japan, Mexico, Turkey, Uganda, the United States and other countries) have been summarized in Table 3.6. Thus, the capsaicin content varies from 0.0% in sweet peppers to as high as 1.86% in Indian chillies which is the highest on record.

Andre and Mile (1975) have developed a new, simple and sensitive TLC method for the determination of capsaicin, even at levels as low as 0.1 μg of capsaicin per spot. This method can readily be used to distinguish the paprika varieties according to their degree of pungency. The International (ISO) sub-committee 34/SC7, which was chaired by Dr J. S. Pruthi at the British Standards Institute in London in October 1968, scanned through the merits and demerits of each method and unanimously recommended (1) Spectrophotometeric or spectrometric method as the ISO *Reference method*; (2) HPLC and TLC method and (3) by Scoville index techniques described in fair detail by Pruthi (1969), as the ISO working methods for the determination of capsaicin in *Capsicums*/chillies.

Major colouring pigments in Capsicums/*chillies*

The colour of spices is important from the point of view of quality as well as economic worth. The colouring pigments present in pepper/paprika have been briefly discussed here (Krishnamurthy and Natarajan, 1970).

Colouring pigments in red pepper

A number of carotenoids have been isolated and characterized from red pepper (*Capsicum annuum*) or the dehydrated product, paprika (Cooper *et al.*, 1962; Curl, 1962; Cholonoky *et al.*, 1963; De la Mar and Francis, 1969). The major red pigment is capsanthin, which was isolated in crystalline form as early as 1927 (Karrer and Jucker, 1950), but the current structure was assigned relatively recently (Barber *et al.*, 1960, 1961). Faigle and Karrer (1961) determined the asymmetry of the carbon-5 position in the five-membered ring of capsanthin and Cooper *et al.* (1962) established that the hydroxyl in the cyclopentane ring is trans to the polyene chain. Capsanthin occurs as the dilaurate in paprika (Phillip *et al.*, 1971). The oxidation degradation sequence for capsanthin in oxygen was worked out by Phillip and Francis (1971a).

Table 3.4 A critical appraisal of the available analytical techniques for the determination of capsaicin in chillies/*Capsicums*

Author(s)	Name of method	Authors critical remarks on each method
Scoville (1912)	Scoville organoleptic test – based on threshold sensory evaluation	Unsatisfactory. Results not reproducible.
Gibbs (1927)	Photometric method	Needs improvement.
Von Fodor (1930, 1931)	VOCl₃ method (vanadium oxy-trichloride)	Colour must be determined within 20–30 seconds; otherwise interference by other colour reactions take place.
Tice (1932)	VOCl₃ method	
Berry and Samways (1937)	Threshold sensory evaluation	Low sensitivity. Lack of reproducibility.
Hayden and Jordan (1941)	Threshold sensory evaluation	Results not reproducible.
Nogrady (1943)	Titration with picric acid and flourescence desorption	
Buchi and Hippenmeier (1948)	Phosphomolybdic acid method	Colorimetric method. Time-consuming.
North (1949)	Phosphostungstic phosphomolybdic acid–vanillin method (Folin-Denis reagent)	Colorimetric method.
Newman (1953)	On Scoville test	Defects of Scoville test pointed out.
Fujita *et al.* (1954)	Paper partition chromatography and colorimetric method	Simple technique.
Schenk (1954)	Ammonium vanadate and HCl method	Photometric method.
Schulte and Krueger (1955)	Diazobenzene sulfonic acid method	Colorimetric method. A satisfactory method.
Spanyar *et al.* (1956a)	Sulfanilic acid method	Authors surveyed the difficulties in earlier method. This method gives considerably accurate and reproducible results.
Spanyar *et al.* (1956b)	Polarimetric titration with *p*-diazobenzene sulfonic acid	Applicable to samples containing capsaicin more than 30 mg/100 g and moisture not above 20%. Mean error less than ± 10%.
Schenk (1957)	VOCl₃ method	Optical inspection and colour comparison. Suitable for rough assessments in pharmacy. A good review of earlier methods presented.
Spanyar *et al.* (1957)	Diazosulfanilic acid method	Applicable to samples containing capsaicin above 5 mg/100 g and moisture not above 20%. Accuracy ± 10%.
Schulte and Krueger (1957)	Photometric method	
Suzuki *et al.* (1957)	Ultraviolet spectrophotometric method	
Waldi (1958)	Paper chromatographic method	The fat extracted from paprika causes elongation of spots and tailing.

(Continued)

Table 3.4 Continued

Author(s)	Name of method	Authors critical remarks on each methods
Spanyar *et al.* (1958)	Sulfanilic acid method, modified	A quick method; rough results obtainable within 4 minutes.
Zitko (1957)	Sulfanilic acid method, modified	
Benedek (1959a)	Ammonium vanadate photometric method	Sensitivity; 0.01 mg of capsaicin.
Spanyar *et al.* (1958)	Photometric method	Satisfactory.
Joint Committee of the Pharmaceutical Society and the Society for Analytical Chemistry (1959)	Column chromatographic purification and spectrometry	A review and improvement of available methods discussed.
Csedo *et al.* (1960)	Review of analytical methods	Authors recommend Spanyar method.
Tiechert *et al.* (1961)	Thin-layer chromatography	Satisfactory.
Brauer and Schoon (1962)	Ultraviolet spectrophotometric method	Satisfactory but time-consuming, involving costly equipment.
Kosuge and Inagaki (1962)	Phosphomolybdate method and colorimetric method	
Joint Committee of the Pharmaceutical Society and the Society for Analytical Chemistry (1964)	Spectrophotometric and modified colorimetric methods	Spectrophotometric and colorimetric methods, recommended.
Heuser (1964)	Thin-layer chromatographic method	Applicable to samples containing at least 0.1 mg of capsaicin. Not sensitive for lower concentrations. Time-consuming.
Csedo and Kopp (1964)	Ammonium vanadate	
Friedrich and Rangoonwala (1965)	Thin-layer chromatography	
Karawya and Balbaa (1967)	Micromethod, diazobenzene sulfonic acid method	
American Spice Trade Association (1968)	Scoville test	Subjective test recommended.
Spanyar and Blazovich (1969)	Thin-layer chromatographic method using ferric chloride and potassium ferricyanide reagent	Reaction not very specific. Colouring pigments interfere. Evaluation to be completedwithin 1–2 minutes after spraying to avoid interference by other pigment compounds.
Blazovich and Spanyar (1969)	Thin-layer chromatographic method	Applicable to oleoresin and other preparations containing large amounts of capsaicin.
Hollo *et al.* (1969)	GLPC method	
Hartman (1970)	GLPC method	
Mathew *et al.* (1971)	TLC method	Satisfactory.
Govindarajan and Ananthakrishnan (1974)	Paper partition chromatography	Simple, inexpensive, adoptable.
Indian Standards Institution (1976)	Scoville test modified	Adoptable method.
Govindarajan *et al.* (1977)	Scoville heat units (organoleptic evaluation)	Modified Scoville test, useful method.

Source: All references are cited in Pruthi (1980, 1999).

Table 3.5 Systematic categorization of analytical techniques for the evaluation of pungency (capsaicin) in *Capsicums*

Techniques/basis	Authors
Threshold sensory (organoleptic) evaluation	Scoville (1912); Berry and Samways (1973); Hayden and Jordan (1941); Indian Standards Institution (1976); Govindarajan *et al.* (1977)
(1) Photometric methods	Gibbs (1927); Joint Committee of the Pharmaceutical Society and the Society for Analytical Chemistry (1959)
(2) VOCl₃ method (vanadium oxytrichloride)	Von Fodor (1930, 1931); Tice (1932); Schenk (1954, 1957); Benedek (1959a); Csedo and Kopp (1964)
(3) Phosphomolybdic acid method	Buchi and Hipenmeier (1948); Kosuge and Inagaki (1962)
(4) Follin-Denis reagent: phospho-tungstic phosphomolybdic–vanillin method	North (1949)
(5) Diazobenzenesulfonic acid method	Schulte and Krueger (1955); Spanyar *et al.* (1957); Karawya and Balbaa (1967)
(6) Sulfanilic acid method	Spanyar *et al.* (1956, 1958); Zitko (1957)
Ultraviolet spectrophotometric methods	Suzuki *et al.* (1957); Brauer and Schoon (1962); Joint Committee of the Pharmaceutical Society and the Society for Analytical Chemistry (1964)
Polarographic titration with *p*-diazobenzene sulfonic acid	Spanyar *et al.* (1956)
Paper chromatography and colorimetry	Fujita *et al.* (1954); Waldi (1958); Govindarajan and Ananthakrishnan (1974)
Thin-layer chromatography	Ticchert *et al.* (1961); Heusser (1964); Friedrich and Rangoonwala (1965); Blazovich and Spanyar (1969); Spanyar and Blazovich (1969); Mathew *et al.* (1971)
Titration with picric acid and fluorescence desorption	Nogrady (1943)
Gas–liquid chromatographic methods	Hollo *et al.* (1969); Hartman (1970)

Source: All references are cited in Pruthi books (1980, 1999).

Colouring parameters in paprika and paprika oleoresin

Column chromatography has been the traditional method for the separation of carotenoids (Davis, 1965). For example, with paprika carotenoids, Sea Sorb 43 is effective, but it does cause some isomerization and oxidation of pigments; hence, a more rapid method is desirable. Paper chromatography based on impregnated papers (Booth, 1962) and papers with suitable fillers (Jensen and Jensen, 1959; Jensen, 1960) have been suggested. Separations are described based on thin-layer systems on alumina (benzene), silica gel G (methylene chloride–ether, petroleum ether–benzene, undecane–methylene chloride and methylene chloride–ethyl acetate), calcium hydroxide (hydrocarbon mixture–methylene chloride), secondary magnesium phosphate (carbon tetrachloride, benzene and petroleum ether–ether), silica gel G mixed with rice starch (*n*-hexane–ether), calcium hydroxide (benzene, benzene–methanol and petroleum ether–benzene) (Demole, 1958, 1959; Stahl *et al.*, 1963; Bollinger, 1965). Stahl *et al.* (1963) reported that not all carotenoid mixtures could be separated with a single solvent and a single absorbent in a thin-layer system. Phillip and Francis (1971b) reported the development of a solvent system

Table 3.6 Variations in capsaicin content in important varieties of red pepper grown in different parts of the world

Variety or type of red pepper	Source (country)	Capsaicin (% w/w)	References
Paprikas	Different sources	0.00–0.10	Kisgyorgy *et al.* (1962), Natarajan *et al.* (1968b); CAL (1969)
Hungarian paprika (ground)	Hungary	0.022–0.03	Pruthi (1969a)
Mambasas var.	Africa	0.60–1.10	Louis Sair (personal communication, 1969)
Red pepper	United States	0.25–0.45	Louis Sair (personal communication, 1969)
Hontakas	Japan	0.40–0.50	Louis Sair (personal communication, 1969)
Santaka	Japan	0.55–0.65	Louis Sair (personal communication, 1969)
Ethiopian chillies	Ethopia	0.35	Louis Sair (personal communication, 1969)
Uganda var.	Uganda	0.85	Suzuki *et al.* (1957)
Turkish	Turkey	0.083	Suzuki *et al.* (1957)
Mexican pequinos	Mexico	0.26	Suzuki *et al.* (1957)
Santana	Japan	0.058	Suzuki *et al.* (1957)
Abyssinian chillies		0.075	Suzuki *et al.* (1957)
Indian chillies	India	1.05–1.59	Deb *et al.* (1963)
		0.60–1.86	Ramanujan and Tewari (1968)

Source: All references are given in Pruthi (1980).

for the thin-layer separation of paprika carotenoids as well as the physico-chemical properties of capsanthin and derivatives. Rosebrook (1968) and Rosebrook (1971) reported the results of systematic collaborative studies on a method for the detection of the extractable colour in paprika and paprika oleoresin. The colour in paprika is extracted with acetone, and absorbance is measured at 460 nm. The method incorporates a standard colour solution as a spectrophotometric check. The method has been recommended for adoption. In a series of papers, Moster and Prater (1952, 1957a,b,c) described the extractable colour, the colour scale, the colour of paprika oleoresin, and the structure of the colouring matter of paprika. Cholonoky *et al.* (1958) reported on the carotenoid pigments in yellow paprika, and Curl (1964) reported on carotenoids of green bell pepper. Thus, the subject of the assessment of colour of *Capsicums* and paparika has been investigated by a number of workers in different countries. Their findings have been condensed in Table 3.7.

Shuster and Lockhart (1954) have reported on the comparative composite colour grading of paprika by visual score, Lovibond units, C-units and absorbancy at 462 nm (Table 3.8).

Colouring pigment in cayenne pepper

Lease and Lease (1956) have reported on the different fractions of red pepper pigments of different varieties of cayenne pepper and paprika as a percentage of total colour of the hexane extract (Table 3.9). Pruthi (1969) reported on the loss in colour (capsanthin and capsorubin) in Hungarian paprika during storage. Kanner *et al.* (1976, 1977, 1978) studied the carotene-oxidizing factors in red pepper. The comparison between the Indian and Zimbabwean paprika varieties for colour production is set out in Table 3.10. All references cited by Pruthi, in his recent books on "Quality Assurance in Spices and Spice Products" (1999), Van Blaricum and Martin (1951) reported the retarding of the loss of red colour in cayene pepper.

Table 3.7 Analytical techniques for colour evaluation of *Capsicums* and factors affecting colour retention therein

Author(s)	Title/principle of method	Remarks
Moster and Prater (1952)	Measurement of extractable colour of *Capsicum*	For routine analysis.
Shuster and Lokhart (1954)	Objective method of colour evaluation, colour grading, absorption spectra	
Moster and Prater (1957a)	Extraction of colour	
Moster and Prater (1957b)	A linear colour scale	
Moster and Prater (1957c)	Measurement of the colour of oleoresins of paprika	
Garcia (1959)	Determination of capsanthin and capsorubin content	
Benedek (1959b)	Determination of colour	
Pohle and Gregory (1960)	Spectrophotometric method, calibration with β-carotene	
American Spice Trade Association (1968)	Photometric method (ASTA)	
Khristova (1961)	Determination of pigments	
Santa Maria and de-Ruiz-Assin (1961)	Methods of analysis of paprika, colour index (transmittance of 0.05 g/100 ml in CCl_4 at 460 nm. 15%)	
Videki and Videki (1961)	American Spice Trade Association method, Spectrophotometric method, compared with Benedek method	Lengthy and expensive although the inaccuracy is negligible. Benedek method preferred.
Serna (1962)	Photometric method measurement of transmission at 460 nm, true colour purity values	Results on 122 samples obtained with spectrophotometry and two colorimeters showed very close agreement with true colour purity.
Sancho and Navarro (1962a)	Variation in colour of factors on carotenoids	
Sancho and Navarro (1962b)	Ultraviolet absorption spectra of carotenoids,	Nine carotenoids obtained by chromatography.
Curl (1962)	Carotenoids of red bell peper (% of different pigments given)	
Sancho (1962)	Photometric method, colour measurement at 450 nm, comparison with standard solution of $K_2Cr_2O_7$ and $CoCl_2$	
Palotas and Koneesni (1964)	Application of ASTA method (No. 10) and comparison with Benedek method	Values of one method are convertible to those of the other.
Benedek (1968)	Improved method for the	
Rosebrook *et al.* (1968a)	determination of extractable colour	
Lantz (1946)	Effect of slicing or slitting on time of dehydration and colour	

(Continued)

Table 3.7 Continued

Author(s)	Title/principle of method	Remarks
Lease and Lease (1956a)	Effect of stage of ripening on the retention of colour	
Lease and Lease (1956b)	Effect of fat-soluble antioxidants	
Lease and Lease (1962)	Effect of drying conditions	
Daoud and Luh (1967)	Method and nature of packaging, laminate and aluminum film pouches	
Chen and Gujmanis (1968)	Autoxidation of extractable colour	
De La Mar and Francis (1969)	Carotenoid degradation in bleached paprika	
Phillip and Francis (1971a)	Nature of fatty acids and capsanthin esters in paprika	
Phillip and Francis (1971b)	Isolation and chemical properties of capsanthin and derivatives	
Rosebrook (1971)	Collaborative study of a method for extractable colour in paprika and parika oleoresin	

Source: All above references are cited in Pruthi (1980).

Table 3.8 Composite colour/grading data on domestic paprika samples[a]

Sample no.	Visual score[b]	Lovibond red[c]	C units[d]	Absorbancy at 462 mμ[e]	Order of grade[f]
1	24.7	17.7	44.6	1.024	1
2	23.9	16.8	41.0	0.950	2
3	21.7	15.6	36.5	0.870	3
4	19.3	15.0	34.0	0.830	4
5	16.1	14.6	32.6	0.752	5
6	12.2	13.7	29.5	0.676	6
7	11.8	13.3	27.7	0.620	7
8	8.2	11.9	23.8	0.540	8
9	6.0	10.8	20.7	0.485	9
10	3.9	9.6	17.4	0.415	10
11	2.2	1.2	14.0	0.342	11
12	1.0	7.0	11.0	0.277	12

Source: Reproduced from Pruthi (1980).

Notes
a From Shuster and Lockhart (1954).
b Average of 10 observers.
c Two-inch cell. Lovibond yellow units at 20–27; concentration 1 : 500; ethanolic extract.
d Two-inch cell. From Lovibond red units.
e Hardy recording spectrophotometer (cell path 1 cm; bandwidth 10 mμ; concentration 1 : 1000; acetone extract).
f An order of 1 signifies the reddest sample. The order was the same for the visual scores, the absorbances, and the C units.

Colouring pigments in paprika

The colour in paprika powder is the principal criterion for assessing its quality. The pigment content of paprika powder can range from 0.1 to 0.8%. The colour value of paprika is usually

Table 3.9 Fractions of pepper pigments of paprika (as a percentage of total colour of the hexane extract)[a]

Capsicum *and paprika* varieties	Free xanthophyll (%)	Ester xanthophyll (%)	Carotene, non-saponiftable (%)	β-Carotene (mg/g of dry pepper)
Louisiana sports I				
New	10.5	81.3	8.1	8.0
Aged (70.8%)[b]	4.9	61.3	10.8	2.9
Long Cayenne 69A				
New	8.8	78.6	12.7	5.0
Aged (44.2%)[b]	6.6	67.8	10.4	1.2
"Seedless" Cayenne				
New	3.9	87.8	8.3	9.2
Aged (35.9%)[b]	6.6	60.6	7.5	1.5
"Stock" Cayenne				
New	7.4	85.4	7.2	4.2
Aged (65%)[b]	4.9	66.0	8.6	1.8
Paprika Carolina				
New	18.5	74.1	7.4	5.7
Aged (35.2%)[b]	11.1	64.7	7.1	0.97
Paprika 1-M-1				
New	12.1	77.2	10.7	13.0
Aged (30.8%)[b]	7.9	51.5	7.2	0.97

Notes

a From Lease and Lease (1956); Pruthi (1999).

b Percentage of total colour left after $11\frac{1}{2}$ months of aging at 25°C.

Table 3.10 Comparison between the Indian and Zimbabwean paprika varieties for colour production

Component/character	Byadagi dabba	Tomato chilli	Zimbabwe type
Colour value (MSD-10)	100,000	115,000	145,000
Capsaicin (%)	1.8	2.0	0.25
Carotenoids (g/kg)	62	63	62
Red pigments (%)	40	45	42
Capsanthin (%)	51	52	50
Xanthophyll (g/kg)	61	62	61
β-carotene (%)	10	8	11
Absorbance ratio	0.965	0.985	0.970

Source: Basic Synthite Chemicals, Kolencherry, Kerala, India and Mathew *et al.* (2000).

expressed in terms of ASTA colour value (American Spice Trade Association). This is the extractable colour present in the paprika. Common paprika ASTA colours present in the industry are 85,100, 120 and 150 (Tainter and Grenis, 1993). The major colouring pigments in paprika are capsanthin and capsorubin, comprising 60% of the total carotenoids. Other pigments are Betacarotene, Zeaxanthin, Violaxanthin, Neoxanthin and Lutein (Anu and Peter, 2000).

Detection of adulteration in *Capsicums*/paprika

Introduction – a global overview

Konecsni (1956–57) studied the microscopic determination of added food-paprika in milled spice-paprika. Mitra *et al.* (1970) made a comparative study of the estimation of starch in 20 samples of chillies by acid hydrolysis and diastase methods separately. They suggest that the Prevention of Food Adulteration ACT (PFA) (Government of India, 1955) specification for the maximum limit of starch content (by the diastase method) in ground chilli may be safely fixed at 4%. Navarao *et al.* (1965) have reported the thin-layer chromatographic determination of synthetic fat-soluble azo dyes in foods, particularly paprika adulterants. Stelzer (1963) reported the identification of synthetic colouring in paprika by thin layer chromatography (TLC). Pure paprika gave very low Rf values. Sudan II coloured and paprika gave Rf values of 0.78.

Schwien and Miller (1967) have reported a method for the detection and identification of dehydrated red beets in *Capsicum*. Dried red beet pulp used as an adulterant in *Capsicums* may be detected by removing the *Capsicum* oil with petroleum ether and isolating the red beet particles from the decolourized spice. The beet particles (which are insoluble in petroleum ether) are identified and confirmed by microscopic examination, paper chromatography and spectrophoto-metric analysis. Todd (1958) described a method for the detection of foreign pungent compounds in oleoresin *Capsicum*, ground *Capsicum* and chilli.

Sacchetta (1960) and Mitra *et al.* (1961) reported the detection of added colours to chilli or *Capsicum* powder by paper chromatography. Datta and Susi (1961) reported the quantitative differentiation between the natural pungent principle (capsaicin) and vanillyl nonamamide in *Capsicum* by infrared spectrophotometric procedures. Sen *et al.* (1973) have reported a quick and reliable TLC method for the detection of a small amount of mineral oil (a prohibited adulterant because of its carcinogenic effect) in chillies. The method can detect mineral oil at levels below 0.1%.

The most likely adulterant or substitute of cayenne pepper is the exhausted *Capsicums* which can be detected by low non-volatile ether extract, which should be at least 15% (Cox and Pearson, 1962). Microscopic examination should reveal adulterants like dried tomato waste from factories (seed and red skins) (Pruthi, 1980).

Detection of oil-soluble coal-tar dyes in chillies (Welcher, 1970)

Reagents

Petroleum ether 40–60°C, Diethyl ether, Benzene, Hexane, Acetic acid.

Procedure

Five grams of the powdered chilli sample is taken in a 250 ml beaker and shaken with 50 ml of 1 : 1 petroleum ether and diethyl ether. The ether extract is filtered and concentrated to 1 ml. The concentrated extract is potted on the 0.2 mm thickness silica gel G plate with the help of glass capillaries. The plate is developed in benzene : hexane : acetic acid in the ratio (40 : 60 : 1) to a 15 cm height. The plate is removed, air-dried and kept in the oven at 100°C for 1 h. The spots of natural colour are changed to pale yellow or fade away while spots of oil-soluble coal-tar dyer remained as such.

Qualitative test for detection of oil-soluble colours

Reagents

(1) 90% Alcohol; (2) Petroleum ether; (3) 40% Stannous chloride.

Procedure

An alcoholic solution of the colour can be obtained by shaking the oil or fat with equal volume of 90% alcohol and washing the alcohol solution with petroleum ether to remove traces of fat. The alcoholic solution is used for further testing. Test two small portions of alcoholic solution by mixing one with equal volume of HCl and the other with equal volume of 40% stannous chloride solution. Common oil-soluble colours show a decrease of red colour or turn blue with acid and are decolourized by stannous chloride.

Detection of vegetable tannin colouring of chilli powder

Often chilli powders of poorer colour are polished with "Ratanjot" i.e. alkanet extract in an oily medium. The oil used is generally of "taramira" seeds. The oil is cheap and has its own pungency to enhance that of chilli powders. The ether extract is treated with 10% of NaOH solution and a violet colour confirms the presence of these tanning agents.

Detection of water-soluble colours in chillies

The residue left after the filtration of ether (for detection of oil-soluble dyes described above) is dried. Then about 100 ml of 80% alcohol is aded and kept overnight. The extract is then filtered and the residue washed with 80% alcohol two to three times so that all the colours are extracted. The filtrate is then concentrated until the alcohol is removed and then sufficient water is added. This process extracts both acidic and basic dyes.

Alternatively, acidulated or alkaline water can also be used for extracting basic or acidic dyes respectively, if no interference is expected in wool dyeing. The method of wool dyeing consists in absorbing the colour in an acid or base from aqueous solution on white wool strands in boiling condition and washing the wool and stripping the dyed wool fibre with an alkaline or acidic eluent. The extract is chromatographed on filter paper (Whatman No. 10) with any of the following solvent systems using control known dyes: (1) Iso-amyl alcohol–ethanol–water–NH_3 (4:4:2:1) or (2) n-butanol–acetic acid–water (2:1:1) or (3) 2% sodium citrate in 5% NH_3 solution.

All three solvents should be tried for unambiguous detection as some colours do not separate in one solvent or can have the same mobility. The current method for the detection of water-soluble colours is by TLC using n-A1203 as the adsorbent. Extract the sample (5–10 g) with 25 ml of water and a few drops of NH_3. Cork the flask and let it stand for 10 min. For extraction: shake, filter and concentrate the extract to 2–3 ml. Spot the sample along with the standard colours on n-alumina plate. Run the plate up to 10 cm in the developing chamber using methanol; NH_3 (9:1) as the developing solvent. Remove the plate and dry.

Detection of lead chromate in chillies, curry powder and turmeric by diphenyl carbazide test for lead chromate (IS: 2446 : 1980)

Lead chromate resembles turmeric powder in colour and solubility in water. These similarities are abused by unscrupulous traders. The hazardous lead chromate is at times employed to highlight the faded colour of old turmeric rhizomes and also to patch up weevil's bores or black mouldy patches on them. At the same time, the element chromium is a constituent of natural turmeric occurring to the extent of 3–5 ppm (Roberts, 1968). Earlier, it was thought that the absence of chromate also confirmed the absence of lead, an assumption subsequently proved false. Several samples of turmeric powder give a negative chromate (CrO_4) test but still have a lead content more than 60 ppm. Thus the chromate test must be negative in turmeric. The procedure is described next.

Reagents

Dilute sulphuric acid (1 : 7 v/v), diphenyl carbazide solution 0.2% w/v in ethyl alcohol (95%).

Procedure

Ash about 2 gm of the material. Dissolve the ash in 4–5 ml of diluted sulphuric acid in a test tube and add 1 ml of diphenylcarbazide solution. The presence of chromate is indicated by violet colour.

Boric acid test for detection of turmeric in chillies

Reagents

Hydrochloric acid (HCl), diluted ethanol, boric acid, ammonium hydroxide.

Procedure

Slightly acidulate the aqueous or dilute ethanolic solution with hydrochloric acid and divide it into two equal parts. To one portion, add solid boric acid. A brown red colouration forms in the presence of turmeric. Compare the colour with that of the paper treated; it should be a similar colour without the addition of boric acid. Dip in another portion of the solution with a strip of turmeric paper and dry on a watch glass over a water-bath. The same brown red colour soon appears. With the addition of a drop of ammonium hydroxide, the colour changes to blue black. This technique is conveniently used for milk system, syrups and the aqueous or alcoholic extract of spices.

Quality evaluation of oleoresins of *Capsicum*, paprika, red pepper, chillies

There is an apparent confusion in the nomenclature of oleoresins of *Capsicums*, chillies, paprika and red pepper, which has been sorted out by the Essential Oils Association of USA (1979) as follows:

Nomenclature of oleoresins of *Capsicums*, red pepper, paprika, chillies and bird chillies

Type of oleoresins of Capsicums	*Botanical name (species)*	*Colour value*	*Colour description*	*Pungency of bite intensity*
Oleoresin *Capsicum*	*Capsicum frutescens*	4,000 max.	Clear red, light amber	Very high bite/ pungency
Oleoresin red pepper	*Capsicum annuum*	240,000 max.	Very deep red colour	High bite/ pungency
Oleoresin paprika	*Capsicum annuum*	40,000– 100,000	Deep red colour	Nil or negligible
Oleoresin chilli	*Capsicum annuum*	4000–10,000	Deep red colour	Highly pungent
Bird chilli	*Capsicum annuum*	2000	Red colour	Highly pungent

Source: Essential Oils Association of USA; Eisvale (1981).

World trade paprika and paprika products

Szabo (1969) has described the manufacture and quality of paprika oleoresin. About 50,000 tonnes of paprika are required for the production of oleoresin and colouring powder. Out of this, a major portion (20,000 tonnes) is produced and utilized in Spain alone (Anu and Peter, 2000). Spain also produces about 500 tonnes of paprika oleoresin. The next major producers of paprika are (1) Hungary, (2) Morocco, (3) Yugoslavia, (4) Czechoslovakia, (5) Chile, (6) Turkey, (7) Portugal, (8) Bulgaria, (9) Romania, (10) Greece, (11) China, (12) Mexico, (13) Israel and (14) South Africa. Paprikas from various countries are known by the names of the respective countries of origin. They include all forms of zero pungent types. In Europe, only two forms are recognized, the long Hungarian type and the round Spanish type. According to the Basic Synthite Chemicals, the paprika exporting firm in Kerala, though 75,000 tonnes of paprika such as Byadagi chillies and 9,000 tonnes of Tomato chillies are produced, very little is available in India for the manufacture of paprika oleoresin. There is an urgent need for the development of paprika varieties from the above two types. It is estimated that 470 tonnes of paprika oleoresin were exported from India (valued at Rs. 55 crores) during 1997–98 (Anu and Peter, 2000; Mathew *et al.*, 2000).

Colour of oleoresin paprika

Definition

Oleoresin paprika is the product obtained by solvent extraction of the powdered dried ripe red pods of paprika *Capsicum annuum* L., with the subsequent removal of the solvent. Oleoresin paprika is evaluated strictly on a unit colour basis. The bulk of the oleoresin paprika has a colour value of 40,000–100,000.

Appearance and odour

A deep red, somewhat viscid liquid, with a characteristic odour of paprika.

Determination of colour value of oleoresin paprika

BY VISUAL METHOD

(1) Measure exactly 1.0 g of oleoresin into a 100 ml volumetric flask. Make up the volume with acetone, and mix and shake well.
(2) Prepare the appropriate dilution with acetone depending upon the approximate colour value of the oleoresin.
(3) *Chart:* Colour value vs dilutions

Colour value of oleoresin	Dilution (%)	Volume of solution used for dilution	Approximate volume of dilute solution to be used (in ml)
100,000	0.01	1 ml/100 ml acetone	8–10
50,000	0.02	1 ml/ 50 ml acetone	8–10
40,000	0.02	1 ml/ 50 ml acetone	10–13
30,000	0.02	1 ml/ 50 ml acetone	15–20
20,000	0.04	2 ml/ 50 ml acetone	10–13
10,000	0.10	5 ml/ 50 ml acetone	8–10

(4) Pipette approximate amount of dilute solution (depending upon its colour value (CV)) into a 100 ml Nessler tube. Bring up the volume to mark with acetone. Compare through the length of the tube with blank. Make small addition until sample and blank match.

(5) Calculation:

$$CV = 100 - (A \times B)/A \times B \times 100$$

Where A is the volume of dilute solution used (in ml) and B is the percentage of dilute solution.

(6) *Blank*: Into a 100 ml Nessler tube, pipette 10 ml of 0.1 N potassium dichromate solution (4.904 gm $K_2Cr_2O_7$ per litre) and 1.0 ml of 0.5 N cobaltous chloride solution (5.948 gm $CoCl_2 \cdot 6H_2O$ per 100 ml) and make up to 100 ml with distilled water.

BY INSTRUMENTAL METHOD

Spectrophotometric assay of oleoresin paprika (Videki and Videki, 1961; Pruthi, 1999):

(1) *Equipment*:

  (i) Spectrophotometer: Beckman Model DU or DK2.
  (ii) 1 cm silica or corex capped cell.

(2) *Procedure*: Weigh 1.0 g of oleoresin into a 100 ml volumetric flask, and make up the volume with acetone. Pipette 1.0 ml of this solution into a second 100 ml volumetric flask and make up the volume with acetone.

Using a tungsten lamp source and a P28 photomultiplier tube, zero the spectrophotometer using acetone as a blank. Measure the absorbance of the 0.01% solution oleoresin at 458 nm. Multiply this reading by 61,000 to obtain the Nesslerimeter colour value.

Solubility in different solvents

Alcohol: Partly soluble with oily separation.
Benzyl benzoate: Soluble in all proportions.
Fixed oils: Soluble in all proportions in most fixed oils.
Glycerine: Insoluble.
Mineral oil: Very slightly soluble.
Propylene glycol: Insoluble.

Determination of residual solvent in oleoresin

By distillation–gas chromatography method (EOA, 1979)

EQUIPMENT AND CHEMICALS

(1) Gas chromatography
(2) "Volatile oils apparatus" for essential oils heavier than water, made according to the ISU specification (J. Am. Pharma. Soc. XVII (4), 346). The neck on the distilling head should be insulated with asbestos (Pruthi, 1999).
(3) Pipettes, volumetric flasks, heaters, etc.
(4) Toluene of purity suitable for this analysis. Purity of the toluene for the purpose of this analysis may be determined by a gas chromatographic analysis using the columns and

conditions prescribed below and injecting the same quantity of toluene as will be injected in the analysis for solvents. If impurities interfering with this test are present in the toluene, they will appear as peaks prior to the toluene and can be removed by fractional distillation.

(5) Column packings: (a) 17% by weight of Ucon 75-H-90,000 on 35–80 mesh chromosorb-W or equivalent support. (b) 20% Ucon LB-135 on 35–80 mesh support. (c) 15% Ucon LV-1715 on 60–80 mesh support. (d) Porapak Q 50–60 mesh.

(6) Potassium carbonate (anhydrous), sodium sulfate anhydrous, a detergent free of volatile compounds, and an anti-foam free of volatile compounds. If volatile compounds are present in the detergent or anti-foam, they may be removed by length boiling of their aqueous solution.

(7) Benzene of sufficient purity for this analysis; purity is to be determined as in the case of toluene.

REFERENCE SOLUTION

(a) Prepare a solution of toluene containing 2,500 ppm of benzene. If the toluene available contains benzene as the only impurity, the benzene level in the toluene can be determined, and sufficient benzene can be added to bring the level to 2,500 ppm. A procedure for determining the benzene content of toluene is as follows.

Procedure: At constant carrier gas flow rate and column temperature, with other instrumental condition constant, determine the area of benzene in a suitable sample size of toluene. Add known amounts of benzene to this toluene, and determine areas for each known amount under the same conditions. Plot the results on a linear graph paper, with the known concentration of benzene as the ordinate and the area as the abscissa. The plot should be linear, and the negative intercept of the ordinate will give the concentration of the benzene in the original toluene.

(b) Prepare a solution containing 0.63% v/w acetone in water.

Procedure: Place 50.0 g of oleoresin, 1.0 ml of solution A, 10 g of sodium sulfate, 50 ml of water, and a small amount of detergent and anti-foam in a 250 ml flask. Attach the distilling head and receiver and collect approximately 15 ml of the distillate. Add 15 g of potassium carbonate to the distillate, cool while shaking, and allow the two phases to separate. All the solvents except methanol will be present in the toluene layer. Draw off the aqueous layer, and place it in a 50 ml flask with boiling chips. Add 1.0 ml of reference solution B. Attach a distilling head and distill off approximately 1.0 ml. This distillate will contain methanol if present in the oleoresin, and acetone as the internal standard.

ANALYSIS

Column: 1/4* o.d. 6–8" long.
Flow rate: 50–80 ml per min.
Temperature: 70–80°C.
Sample size: 15–20 μl for hotwire detector.

(1) Ucon 90,000: Use this column for the separation of acetone and methanol from their aqueous solution. Use it for the separation and analysis of hexane, acetone and trichloroethylene in the toluene layer from the distillate. Elution order on this column is acetone, methanol, water; hexane, acetone, isopropanol plus methylene chloride, benzene, trichloroethylene, ethylene dichloride plus toluene.

(2) Ucon LB-135: Use this column for the separation of ethylene chloride, isopropanol and ethylene dichloride. Elution order on this column is: hexane plus acetone, ethylene chloride, isopropanol, benzene, ethylene dichloride, trichloroethylene, toluene.

(3) Ucon LB-1715: This is the best general purpose column, except for the determination of methanol. Elution order for this column is hexane, acetone, benzene, ethylene dichloride, toluene.

COMPUTATIONS

(1) *Calibration of the instrument*: Determine the response of the detector for known ratios of solvents by injecting known mixtures of solvents and benzene in toluene. The levels of the solvents and benzene in toluene should be of the same magnitude as present in the analysis. Calculate the areas of the solvent in relation to benzene. Compute the calibration factor C of the detector as follows:

$$C \text{ (solvent)} = \frac{\text{wt\% solvent}}{\text{wt\% benzene}} \times \frac{\text{area of benzene}}{\text{area of solvent}}$$

The recovery of the various solvents from the oleoresins in relation to the recovery of benzene by the distillation procedure is:

hexane 52%; methylene chloride 87.5%;
acetone 85%; trichloroethylene 113%;
isopropanol 100%; ethylene dichloride 102%; methanol 87%.

(2) Calculation of residual solvent in oleoresin: The level of the internal standards, related to the 50 g of oleoresin used, is: Benzene 43.4 ppm and acetone 100 ppm. The level of residual solvent using benzene as an internal standard is computed as follows:
     Residual solvent ppm $= 43.4 \times C$ (solvent) $\times 100/(\%$ recovery of solvent $\times$ area of solvent/area of benzene. The level of residual methanol using acetone as an internal standard is:
Residual methanol ppm $= 100 \times C$ (methanol)$/0.87 \times$ area of methanol/area of acetone.

By gas chromatography head-space (alternate) method (EOA 1979)

(1) *Principle*: an oleoresin sample is diluted with diethyl phthalate and equilibrated at 60°C with its head-space in a flask sealed with a septum. A sample of head-space vapour is assayed by gas chromatography. The residual solvent content of the oleoresin can be obtained from that of the vapour by use of a calibration chart.
(2) *Apparatus*: (a) Bottles, glass, 2 oz bottle equipped with septum caps; (b) injection syringe, 2.5 ml, gas tight; (c) Water-bath, set the thermostat at 60°C; (d) Gas chromatograph (GC) with flame ionization.
(3) *Conditions*:

(a) Column: Stainless steel, 1 m $\times$ 4 mm (i.d.).
(b) Packing: Poropak Q (or other suitable column packing)
(c) Column temperature: 150°C, isothermal.
(d) Injection port temperature: 220°C.
(e) Detector: Flame ionization

Hydrogen flow rate: 50 ml/min.
Air-flow rate: 500 ml/min.

(f) Carrier gas: Nitrogen, flow rate : 40 ml/min.
(g) Chart speed: 30 cm/h.
(h) Sample size: 2.5 ml.

(4) *Reagents*:

  (a) Residual solvent standard (RS) essentially single peak by GC.
  (b) Diethyl phthalate (DEP) essentially single peak by GC.

(5) *Procedure:*

  (a) Preparation of solutions:

    (a-1) Sample solution: Weigh 10 g of the oleoresin to be tested into the 2 oz bottle and make up to 20 g with DEP. Seal with a septum cap.

    (a-2) Calibration solution: Weigh 1.0 g of RS standard in a 100 ml flask, and make up to 100 g with DEP to obtain a standard solution. Into a series of 2 oz bottles, weigh 1.0, 2.0, 3.0, 4.0 and 5.0 g of standard solution and make up to 20 g with DEP. Seal the bottles with septum caps. These calibration solutions will contain respectively, 10, 20, 30, 40 and 50 ppm of RS.

  (b) Calibration and determination:

    (b-1) Immerse the bottles up to the necks in a water-bath thermostated at 60°C. After 15 min remove the bottles, mix by whirling, and return to the water-bath for a further period of 2 h.

    (b-2) Using the 2.5 ml syringe, inject a sample of the vapour in each bottle onto the GC column, as follows: draw 2.5 ml of air into the syringe and insert the needle into the bottle. Discharge the syringe contents into the bottle and withdraw 2.5 ml of vapour from the bottle. Discharge the syringe contents back into the bottle, again withdraw, and repeat these operations once more. Leave the needle in the bottle for 30 s, withdraw the needle, and inject the syringe contents into the GC column.

  (c) Calculation:

    (c-1) Determine the areas of the RS peaks from the gas chromatograms of the calibration samples and plot in a graph the corresponding concentration of the RS. A linear calibration curve passing through the origin should be obtained.

    (c-2) Determine the area of the RS peak from the gas chromatogram of the oleoresin sample and read from the calibration curve the concentration of RS expressed in ppm, correcting, if necessary for attenuation.

Quality of oleoresin Capsicum

The oleoresin *Capsicum* is obtained by solvent extraction of powdered dried ripe fruit of *Capsicum frutescens* (bird chillies). Stahl (1965) has presented a critical review of the available methods of quality assessment of oleoresins.

Appearance and odour

A clear red, light amber or dark red, somewhat viscid liquid with a characteristic odour and very high bite.

Scoville units

The method is the same as given for chilli oleoresin (min Scoville units 240,000).

Odour value

Four thousand units maximum. The paprika method is the same as for paprika oleoresin.

Residual solvent

As described here:

(a) Alcohol: Partly soluble with oily separation and/or sediment.
(b) Benzyl benzaoate: Soluble in all proportions.
(c) Fixed oil: Soluble in all proportions in most fixed oils.
(d) Glycerine: Insoluble.
(e) Mineral oil: Insoluble.
(f) Propylene glycol: Insoluble.

Quality of oleoresin red pepper

The oleoresin red pepper is obtained from *Capsicum annuum* L. or its hybrid pepper, known as "*Louisiana sport*".

Appearance and odour

A deep red coloured liquid with a characteristic odour and high bite.

Scoville heat units

24,000 max.

Residual solvent:

As described here:
(a) Alcohol: Partly soluble with oil, separation or sediment.
(b) Benzyl benzoate: Soluble in all proportions.
(c) Fixed oils: Soluble in all proportions in most fixed oils.
(d) Glycerine: Insoluble.
(e) Propylene Glycol: Insoluble.

Quality of oleoresin chilli

Oleoresin chilli is extracted from chillies (the fruit of red pepper, *Capsicum annuum* L. or *Capsicum frutescens* L.) using approved food grade solvent and subsequent careful removal of the residual solvent by distillation. In addition to the intense pungency caused by the capsaicin and the small quantities of allied alkalloids, the oleoresin chilli has dark red carotenoid pigments. The oleoresin should be free from rancid flavour.

Determination of the colour value of oleoresin chilli

(a) *Principle*: To make the estimation completely objective, instrumental analysis is employed. The colour of a specified dilution is estimated at 458 nm and multiplied by the appropriate factor (ISI, 1979; ISO, 1989).
(b) *Apparatus*: (1) 100 ml volumetric flask; (2) 1 ml pipette; (3) Spectrophotometer or a Spectronic photo-electric colorimeter with adjustable wavelength.
(c) *Reagent*: Acetone.
(d) *Procedure*: Weigh 1.0 g of oleoresin into a 100 ml volumetric flask and make up the volume with acetone. Using a tungsten lamp source and acetone as the blank, measure the absorbance level of the 0.01% solution of oleoresin at 458 nm. Multiply this reading by 61,000 to obtain the colour value.

Determination of Scoville heat units in oleoresin chilli

(a) *Principle*: This is a method based on sensory evaluation. Although subjective, industry and trade still employ this method. Oleoresin solutions variably diluted with sugar solution are tested in increasing concentration. The highest dilution at which a pungency is first detected is taken as the measure of the heat value (ISO, 1977; ISI, 1976).

(b) *Reagents*: 5% sugar solution 95% alcohol.

(c) *Procedure*:

    (a) 0.2 g of the oleoresin is weighed into a 50 ml volumetric flask. 95% alcohol is added up to the volume, the material is shaken and allowed to settle.

    (b) To 140 ml of 5% sugar solution, add 0.15 ml of alcoholic solution. Swallow 5 ml of this solution after thoroughly shaking it; and

    (c) A bite or stinging sensation in the throat which is just perceptible at this dilution is equal to 240,000 Scoville units. If bite is very strong, a further dilution is necessary. The following dilutions are equal to the corresponding Scoville heat units when 5 ml of swallowed causes a perceptible pungent sensation:

        (i) 15 ml of sugar solution, 30 ml of solution (b) = 360,000

        (ii) 20 ml of sugar solution, 20 ml of solution (b) = 480,000

        (iii) 30 ml of sugar solution, 15 ml solution (b) = 720,000.

A panel of five members may be used. Three out of the five members should find the solution to be pungent.

Determination of capsaicin content

The pungency of chilli is caused by capsaicin and traces of allied pigments. Therefore, pungency can be more objectively evaluated by estimating the capsaicin content. For this purpose, two methods are currently used and are equally suitable (Govindarajan *et al.*, 1977; ISI, 1976; ISO, 1988).

TLC METHOD FOR CAPSAICIN DETERMINATION

(a) *Principle*: The capsaicin and allied pigments may be separated from other constituents by TLC on a silica gel. The pungent principle can react with Folin-Dennis reagent to give a blue complex which may be estimated colorimetrically. The HPLC and spectrometric methods have been described earlier for the determination of capsaicin content of chillies (whole and ground). The presence of much simpler TLC/PPC methods are briefly described below:

(b) *Apparatus*: (1) Complete TLC set; (2) Spectrophotometer or colorimeter capable of measuring in the region of 660–669 nm; (3) 10 ml measuring jar with stopper; (4) Test-tubes; (5) Graduated 10 ml pipettes (graduated at 0.1 ml).

(c) *Reagents*:

    (1) Silica gel (TLC grade);

    (2) Developing solvent – Mixture of 80 ml of benzene and 5 ml of methanol;

    (3) Phosphomolybdic–Phosphotungstic acid (Folin-Dennis Reagent)-Reflux for 2 h, 750 ml water, 100 g sodium fungstate, 20 g phosphomolybdic acid and 50 ml of phosphoric acid. Cool the mixture and dilute to 1 L with water.

    (4) Vanillin – Analytical grade.

    (5) Standard aqueous vanillin solution – The standard aqueous solution of vanillin should contain 0.01 mg/ml. This standard solution (1–3.5 ml) is made to react with 0.5 ml of

Folin-Dennis reagent and 1 ml of saturated aqueous solution of sodium carbonate, keeping the total volume to 5 ml by the addition of distilled water.

(d) *Procedure*:

(1) Preparation of TLC plates: Mix 30 g of silica gel containing calcium sulphate as a binder. Mix with 60 ml of distilled water. Pour the slurry into a TLC spreader adjusted to a thickness of 150 μm and spread over five to six glass plates of 20 × 20 cm. Air dry the plates for 4–5 h, and later activate them by drying them in a desiccator.

(2) Estimation: Weigh 5 g of oleoresin in to a 10 ml stoppered measuring jar and make up the volume with acetone. Spot 10 μl (5–20 μl depending upon the capsaicin content) of this solution on TLC plate and pour developing solvent into an all glass chamber which has been thoroughly saturated with the vapours of the solvent. Expose the plate for about half an hour to free it of the solvent and lightly spray with phosphomolybdic–phosphotungstic reagent. Mark the clear blue capsaicin spot (Rf 0.16) with a small stainless steel scooper or spatula in the form of a circle, enclosing both the spot and an area of 0.25 cm beyond the spot. Scoop out the silica gel containing capsaicin in the marked area into a clean butter paper and transfer it into a test-tube. Add 3.5 ml of distilled water and shake it well. Pipette out 0.5 ml of phosphomolybdic–phosphotungstic reagent into the tube and mix well. After 3 min add 1 ml of a saturated aqueous sodium carbonate solution. Mix thoroughly for 5 min and set in aside for 1 h. Prepare a reagent blank using silica gel layers from a blank area in the plate. Centrifuge the tubes (or filter) at the end of one hour to separate solids. Read optical density in a spectrophotometer at 725 nm (or in a colorimeter using a glass filter in 660–690 nm region). Determine the amount of capsaicin (X) in the spot from a standard graph of known concentrations of pure vanillin vs optical density. (Multiply by a factor of 2 to correct for the difference in the molecular mass of capsaicin and vanillin.)

(3) Preparation of standard graph: Different volumes of a standard aqueous solution of vanillin are made to react with Folin-Dennis reagent and the optical density is determined. Draw a standard graph with the mass of vanillin against optical density (Govindarajan and Ananthakrishna, 1970).

(4) *Calculation*:

Capsaicin (%) = X × 1000 × 200/r

where X is the number of grams of capsaicin (vanillin × 2) in the spot, and r is the number of μl of 50% acetone solution of oleoresin.

(B) PAPER CHROMATOGRAPHIC METHOD FOR CAPSAICIN DETERMINATION

(a) *Apparatus*:

(1) Chamber for ascending paper strip chromatography, complete with accessories.
(2) Spectrophotometer or colorimeter with narrow band filter.

(b) *Reagents*: (1) Ethyl acetate solvent of capsaicin or dilution of total extracts; (2) Methanol (AR); (3) Boric acid (AR); (4) Potassium chloride; (5) Buffer-A solution containing 3.1 g of boric acid and 3.7 g of potassium chloride in 100 ml of distilled water adjusted to pH 9.6 with 1 N sodium hydroxide; (6) *Methanol Buffer* pH 9.6 (60 : 40 v/v) and developing solvent; (7) Gibba reagent 0.1% 2,6-dicholro-*p*-benzoquinone-4-chloromine in acetone as the chromogenic reagent.

(c) *Procedure*:

    (1) Dissolve the oleoresin or extractives in ethyl acetate to give a solution containing about 2.5 g/L of capsaicin. Apply an amount of the solution containing 10–50 μg of capsaicin (that is 4–20 μl) as a thin streak covering the entire width of a 2 × 20 cm strip of Whatman no. 3 paper. Develop by ascending chromatography in methanol buffer solvent until the solvent front has moved up about 15 cm, usually in about 1 h. Dry the strip in air (Govindarajan and Ananthakrishna, 1974, ISI, 1979).

    (2) Pass the dried paper strip uniformly through Gibbs reagent and dry it in air. Capsaicin becomes faintly visible as a blue spot near the solvent front. Develop the colour further by spraying it with buffer (pH 9.6) lightly on both sides of the paper. Transfer the strip to a dark cupboard for 30 min for the full development of colour and drying.

    (3) Excise the blue spot and elute it with methanol in the dark. Make up the eluted colour to a convenient volume (5–10 ml) and measure at 615 nm in a spectrophotometer against a reagent blank eluted from the corresponding area of strip run simultaneously without the sample. Calculate the percentage of capsaicin from a standard graph as given in Section "TLC method for Capsaicum Determinants" or by using E 1% cm = 640 at 615 nm.

(d) Estimation of capsaicin content in *Capsicum*: Prepare the oleoresin or total extract of the sample by Soxhlet extraction of the powdered *Capsicum* with a suitable solvent (ethyl acetate) for 2.5–3 h. After removal of the solvent, weigh the residue and proceed as in "TLC method for Capsaicum Determinants".

(e) Standard curve and E 1%/1 cm for pure capsaicin: Prepare a solution of pure capsaicin in ethyl acetate (2 g/L). Spot 10, 20, 30, 40 and 50 μg of capsaicin (5–25 Ul) on strips of Whatman no. 3 paper. Develop and visualize the capsaicin spot elute it, and proceed further, as usual, for the estimation of capsaicin.

Development of International (ISO) Standards for spices including chillies and their methods of testing

Organizational set up and objectives of International Organization for Standardization

The International Organization for Standardization (ISO) is the specialized international agency for the development of international standards. The ISO has 191 technical committees covering different commodities traded on an international level. The membership of the ISO, at present, comprises 89 countries. The ISO aims to promote the development of standardization with a view to facilitate the international exchange of goods and services and to develop cooperation in the sphere of intellectual, scientific, technological and economic activities. The results of the ISO technical work are published as international standards (ISO standard).

One of the ISO technical committee, ISO/TC 34, deals with agricultural food products. This technical committee has 15 sub-committees dealing with various food commodities, such as fruits and vegetables, milk and milk products, meat and meat products, spices, etc. One of the sub-committees (ISO/TC 34/SC 7) deals with spices and condiments. The Indian Standards Institution (now called the Bureau of Indian Standards, BIS), New Delhi, holds the ISO/TC

Table 3.11 International (ISO) standards for *Capsicums*, chillies, paprika products and methods of testing

ISO standard no.	Title of ISO specification	Total pages
ISO standards for whole and ground Capsicums, etc:		
ISO 972: 1997	Chillies and *Capsicums*, whole or ground (powdered) – specification	
ISO 7540: 1984	Ground (powdered) paprika (*Capsicum annuum* L.) – Specification	4
ISO 3588: 1977	Determination of degree of fineness of grinding – hand seving method (reference method)	2
ISO 7542: 1984	Ground (powdered) paprika (*Capsicum annuum* L.) – Microscopical examination	8
ISO 3513: 1995	Chillies – determination of Scoville index	
ISO 7541: 1989	Ground (powdered) paprika – determination of total natural colouring matter content	3
ISO 7543-1: 1994	Chillies and chilli oleoresins – determination of total capsaicinoid content, Part 1: Spectrometric method	
ISO 7543: 1993	Chillies and chilli oleoresins – determination of total capsaicinoid content, Part 2: Method using High Performance Liquid Chromatography (HPLC method)	
International (ISO) methods of test spices and spice-products:		
ISO 2825: 1981	Preparation of a ground sample for analysis	
ISO 948: 1980	Method of sampling	
ISO 927: 1982	Determination of extraneous matter	
ISO 1298: 1982	Determination of moisture content – entrainment method	
ISO 941: 1980	Determination of cold water-soluble extract	
ISO 6571: 1984	Determination of volatile oil content	
ISO 928: 1997	Determination of total ash	
ISO 1108: 1992	Determination of non-volatile ether extract	
ISO 930: 1997	Determination of acid-insoluble ash	

Source: Compiled by Pruthi (1999).

34/SC 7 Secretariat, and the Director of Agriculture and Food Division of BIS is the permanent Secretary of SC 7, while the Chairman is elected at each sub-committee meeting which is generally held once in every two years by rotation in different member countries. Generally, the Indian representative is invited to be Chairman.

The membership of this Spices sub-committee (SC 7) has 18 participating members, 23 observer members and 6 international bodies having category "A" liaison and category "B" liaison. Under the rules of procedure of ISO, every participating member has an obligation to vote and attend meetings of ISO/TC 34/SC 7. The observer and liaison members receive all the technical documents for comments but have no voting rights, but they can attend meetings. A list of the ISO standards published on chilli (whole and ground) and chilli products is given below (Table 3.11).

These standards are subject to a periodical review (every five years) under the working procedure of the ISO (with a view to keeping the standards up-to-date).

National standards for *Capsicums* and chilli products

The following three bodies are responsible for formulating Indian National Standards:

(A) *The Agmark standards (Directorate of Agricultural Marketing and Inspection (DMI) Government of India under the aegis of the Agricultural Marketing Advisor to Government of India*: This is the oldest (1935) standard formulating body in the country under the APGM Agricultural Produce Grading and Marking Act, 1937. The DMI laid down numerous Agmark standards for over 20 spices including nine grade standards for the most important commercial varieties of dried ripe, whole red chillies. These standards encompass "stalkless chillies" (mainly for export to USA), and "clipped chillies" (without calyces) produced in India, and "red chilli powder". The eight standards for whole chillies cover important grade designations and physical quality parameters, while the chilli powder standard covers specifications for chemical factors like moisture, total ash, ash-insoluble in HCl, crude fibre and non-volatile ether extract. This standard also covers physical quality factors like freedom from extraneous matter, damage, freedom from insect and fungal contamination or mould growth, added colouring and flavouring matter, oil and preservatives, etc. The nine Agmark standards are described in schedules I–V, V A, B and C and on the section covering powder.

(B) *The Indian (IS) Standards*: Now called the Bureau of Indian Standards (BIS), it came into being in 1944–45. Since then, it has laid down standards as well as Methods of Testing for important spices, including the three standards for chilli (whole and ground) and one for chilli oleoresin. These are listed below, along with their code numbers:

Standards published	Title of standard
Chillies (whole and ground)	
(1) IS: 2322-1984	Chillies, whole (first revision)
(2) IS: 2445-1984	Chillies, powder (first revision)
(3) IS: 2322-1998	Chillies (whole and ground) combined
Chilli oleoresin	
(4) IS: 13663-1993	Oleoresin, chillies specification

Source: Bureau of Indian Standards, Director, Department of Food and Agriculture, BIS, New Delhi.

(C) *PFA standards/specifications*: The Directorate General of Health Services (Ministry of Health, Government of India), has under the Prevention of Food Adulteration Act (PFA), 1955, laid down the following two PFA standards/specifications for chillies (whole) and chilli (ground or chilli powder).

PFA specification of chillies (whole and powder)

(1) A. 05.05 – chillies (*Lal Mirchi*) whole: The dried ripe fruits or pods of *C. annuum/C. frutescens* L. The proportion of extraneous matter including calyx pieces, loose tops, dirt, lumps of earth, stones shall not exceed 5.0% by weight. The pods shall be free from extraneous colouring matter, coating of mineral oil and other harmful substances. The amount of insect-damaged matter shall not exceed 5.0% by weight.

(2) A. 05.05.01 – chilli (*Lal Mirchi*) powder: The powder is obtained by grinding clean dried chilli pods of *C. frutescens* L./*C. annuum*. The chilli powder shall be dry, free from dirt, mould growth, insect infestation, extraneous matter, added colouring matter (and flavouring matter). The chilli powder may contain any edible oil to a maximum limit of 2% by weight under a label

Agmark grade specifications for chillies (whole)
Schedule 1: Sannam chillies

Grade designations and definitions of quality of chillies commercially known as Sannam variety and harvested in January–August – special characteristics

Grade desig-nations	Trade name	Length (in cm)	Colour	Maximum limits of tolerances						General characteristics
				Damaged and discoloured pods (%)	Pods without stalk (%)	Moisture (%)	Loose seed (%)	Foreign matter (%)	Broken chillies (%)	
SS	Sannam special	5 and above	Light red shining	2.00	2.00	11.50	1.00	1.00	5.00	Chillies shall (a) be the dried ripe fruits belonging to the species C. annum;
SG	Sannam general	Below 5 and above 3	Light red shining	4.00	3.00	11.50	2.00	2.00	7.00	(b) have shape, pungency and seed contents normal to the variety;
SF	Sannam fair	- do -	Dull red	6.00	4.00	11.50	2.00	2.00	7.00	(c) be free from mould or insect damage and be in sound condition and fit for human consumption; (d) be current year's crop and should be free from extraneous colouring matter, oil and other harmful substances.

Notes

Basis of quantitative determination: All determinations and percentages shall be reckoned on the basis of the total weight of representative samples.

Length: The tolerance specified in column 3 shall be based upon the average length of 20 fruits selected at random. The measurement will be taken from the tip of the fruit to the pedicel point (where the stalk is attached).

Discoloured pods: Pods having brown, black, white and other coloured patches will be considered as discoloured pods.

Foreign matter: All extraneous matter including calyx pieces and loose stalks will be treated as foreign matter. For accidental errors a tolerance is permissible up to 5.0% in excess of the tolerance specified under column 3 in respect of S.S. and S.G. and S.F. grades. For accidental errors a tolerance of 0.5% under column 5 and 1.0% under column 6 is permissible for all three grades.

Moisture: For accidental errors, a tolerance of 0.5% for moisture content will be allowed over and above 11.50% only.

Schedule II: Sannam chillies

Grade designations and definitions of quality of chillies commercially known as Sannam variety and harvested in September–December – special characteristics

Grade designations	Trade name	Length (in cm)	Colour	Maximum limits of tolerance						General characteristics
				Damaged and discoloured pods (%)	Pods without stalk (%)	Moisture (%)	Loose seed (%)	Foreign matter (%)	Broken chillies (%)	
SS	Sannam special	5 and above	Light red shining to light red dull	5	10.00	11.50	3.00	1.00	0.50	Chillies shall (a) be the dried ripe fruits belonging to the species *Capsicum annuum*;
SG	Sannam general	Below 5 and above 3	- do -		20.00	11.50	3.00	1.00	2.00	(b) have shape, pungency and seed contents normal to the variety; (c) be free from mould or insect damage and be in sound condition and fit for human consumption; (d) be current year's crop and shall be free from extraneous colouring matter, oil and other harmful substances.

Notes

Basis of quantitative determination: All determinations and percentages shall be reckoned on the basis of the total weight of representative samples.

Length: The tolerance specified in column 3 shall be based upon the average length of 20 fruits selected at random. The measurement will be taken from the tip of the fruit to the pedicel point (where the stalk is attached).

Discoloured pods: Pods having brown, black, white and other coloured patches will be considered as discoloured pods.

Foreign matter: All extraneous matter including calyx pieces and loose stalks will be treated as foreign matter. For accidental errors a tolerance of 0.5% under column 5 is permissible for both the grades.

Moisture: For accidental errors, a tolerance of 0.5% for moisture content will be allowed over and above 11.50% only.

Schedule III: Mundu chillies

Grade designations and definitions of quality of chillies commercially known as Mundu variety and harvested in September–December – special characteristics

Grade designations	Trade name	Length (in cm)	Colour	Maximum limits of tolerance						General characteristics
				Damaged and discoloured pods (%)	Pods without stalk (%)	Moisture (%)	Loose seed (%)	Foreign matter (%)	Broken chillies (%)	
MS	Mundu special	Not exceeding 2.5 cm	Deep red shining	2.0	5.0	11.50	1.00	1.00	1.00	Chillies shall (a) be the dried ripe fruits belonging to the species *C. annuum*; (b) have shape, pungency and seed contents normal to the variety; (c) be free from mould or insect damage and be in sound condition and fit for human consumption; (d) be current year's crop and shall be free from extraneous colouring matter, oil and other harmful substances.
MG	Mundu	- do -	- do -	4.0	10.0	11.50	1.00	1.00	1.00	

Notes

Basis of quantitative determination: All determinations and percentages shall be reckoned on the basis of the total weight of representative samples.
Length: The tolerance specified in column 3 shall be based upon the average length of 20 fruits selected at random. The measurement will be taken from the tip of the fruit to the pedicel point (where the stalk is attached).
Discoloured pods: Pods having brown, black, white and other coloured patches will be considered as discoloured pods.
Foreign matter: All extraneous matter including calyx pieces and loose stalks will be treated as foreign matter. For accidental errors a tolerance of 0.5% under column 5 is permissible for both the grades.
Moisture: For accidental errors, a tolerance of 0.5% for moisture content will be allowed over and above 11.50% only.

Schedule IV: Rari Patarki or Patli chilli

Grade designations and definitions of quality of chillies commercially known as Rari (Patarki or Patli) variety – special characteristics

Grade desig-nations	Trade name	Length (in cm)	Colour	Maximum limits, percentage by weight						General characteristics
				Damaged and discoloured pods (%)	Pods without stalk (%)	Moisture (%)	Loose seed (%)	Foreign matter (%)	Broken chillies (%)	
RS	Rari special	8 and above	Bright red	1.0	2.00	11.50	1.0	1.0	5.0	Chillies shall (a) be the dried ripe fruits belonging to the species *Capsicum annuum*; (b) have shape, pungency and seed contents normal to the variety; (c) be free from mould or insect damage and be in sound condition and fit for human consumption; (d) be current year's crop and shall be free from extraneous colouring matter and other harmful substances.
RG	Rari general	Below 8 and above 6	- do -	2.0	2.0	11.50	1.0	1.0	5.0	

Notes

Basis of quantitative determination: All determinations and percentages shall be reckoned on the basis of the total weight of representative samples.

Length: The tolerance specified in column 3 shall be based upon the average length of 20 fruits selected at random. The measurement will be taken from the tip of the fruit to the pedicel point (where the stalk is attached).

Discoloured pods: Pods, having brown, black, white and other coloured patches will be considered as discoloured.

Foreign matter: All extraneous matter including calyx pieces and loose stalks will be treated as foreign matter. For accidental errors a tolerance is permissible up to 5% in excess of the tolerance specified in column 3 under R.S. and R.G. grades. For accidental errors a tolerance of 0.5% under column 5 and 1.0% under column 6 is permissible for both the grades.

Moisture: For accidental errors, a tolerance of 0.5% for moisture content will be allowed over and above 11.50% only.

Schedule V: Gospurea chillies

Grade designations and definitions of quality of chillies commercially known as Gospurea variety – special characteristics

| Grade desig-nations | Trade name | Length (in cm) | Colour | Maximum limits, percentage by weight | | | | | | General characteristics |
				Damaged and discoloured pods (%)	Pods without stalk (%)	Moisture (%)	Loose seed (%)	Foreign matter (%)	Broken chillies (%)	
GS	Gospurea special	5 and above 5	Bright red	2.0	2.0	11.50	1.0	1.0	5.0	Chillies shall (a) be the dried ripe fruits belonging to the species C. *annuum*; (b) have shape, pungency and seed contents normal to the variety; (c) be free from mould or insect damage and be in sound condition and fit for human consumption; (d) be current year's crop and shall be free from extraneous colouring matter and other harmful substances.
GG	Gospurea general	Below 5 and above 3	- do -	3.0	2.0	11.50	1.0	1.0	5.0	

Notes

Basis of quantitative determination: All determinations and percentages shall be reckoned on the basis of the total weight of representative samples.

Length: The tolerance specified in column 3 shall be based upon the average length of 20 fruits selected at random. The measurement will be taken from the tip of the fruit to the pedicel point (where the stalk is attached).

Discoloured pods: Pods, having brown, black, white and other coloured patches will be considered as discoloured.

Foreign matter: All extraneous matter including calyx pieces and loose stalks will be treated as foreign matter. For accidental errors, a tolerance of 0.5% under column 5 and 1.0% under column 6 is permissible for both the grades.

Moisture: For accidental errors, a tolerance of 0.5% for moisture content will be allowed over and above 11.50% only.

Schedule VI: other chillies

Grade designations and definitions of quality of chilli not covered by Schedules I, II, III, IV and V of these rules and produced in India – special characteristics

Grade designations	Colour	Maximum limits of percentage of weight						General characteristics
		Damaged and discoloured red pods (%)	Pods without stalk (%)	Moisture (%)	Loose seed (%)	Foreign matter (%)	Broken chillies (%)	
Special	Characteristic of the variety	2.0	2.0	11.5	1.0	1.0	3.0	Chillies shall (a) be the dried ripe fruits belonging to the species *C. annuum*; (b) have the characteristic shape, colour, length, pungency and seed contents normal to the variety;* (c) be free from mould or insects and be in sound condition and fit for human consumption; (d) be of current year's crop and free from extraneous colouring matter, oil and any other harmful substance.
General	- do -	2.0	3.0	11.5	2.0	1.5	5.0	
Standard	Characteristic of the variety. Dull shade up to 50% of the pods permissible	6.0	4.0	12.5	3.0	2.0	7.0	

Notes

Discoloured pods: Pods having brown, black, white and other coloured patches.

Foreign matter: All extraneous matter including calyx pieces and loose stalks will be treated as foreign matter. A tolerance is permissible up to 0.5% in excess of the tolerance specified under column 7 in respect of Special and General grades. For accidental errors a tolerance of 0.5% under column 3 and 1.0% under column 4 is permissible for the grades Special and General.

Moisture: A tolerance of 0.5% for moisture content will be allowed in Special and General grades only.

* Variety: The name of the variety shall be separately stamped on the grade designation label.

Schedule VII: Stalkless chillies

Grade designations and definitions of quality of chillies (stalkless) produced in India – special characteristics

Grade designation	Trade name	Pods with stalk (%)	Pods with calyx (%)	Maximum limits of tolerance			General characteristics
				Moisture (%)	Loose seed (%)	Foreign matter (%)	
Chillies general (stalkless)	Stalkless chillies	1.00 by count	5.00 by count	11.00	5.00	0.5	Chillies shall (a) be the dried ripe fruits, belonging to the species *C. annuum*; (b) be free from visible mould or insects and be in sound condition and fit for human consumption; (c) be of one year's crop and free from extraneous colouring matter, and any other harmful substance.

Notes

Foreign matter: All extraneous matter including calyx pieces and loose stalks will be treated as foreign matter. A tolerance is permissible up to 0.25% in excess of tolerance specified under columns 3 and 7 and 0.5% in column 4.

Moisture: A tolerance of 0.5% for moisture content will be allowed over and above the stipulated limit.

Stalkless chillies: Chillies from which the calyx together with the stalk is removed.

Schedule VIII: Clipped chillies

Grade designations and definitions of quality of chillies (clipped) produced in India – special characteristics

Grade designation	Trade name	Pods with calyx (%)	Maximum limits of tolerance			General characteristics
			Moisture (%)	Loose seed (%)	Foreign matter (%)	
Chillies general (clipped)	Clipped chillies	1.00 by count	11.00	3.00	0.5	Chillies shall (a) be the dried ripe fruits, belonging to the species *Capsicum annuum*; (b) be free from visible mould or insects and be in sound condition and fit for human consumption; (c) be of one year's crop and free from extraneous colouring matter, oils and any other harmful substance.

Notes

Foreign matter: All extraneous matter including calyx pieces and loose stalks will be treated as foreign matter. A tolerance is permissible up to 0.25% in excess of tolerance specified under columns 3 and 6.

Moisture: A tolerance of 0.5% for moisture content will be allowed over and above the stipulated limit.

Clipped chillies: Chillies having stalks clipped from the very base without having calyx.

Agmark grade specifications for chilli powder

Grade designation and definition of quality of powdered chillies

Grade designation	Total ash, percentage (by weight) maximum	Special characteristics (physico-chemical)				General characteristics
		Ash insoluble in HCl, percentage (by weight) maximum	Crude fibre, percentage (by weight) maximum	Non-volatile ether extract, percentage (by weight) maximum	Moisture, percentage (by weight) maximum	
Standard	8.00	1.25	30.00	12.00	10.00	Powdered chillies shall be the product obtained by grinding pure, clean, dried, ripe fruits of the genus *Capsicum* only and shall be free from extraneous matter. It shall also be free from damage by insect infestation and/or fungus contamination, mould growth, added colouring matter and preservatives and other foreign substances or substitutes and from any extraneous or undesirable odour or flavour.

declaring the amount and the nature of the oil used. The chilli powder shall conform to the following physico-chemical standards:

Moisture:	Not more than 12.0% by weight.
Total ash:	Not more than 8.0% by weight.
Ash insoluble in dilute HCl:	Not more than 1.3% by weight.
Non-volatile ether extract:	Not less than 12.0% by weight.
Crude fibre:	Not more than 30.0% by weight.

Indian standard for chilli, whole and ground (powdered) specification (IS 2322: 1998 – 2nd revision)

Description of chillies (whole and ground)

(1) The chilli shall be dried, ripe fruits or pods of the species *C. frutescens* L. and *C. annuum* L. The chilli may be with or without stalk.

(2) The ground chilli is the product obtained by grinding the whole chilli without stalk and without any added matter.

Grades

(1) Whole chilli may be graded on the basis of length, colour and characteristics like moisture, extraneous matter, damaged and discoloured pods, pods without stalks, loose seeds and broken chillies.

(2)   Grade designation, trade name and their requirements are given in Table 3.12.

Requirements

ODOUR AND FLAVOUR

Chilli shall have a characteristic pungent taste. It shall be free from foreign taste and flavour including rancidity and mustiness.

FREEDOM FROM MOULDS, INSECTS, ETC.

Chilli, whole or ground (powdered), shall be free from living insects and shall be practically free from mould growth, dead insects, insect fragments and rodent contamination, which may be visible to the naked eye (corrected, if necessary, for abnormal vision), or by using the required magnifying instrument. If the manification exceeds ×10, this fact shall be mentioned in the test report.

COLOUR

The whole chilli and powder may vary in colour from dark blackish red to orangish yellow according to the variety. The colour of the graded chilli shall be as given in Table 3.12 for particular grade.

LENGTH

The length of the ungraded whole chilli shall be agreed between the purchaser and the supplier. The length of the graded chilli shall be as given in Table 3.12 for the particular grade. The measurement shall be taken from the tip of the fruits to the pedicel point (where the stalk is attached) and the average length of 20 fruits selected at random shall be reported.

EXTRANEOUS MATTER

Extraneous matter includes all vegetable matter other than the whole chilli; but including calyx pieces and loose stalks, and mineral matter such as sand, soil and earth. The proportion of extraneous matter in whole ungraded chilli when tested by the method given in IS 2322 shall not be more than 4% (m/m). In case of graded chilli the extraneous matter shall be as given in Table 3.12 for the particular grade.

DAMAGED AND DISCOLOURED PODS

These include pods damaged due to insect attack or diseases and which have developed brown, black, white and other coloured patches. The proportion of such damaged and discoloured pods in "ungraded whole chilli" shall be not more than 5% by mass when determined by physical separation. In the case of "graded whole chilli", the damaged and discoloured pods shall be as given in Table 3.12 for the particular grade of chillies.

PODS WITHOUT STALKS

In the case of "ungraded whole chilli", "the pods without stalks" shall not be more than 10% by mass when determined by physical separation. In case of "graded whole chilli", "the pods without stalks" shall be as given in Table 3.12 for the particular grade of chilli.

Table 3.12 Grade designations of whole chilli and their quality requirements

Grade	Trade	Length	Colour	Maximum limits, percent (m/m)						
				Damaged and dis- coloured pods	Pods without stalk	Moisture	Loose seed	Extraneous matter	Broken chilli	Pods with calyx
(1)	(2)	(3)	(4)	(5)	(6)	(7)	(8)	(9)	(10)	(11)
SS1	Sannam special	5 and above	Light red shining	2.0	2.0	11.5	1.0	1.0	5.0	—
SG1	Sannam general	Below 5 and above 3	Light red shining	4.0	3.0	11.5	2.0	2.0	7.0	—
SP1	Sannam fair	- do -	Dull red	6.0	4.0	11.5	2.0	2.0	7.0	—
SS2	Sannam special	5 and above	Light red shining to light red dull	5.0	10.0	11.5	3.0	1.0	0.5	—
SG2	Sannam general	Below 5 and above 3	- do -	5.0	20.0	11.5	3.0	1.0	2.0	—
MS	Mundu special	Not exceeding 25	Deep red shining	2.0	5.0	11.5	1.0	1.0	1.0	—
MG	Mundu general	- do -	- do -	4.0	10.0	11.5	1.0	1.0	1.0	—
RS	Rari special	8 and above	Bright red	1.0	2.0	11.5	1.0	1.0	5.0	—

RG	Rari general	Below 8 and above 6	- do -	2.0	2.0	11.5	1.0	1.0	5.0	—
GS	Gospurea special	5 and above	- do -	2.0	2.0	11.5	1.0	1.0	5.0	—
GG	Gospurea general	Below 5 and above 3	- do -	3.0	2.0	11.5	1.0	1.0	5.0	—
USS	Special	—	Characteristics of the variety	2.0	2.0	11.5	2	1.0	4.0	—
USG	General	—	- do -	4.0	3.0	11.5	3	2.0	6.0	—
USF	Fair	—	- do -	6.0	4.0	11.5	3	2.1	8.0	—
SL	Stalkless chillies	—	—	*	*	11.0	5.0	0.5	—	5.0
CC	Clipped	—	—	*	*	11.0	3.0	0.5	—	—

Source: Indian standard: IS 2322 : 1998 (2nd revision) – Courtesy BIS.

Notes

1 For accidental errors in the case of extraneous matter, a tolerance is permissible up to 5.0% in excess of the tolerance specified in respect of the Grades SS1, SG1, SF1, SS2, MS, MG, RS, RG, GS and GG.

2 A tolerance of 0.5% in respect of grade USS and USG in excess of the tolerance specified for extraneous matter is permissible. A tolerance of 0.25% in excess of the tolerance specified in respect of pods with stalk and extraneous matter and 0.5% for pods with calyx for grade SL is permissible.

3 For accidental errors in the case of damaged and discoloured pods, a tolerance of 0.5% and in the cased of pods without stalks, a tolerance of 1.0% is permissible in respect of the Grades SS1, SG1, SF1, SS2, SG2, MS, MG, RS, RG, GS, GG, USS, USG, USF.

4 For accidental errors, except for the grade USF, a tolerance of 0.5% over and above the tolerance specified for moisture is permissible.

* Pods with stalk – 1.0% by count.

LOOSE SEEDS

The proportion of loose seeds in ungraded whole chilli shall not be more than 3% by mass when determined by separating the loose seeds and following the method given in Para 4 of IS 1797 (Indian Standard). In the case of graded chillies, the proportion of loose seeds shall be as given in Table 3.12 for the particular grade of chilli.

BROKEN PODS

The proportion of broken pods in "ungraded whole chilli" shall be not more than 8% by mass when determined by separating the broken pods and following the method given in Para 4 of IS, 1797. In the case of "graded whole chillies", the proportion of broken pods shall be as given in Table 3.12 for the particular grade.

MOISTURE

The moisture content in "ungraded whole chilli" and "ground (powdered) chilli" shall not be more than 12% and 10% by mass, respectively, when determined in accordance with the method given in Section 9 of IS 1797. In case of graded whole chillies, the moisture content shall be as given in Table 3.13 for the particular grade.

FINENESS

The chilli (powdered) shall be ground to such a fineness that all of it passes through a 500 μm IS sieve and nothing remains on the sieve.

CHEMICAL REQUIREMENTS

The chillies ground (powdered) shall also comply with the requirements given in Table 3.13.

SCOVILLE INDEX

When tested in accordance with the method given in IS 8104, the Scoville index in chilli, whole and ground (powdered) chilli shall not be less than 24,000. Chilli of lower Scoville index than that specified may be supplied as agreed between the purchaser and the supplier.

Table 3.13 Physico-chemical requirements for chillies, ground (powdered)

Characteristics	Requirements	Methods of test (IS, 2322: 1998)
Total ash, percent by mass, (maximum)	8.0	6
Acid insoluble ash, percent by mass, (maximum)	1.3	8
Crude fibre, percent by mass, (maximum)	30.0	13
Non-volatile ether extract, percent by mass, (minimum)	12.0	14

Source: IS-2322:1998 (BIS) (2nd revision).

Note
The characteristics shall be tested on dry basis.

FREEDOM FROM ADDED COLOUR AND THE LIKE

The pods shall be free from extraneous colouring matter, coating of mineral oil and other harmful substances. The ground chillies shall be free from extraneous colouring matter and flavouring matter. However, it may contain any edible oil upto a maximum limit of 2% by mass. The amount and name of the oil used, however, shall be declared on the label.

HYGIENIC CONDITIONS

The chilli, whole and ground shall be processed and packed under hygienic conditions (see IS 2322: 1998).

Sampling

Representative samples shall be drawn and conformity of the product to the requirements of this standard shall be determined in accordance with the methods given in IS 13145.

Packing and marking

PACKING

Chillies, whole or ground, shall be packed in any clean, sound and dry container made of metal, glass, food-grade polymers, wood or jute bags. The wooden boxes or jute bags shall be suitably lined with moisture-proof lining which does not impart any foreign smell to the product. The packing material shall be free from any fungal or insect infestation and should not impart any foreign smell. Each container shall be securely closed and sealed. Virakamath (1964) studied the packaging of chillies and other spices.

MARKING

The following particulars shall be marked or labelled on each container/bag:

(a)   Name and address of the manufacturer or packer;
(b)   Name of the material (whole or ground);
(c)   Trade name or brand name, if any;
(d)   Grade designation (in case of whole)
(e)   Batch or code number;
(f)   Net mass;
(g)   Best before ... (month/year);
(h)   Year of the harvest (in case of whole);
(i)   Month and year of packing (in case of ground); and
(j)   Any other marking as required under the "Standards of Weights and Measures (Packaged Commodities) Rules", 1977, and the Prevention of Food Adulteration Act, 1954, and the rules framed thereunder.

Indian standard for chilli oleoresin specification

Requirements

DESCRIPTION

The chilli oleoresin shall be obtained by solvent extraction of the dried ripe fruits of *C. annuum* L. or *C. frutescens* L., with the subsequent removal of the residual solvent.The material shall be a red viscous liquid with the characteristic odour of chilli, high pungency and deep red colour. The material shall be free from rancidity. It shall be free from any added colour. The material shall also comply with requirements given in Table 3.14.

Packing

(1) The material shall be supplied in tightly closed glass, pure aluminium or tinplate containers which shall be nearly full with the material under reference.
(2) The material shall be protected from light and stored in a cool, dark and dry place.

Marking

The container shall be marked with the following:

(a) Name of the material;
(b) Manufacturer's name and trade mark, if any; and
(c) Net mass of the material when packed.

Sampling

Representative samples of the material sufficient to give a composite sample for all the determinations shall be drawn from the containers selected from the lot. Tests for all the characteristics shall be conducted on the composite sample.

Tests

Tests shall be conducted as prescribed in Table 3.14.

QUALITY OF REAGENTS

Unless specified otherwise, pure chemicals (chemicals that do not contain impurities which affect the results of analysis) and distilled water (see IS 1070: 1992) shall be employed in tests.

Table 3.14 Physico-chemical requirements for chilli oleoresin

Characteristics	Requirement
Colour value	4,000–20,000
Scoville heat units (min)	240,000
Capsaicin content, percent by mass (min)	1.5
Residual solvent, mg/kg (max)	—
Hexane	25
Acetone/Ethylene dichloride/trichloroethylene	30
Methanol/isopropanol	30

Sources: Indian specification (IS : 7543-1 : Part I, 1993; IS : 7543 : 1994 : Part II).

Determination of colour value

To make the estimation completely objective, instrumental analysis is employed. Colour of a specified dilution is estimated at 458 nm, the absorbance reading multiplied by a dilution factor and the result expressed as Nesslerimetric colour value.

Apparatus

Volumetric flasks (100 ml), 1 ml Pipette, Spectrophotometer or a colorimeter with adjustable wavelength.

Reagents

Acetone.

Procedure

Measure 1.0 g of oleoresin into a 100 ml volumetric flask, and make up the volume with acetone. Pipette 1.0 ml of this solution into a second 100 ml volumetric flask and make up the volume with acetone. Using a tungsten lamp source and acetone as the blank, take the absorbance of the 0.01% solution of oleoresin at 458 nm. Multiply this reading by 61,000 to obtain the colour value.

Determination of Scoville heat unit

This is a method based on sensory evaluation. Oleoresin solutions variably diluted with sugar solution are tasted in increasing concentration. The highest dilution at which a pungency is just detected is taken as a measure of the heat value.

Reagents

Sugar solution 5%, Alcohol 95%.

Procedure

(1) Accurately measure 0.2 g of the oleoresin into a 50 ml volumetric flask. Add alcohol up to the volume mark and shake the contents and allow to settle.

(2) Add 0.15 ml of this alcoholic solution to 140 ml sugar solution. Swallow after thoroughly shaking 5 ml of this solution.

(3) A biting or stinging sensation in the throat which is just perceptible at this dilution is equal to 240,000 Scoville units. If bite is very strong, a further dilution of this solution, is necessary for a perceptible pungent sensation:

    (a)   15 ml of sugar solution, 30 ml solution (2) = 360,000 Scoville units
    (b)   20 ml of sugar solution, 20 ml solution (2) = 480,000 Scoville units
    (c)   30 ml sugar solution, 15 ml solution (2) = 720,000 Scoville units.

The panel for evaluation shall consist of five members, and three members out of five shall find the solution to be pungent.

Concluding remarks

In this chapter, various aspects of chilli standardization and quality control have been discussed. Some aspects have included the importance and advantages of the standardization of *Capsicums*, quality evaluation of chillies and chilli products (notably oleoresin), variations in chemical composition, major principles or constituents of *Capsicum*/pepper/chilli, determination of pungent principles (constituents) like capsaicinoids and colouring pigments (β-carotene, capsanthin and capsorubin) in all types of chillies, or *Capsicum* and paprika, detection of adulteration in chillies or *Capsicum* or chillies, quality/colour evaluation of chillies and oleoresins of different types, national and international (ISI) standard for chillies (whole and powder) and oleoresin, development of the international standards for chillies, development of national (Agmark, ISI and PFA) standards for different commercial varieties of chillies, and ISI standard for chilli oleoresin.

There is a need for the development of improved high yielding varieties of paprika that are rich in attractive red pigments and which have higher drying yield and disease resistance. There is also a need to develop better, simpler and quicker methods of quality evaluation of chillies and chilli oleoresin in conjunction with better sensory evaluation techniques. There is a great potential for the development and export of paprika and other export varieties of *Capsicum*. There is an estimated world demand for about 50,000 tonnes, of which Spain alone utilizes 20,000 tonnes annually (Anu and Peter, 2000). Greater emphasis may also be laid on quality production of chilli and paprika. Oleoresin, including tailored oleoresins attempt, to meet the varying demands of the various types of food industries and the pharmaceutical industry.

It may also be noted that the paprika oleoresin is one of the most celebrated and most popular natural colourants for different categories of processed foods. Hence the greater need for emphasis on intensive R&D work, notably in developing countries like India, which produces the largest quantity (about 900,000 tonnes) of *Capsicums*/chillies annually.

References

American Spice Trade Association (1968) *Official Analytical Methods for Spices*, 2nd edn, ASTA, New York, USA.

American Spice Trade Association (1977) The National Constituents of Spices. ASTA Research Committee Report, File 1977. Englewood, New Jersey, 07632, USA.

Andre, L. and Mile, L. (1975) Analysis of capsaicin content of Paprika. *Acta Aliment Acad. Sci.*, Hung., 4, 113.

Anu, A. and Peter, K. V. (2000) The Chemistry of Paprika. *Indian Spices*, 37(2), 15–18.

Association of Official Analytical Chemists (AOAC) (1990) *Official and Tentative Methods of Analysis*, AOAC, Arlington, Virgina, USA.

Bajaj, K. L., Gurdeep Kaur and Soach, B. S. (1978) Varietal variations in capsaicin content *C. annum* Linn. fruits. *Veg. Sci.*, 5(1), 23–29.

Bajaj, K. L., Gurdeep Kaur and Sooch, B. S. (1980) Varietal variations in some important chemical constituents in chilli fruits. *Veg. Sci.*, 7(1), 48–54.

Barber, M. S., Jackman, L. M., Warren, J. K. and Weedon, B. C. I. (1960) Structure of the paprika ketones. *Proc. Chem. Soc.*, Lond., 19.

Barber, M. S., Jackman, L. M., Warren, J. K. and Weedon, B. C. I. (1961) Carotenoids and related compounds 9. The structure of capsaicin and capsorubin. *J. Chem. Soc.*, 4019.

Barnyai, M. and Szabolis, J. (1976) Determination of the red pigment in paprika products. *Acta Aliment Acad. Sci.*, Hung., 5, 87.

Bollinger, H. R. (1965) *Thin Layer Chromatography* (Stahl, E. ed.). Academic Press Inc; New York, USA.

Bollinger, H. R. (1968) *Thin Layer Chromatography* (E. Stahl, ed.). Academic Press, New York, USA, p. 124.

Booth, V. H. (1962) A method for separating lipid components of Leaves. *Biochem. J.*, 84, 444.

Buttery, R. C., Selfort, R. M., Gudanini, D. G. and Ling, L. C. (1969) Characterization of some volatile constituents of Bell Peppers. *J. Agric. Food Chem.*, 17, 1322–1327.

Cholonoky, L. and Szabolcs, J. (1960a) Structure of capsorubin. *Experienta*, 16, 183.

Cholonoky, L. and Szabolcs, J. (1960b) Structure of colouring matter of paprika. *Acta Chim. Acad. Sci.*, Hung., 22, 117.

Cholonoky, L., Gyorgyfy, K., Magy, E. and Panelzel, M. (1958) Investigation of carotenoid pigments III. Pigments of the yellow paprika. *Acta. Chem. Acad. Sci.*, Hung., 16, 227.

Cholonoky, L., Szaboles, J., Cooper, R. D. G. and Weedon, B. C. L. (1963) The structure of krypto-capsin. *Tetrahydron Lett.*, 19, 1257.

Cooper, R. D. G. *et al.* (1962) Stoichemistry of capsorubin and syntheses of the optically active pigments. *Proc. Chem. Soc. Lond.*, 25.

Cox, H. E. and Pearson, D. (1962) *Chemical Analysis of Goods*. Chem. Pub. Co., New York, USA.

Curl, A. L. (1962) The carotenoids of red bell peppers. *J. Agric. Food Chem.*, 10, 504.

Curl, A. L. (1964) The carotenoids of green bell peppers. *J. Agric. Food Chem.*, 12, 522.

Datta, P. R. and Susi, H. (1961) Quantitative differentiation between the natural capsaicin (8-methyl-*N*-vanilly-l-nomenamide) and vanillyl-n-nondynamide. *Anal. Chem.*, 35, 148.

Davis, B. H. (1965) Analysis of carotenoid pigments. In: *Chemistry and Biochemistry* (T. W. God, ed.).

De la Mar, R. and Francis, F. J. (1969) Carotenoid degradation in bleached paprika. *J. Food Sci.*, 34, 96.

Demole, E. (1958–59) Application of absorption micro-chromatograph in thin layers. *J. Chromatography*, 1(1), 24.

Directorate General of Health Services (DGHS), Min. of Health, Govt. of India (1986) Prevention of Food Adulteration Act, 1954. Rules amended upto 1985.

Directorate of Marketing and Inspection (DMI), Govt. of India (1982) APGM Act 1937 with Rules amended upto 31.12.1979. *Marketing Series* No. 192. *Compendium*. Vol. I, pp. 1–431, 5th edn.

Directorate of Marketing and Inspection (DMI), Govt. of India (1985). APGM Act 1937 with Rules amended upto 31.3.1985.

Directorate of Marketing and Inspection, Govt. of India (1982) APGM. Act 1937 with Rules amended upto 31.12.1979. Marketing series No. 192. *Compendium*. Vol. I, pp. 1–431 (5th edn).

Directorate of Marketing and Inspection, Govt. of India (1985) APGM Act 1937 with Rules amended upto 31.3.1985. Marketing series No. 193. *Compendium*. Vol. II, pp. 1–192.

Eisvale, R. J. (1981) *Oleoresins Handbook*, 3rd edn, Dodge & Elcott Inc., New York, USA.

Essential Oil Association of America (EOA) USA (1979) EOA Book of Standards and specifications. EOA. of USA. No. 60, 42nd Street, NY 10017, USA.

Faigle, H. and Karrer, S. (1961) Constitution and configuration of capsanthin and capsorubin. *Helv. Chim. Acta.*, 44, 1257.

Friedrich, H. and Rangoonwala, R. (1965) Separation of capsaicin and non-anoic-acid manillyl amide. *Natiowisenchafen*, 52, 514.

Govindarajan, V. S. (1986) Capsicum – production technology, chemistry, technology, standards and world trade. *CRC Crit. Rev. Food Sci.*, 23(3), 207–288.

Govindarajan, V. S. and Ananthakrishna, S. M. (1970) Observations on the separations of capsaicin from capsicum and its oleoresin. *J. Food. Sci. Technol.*, 7, 212.

Govindarajan, V. S. and Ananthakrishna (1974) Paper chromatographic determination of capsaicin. *Flavour Indus.*, 3, 176–178.

Govindarajan, V. S., Shanthi Narasimhan and Dhanaraj, S. (1977) Evaluation of spices and oleoresins. II. Pungency of *Capsicum* by Scoville Heat Units. *J. Food. Sci. Technol.*, 14, 28–34.

Govinda Reddy, K., Lalitha Kumari, A. and Sri Bavaji, J. N. (1999) Variations in the biochemical constituents in chilli varieties (Indian). *Indian Spices*, 36(1), 13–15.

Indian Standards IS: 1797 (1985) Methods of Test for Spices and Condiments. Indian Standards Institution, New Delhi.

Indian Standards IS: 7543 (1994) Quality Standards for Chilli Oleoresin. Part I.

Indian Standards IS: 8104 (1976) Methods of Test for Pungency of Chillies by Scoville Heat Units.

Indian Standards IS: 13145 (1993) Spices and Condiments – Methods of Sampling and Test (First Revision).

Indian Standards IS: 13663 (1993) Oleoresins Chillies (Capsicums).

Indian Standards IS: 2446 (1980) Detection of lead chromate in chillies, curry powder and turmeric.

Indian Standards Institution (now BIS), New Delhi (1976) Pungency of chillies by Scoville Heat Units. IS: 81704, 1976.

Indian Standards Institution (BIS) New Delhi (1993) Indian Standards IS: 13663, 1993.

Indian Standards Institution (BIS) New Delhi (1998) Chilli – Whole and Ground (Powdered). Indian Standards, (Hormonised) IS: 2322, 1998, pp. 1–4 (Formal details, Dec. section Indian Standards).

International Organization for Standardization (ISO), Budapest (1972–97) International Standards for Chillies, products and their Methods of Test.

International Organization for Standardization (ISO) (1982) Spices and Condiments – Nomenclature. ISO/R-676-1982.

James Varghese (1999) Scanning paprika oleoresin and chilli colour specifications – quality parameters. *Indian Spices*, **36**(1), 7–16.

Jensen, A. (1960) Chromatographic separation of carotenes and other chloroplasts on paper containing aluminium oxide. *Acta Chem. Scand.*, **14**, 205.

Jensen, A. and Jensen, S. L. (1959) Quantitative paper chromatography of carotenoids. *Acta Chim. Scand.*, **13**, 1863.

Joint Committee for Pharmaceutical Society and Society for Analytical Chemistry on Methods of Assay of Crude Drugs (1959) Recommended methods for determination of capsaicin content of *Capsicum* and its preparations. *Analyst*, Lond., **84**, 603.

Joint Committee for Pharmaceutical Society and Society for Analytical Chemistry on Methods of Assay of Crude Drugs (1964) The determinations of capsaicin content of *Capsicum* and its preparations. *Analyst.*, Lond., **89**, 377 (Second Panel Report).

Kanner, J., Mendel, J. and Budowski, R. (1976) Carotene oxidising factors in red pepper fruits I. Ascorbic acid. *J. Food Sci.*, **41**, 183.

Kanner, J., Mendel, J. and Budowski, R. (1977) Carotene oxidising factors in red pepper fruits, Peroxidase activity. *J. Food Sci.*, **42**, 154–184.

Kanner, J., Mendel, J. and Budowski, R. (1978) Carotene oxidising factors in red pepper fruits Oleoresin – Cellulose solid made. *J. Food Sci.*, **43**, 709.

Karrer, P. and Jucker, E. (1950) *Carotenoids*, Amer. Elsevier, New York, p. 241.

Konecsni, I. (1956–57) Studies on the microscopic determination of added food paprika in milled spice-paprika. *Kuloslerryomat, Ommi*, **4**, 439.

Kosuge, S., Inagaki, Y. and Okumura, H. (1965) Pungent principles in *Capsicum. Nippon Nogei Kagaki Kaishi*, **35**, 540.

Kosuge, S. and Furuta, P. (1970) Studies on pungent principles of *Capsicums*. Chemical composition of pungent principles. *Agri. Biol. Chem.*, **34**, 268.

Krishnamurthy, K. and Natarajan, C. P. (1970) Colour and its changes in chillies. *Indian Food Packer*, **27**(1), 39.

Kulka, K. (1967) Aspects of functional groups and flavour. *J. Agri. Food Chem.*, **15**, 48.

Lease, J. B. and Lease, E. J. (1956) Factors affecting the retention of red colour in Pepper. *Food Technol.*, **10**, 268.

Mathew, P. A., Peter, K. V. and Zacharia, J. (2000) Production and export potential of paprika. *Spice India*, **13**(10), 13–18.

Mauriya, K. R., Jha, R. C. and Choudhari, M. L. (1983) Physico-chemical qualities of some varieties of chilli. *Indian Cocoa Areca and Spices J.*, **11**(4), 120–121.

Mitra, S. N., Sengupta, P. M. and Roy, B. R. (1961) Detection of oil-soluble coal-tar dye in chilli capsicing. *J. Proc. Inst. Chem.*, Calcutta, **33**, 69.

Mitra, S. N., Sengupta, P. M. and Sen, A. R. (1970) A comparative study of starch estimation in chillies (Mirch). *J. Inst. Chem.*, **42**, Part I, 15.

Moster, J. B. and Prater, A. N. (1952) Colour of *Capsicum* spice – measurement of the extractable colour. *Food Technol.*, **6**, 459.

Moster, J. B. and Prater, A. N. (1957a) Colour of *Capsicum* spice – measurement of the extractable colour II. Extraction of colour. *Food Technol.*, **11**, 146.

Moster, J. B. and Prater, A. N. (1957b) Colour of *Capsicum* spice – measurement of the extractable colour III. A linear colour scale. *Food Technol.*, 11, 222.

Moster, J. B. and Prater, A. N. (1957c) Colour of *Capsicum* spice – measurement of the extractable colour IV. Oleoresin paprika. *Food Technol.*, 11, 226.

Navarao, S., Orunuo, A. and Oosta, F. (1965) Thin layer chromatography determination of synthetic dyes in Foods I. Fat-soluble azo dyes in paprika. *An. Bromatol.*, 17, 269.

Nelson, E. K. (1919) The constitution of capsaicin – the pungent principle of *Capsicum. J. Amer. Chem. Soc.*, 41, 1115.

Nelson, E. K. and Dawson, L. E. (1923) The constitution of capsaicin – the pungent principle of *Capsicum. J. Amer. Chem. Soc.*, 45, 2179.

Newman, A. A. (1953) Natural synthetic pepper flavonoids. *Chem. Prod. Chem. News*, 16, 413.

Parvathi, S. and Aminuddin, Y. (2000) Chilli – an economically profitable crop of Malaysia. *Spice India*, 13(7), 18–19.

Phillip, T. and Francis, F. J. (1971a) Oxidation of *Capsicum. J. Food. Sci.*, 36, 96.

Phillip, T. and Francis, F. J. (1971b) Isolation and chemical properties of capsanthin and derivatives. *J. Food Sci.*, 36, 1823.

Phillip, T., Nawar, W. W. and Francis, F. J. (1971) The nature of fatty acids and capsanthin esters. *J. Food Sci.*, 36, 98.

Pruthi, J. S. (1969) International collaborative studies on the pungency (capsaicin) and colour (capsanthin) content of Hungarian ground red paprika. Central Agmark Lab., DMI, Govt. of India, Nagpur, Tech. Commun. No. CAL/30/69.

Pruthi, J. S. (1970a) International collaborative work on the determination of capsanthin and capsaicin content in ground paprika (powder). ISO/TC 34/SC-7 DOC. No. 227E. International Organization in Standardization (ISO), Budapest, Hungary.

Pruthi, J. S. (1970b) Packaging of "Spices and Condiments". *J. Packaging*, India, 3(1), 11.

Pruthi, J. S. (1971) Packaging of "Spices and Condiments". *J. Packaging*, India, 4(1), 1.

Pruthi, J. S. (1980) *Spices and Condiments – Chemistry, Microbiology and Technology.* Acad. Press Inc., New York, USA, pp. 1–450 (Reprinted several times in USA on popular demand).

Pruthi, J. S. (1989) Spice Extractives (essential oils and oleoresins including *Capsicum* oleoresin): Present Scenario and Prospects. An invited status paper (sponsored by the International Society for Essential Oils, Flavours and Fragrances presented at the 11th International Congress. *Proc.* Vol. VI, pp. 217–243.

Pruthi, J. S. (1993) *Major Spices of India: Crop Management and Post Harvest Technology.* Second reprint in 1998, pp. 1–514 + XXII. Special Chapter on Capsaicins/chillies (with References) Indian Council of Agri. Research, DIPA, New Delhi, pp. 180–243.

Pruthi, J. S. (1999) *Quality Assurance in Spices and Spice Products – Modern Methods of Analysis*, 1st edn, Allied Publishers Ltd, New Delhi, pp. 1–576 + XXVI.

Purseglove, J. W., Brown, E. G., Green, C. L. and Robins, S. R. J. (1981) *Spices.* 1st edn, Longman Inc., New York, USA, p. 331, (chillies/capsicum).

Roberts, L. A. (1968) as chromatographic determination of methylene chloride, ethylene dichloride and trichloroethylene residues in spice oleoresins. *J. Assoc. Off. Anal. Chem.*, 51, 825.

Rogers, J. A. (1966) Advances in flavour and oleoresin chemistry, a chapter cited in *Flavour Chemistry* by Gold 1966. *Adv. Chem. Ser.*, 56, 203.

Rosebrook, D. D. (1968) An improved method for determination of extractable colour in capsicum. *J. Assoc. Off. Anal. Chem.*, 51, 637.

Rosebrook, D. D. (1971) Collaborative study of a method for the extractable colour in paprika oleoresin. *J. Assoc. Off. Anal. Chem.*, 54, 37.

Sacchetta, R. A. (1960) Paper chromatography of red paprika powder. *Rev. Assoc. Bioquim Argent*, 25, 187.

Schwien, W. G. and Miller, B. J. (1967) Detection and identification of dehydrated red beets in *Capsicum* spices. *J. Assoc. Off. Anal. Chem.*, 50, 223.

Scoville, W. L. (1912) Note on *Capsicum. J. Amer. Pharm Assoc.*, 1, 453.

Sen, A. R., Sengupta, P., Ghose Dastidar, N. and Mathew, T. V. (1973) Chromatographic detection of small amounts of mineral oil in chillies. *Res. Indust.*, 10, 97.

Shuster, H. I. and Lockhart, E. E. (1954) Development and application of objective methods of quality evaluation of spices. I. Colour grading of *Capsicums*. *Food Res.*, 19, 472.

Spices Board, Govt. of India, Men of Commerce, Cochin (2000) Annual Review of Spices Export (whole, ground and processed prod.), a regular annual publication and *Spices Markets (Weekly)*, 12(44), 1–3.

Stahl, W. H. (1963) Critical review of methods of analysis of oleoresins (quality assessment). *J. Assoc. Off. Anal. Chem.*, 48, 515.

Stelzer, H. (1963) Identification of synthetic colouring in paprika. *Nutr. Bromatol. Texicol.*, 2, 177.

Suzuki, T. and Iwai, K. (1984) Constituents of red pepper species – chemistry biochemistry, pharmacology and food science of the pungency principle of *Capsicum* species, In: *The Alkaloids* (A. Brossi, ed.), Vol. XXIII Acad. Press, Orlando, FL, Chap. 4.

Szabo, P. (1969) Manufacture of paprika oleoresin. *Kenserve Parikarp*, 5, 7.

Szabo, P. (1970) Production of paprika oleoresin. *Amer. Perfum Cosmet.*, 85, 39.

Tainter, D. R. and Grenis, A. T. (1993) *Spices and Seasonings – A Food Technology Hand Book*, VCH Publishers, New Delhi, p. 45.

Thresh, L. T. (1846) Isolation of capsaicin. *Pharm. J.*, 6, 941.

Todd, P. H. Jr. (1958) Species: detection of foreign pungent compounds in paprika, oleoresin, ground *Capsicums* and chilli (whole and ground). *Food Technol.*, 12, 468.

USDA. (1977) Composition of Foods, Spices and Herbs. *USDA. Agr. Handbook*, 8–12 Jan., 1977.

Van Blaricum, L. O. and Martin, J. A. (1951) Retarding the loss of red colour in cayenne pepper with oil-soluble antioxidants. *Food Technol.*, 5, 337.

Videki, L. and Videki, I. (1961) Determination of pigmency content of paprika by spectrophotometric (ASTA) method. *Elemiservezsaglati Kozl*, 7, 103.

Virakamath, C. S. (1964) Studies on the packaging of spices – Dry chillies (*C. annuum*). *Indian Food Packer*, 18(4), 4.

Welcher, F. J. (ed.) (1970) Detection of oil-soluble coal-tar dyes in chillies by thin-layer chromatography (YLC). In: *Standard Methods for Chemical Analysis*, Vol. IIIA, 6th edn, pp. 743–747.

4 Pungency principles in *Capsicum* – analytical determinations and toxicology

P. Manirakiza, A. Covaci and P. Schepens

Capsaicinoids are a natural group of alkaloids responsible for the pungency and hotness of the *Capsicum* fruits. Increasing interest in both their role in food additives and their pharmacological properties has led many researchers to conduct more investigations. This chapter deals with their biochemistry, toxicology and analytical methods for determination. Human risk of intoxication with these compounds was reported to be low. Chromatographic analytical methods, including liquid and gas chromatography, were found to be the most convenient for their determination, while spectrophotometric methods were advised for total capsaicinoid measurement.

Introduction

Capsicum fruits in different forms are intensively used in many parts of the world as additives in food for their very strong pungency and aroma, as well as for their pharmacological and therapeutic effects as a stimulant and counter irritant (Govindarajan, 1986; Manirakiza *et al.*, 1999). Paprika normally grows in tropical and temperate regions, and has more recently been introduced in the southern parts of US. It is used as a table spice, as well as in the food and meat processing industry in Europe and in Japan. Several varieties of peppers are commercialized all over the world and are generally considered to be a balanced source of essential nutrients, containing proteins, minerals, vitamin C, sugars and lipids. Several natural pigments and aromas are contained in sweet peppers varieties (Govindarajan, 1986).

Chemical composition of *Capsicum* fruit

In addition to water which represents more than 90% of the *Capsicum* fruit, fibre, pectin, glucose, starch and fructose represent the main components (Table 4.1). Among the high number of natural chemical group components occurring in the *Capsicum* species, carotenoids are responsible for the colour in peppers (α-, β-, γ-, δ-carotenes). Together with xanthophylls, they are not only precursors of vitamin A (retinol), but are also reported to contribute to cancer and cardiovascular diseases prevention (Machlin, 1995; Muller, 1997). In most of the *Capsicum* species (*Capsicum annuum, Capsicum frutescens, Capsicum chinense*), maturation corresponds to changes in carotenoids, flavonoids, chlorophylls, phenolic acid, ascorbic acid and total soluble reducing equivalent (Jurenitsch *et al.*, 1979; Markus *et al.*, 1999; Howard *et al.*, 2000). When the *Capsicum* fruits start the ripening process, chlorophylls disappear in favour of carotenoids of red varieties with the occurrence of related xanthophylls, cryptoxanthin, zeaxanthin, epoxide antheraxanthin, luteoxanthin and new synthesis of the characteristic red carotenoids, cryptocapsin, capsanthin (Markus *et al.*, 1999). These biochemical changes correspond with structural

Table 4.1 Chemical composition of *Capsicum annuum* L.

Compounds	
	Mean ± SD (g/100 g fresh fruit)
Water	91 ± 0.6
Glucose	0.85 ± 0.1
Fructose	0.75 ± 0.1
Sucrose	nd
Starch	0.81 ± 0.2
Fibre	2.2 ± 0.3
Pectin	0.73 ± 0.1
	Mean ± SD (mg/100 g fresh fruit)
Citric acid	28 ± 12
Fumaric acid	1.1 ± 0.4
Malic acid	208 ± 18
Oxalic acid	140 ± 24
Quininic acid	183 ± 62
Vitamin C	24 ± 12
Chlorophyll a	7.9 ± 2
Chlorophyll b	3.4 ± 0.6
All-*trans*-lutein	1.4 ± 0.3
All-*trans*-β-carotene	0.92 ± 0.4

Source: Lopez-Hernandez *et al.*, 1996.

Note
nd not detected.

Table 4.2 Occurrence of carotenoids in *Capsicum* fruits varieties

Carotenoids	TLC colour	Capsicum fruit varieties						
		Y	G	R	YB	RB	CG	RY
Hydrocarbon								
Phytoene	—	−	+	+	+	+	+	−
Phytofluene	—	−	+	−	−	+	+	+
ζ-Carotene	—	−	+	+	+	−	+	−
β-Carotene	yellow	+	+	+	+	+	+*	+
α-Carotene	yellow	+	+	−	+	+	−	−
β-Carotene epoxide	yellow	−	−	−	+	−	+	−
Monols								
Hydroxy-α-carotene	yellow	−	+	−	+	+	−	−
Cryptoxanthin	—	+	+	−	+	+	−	+
Cryptoxanthin epoxide	—	−	−	−	+	+	−	−
Capsolutein	—	−	−	−	+	+	−	−
Diols/epoxides								
Zeaxanthin	orange	+	+	+	+	+	+	+
Antheraxanthin	yellow	+	+	+	+	+	+	+
Violaxanthin	yellow	+	+	+	+	+	+	+
Mutatoxanthin	yellow	+	+	+	+	+	+	+
Luteoxanthin	yellow	−	+	−	+	+	−	−
Lutein**	orange	+	+	−	+	−	−	−
Neoxanthin	yellow	+	+	+	+	+	+	+
Red keto carotenoids								
Cryptocapsin	—	−	−	+	−	+	−	−
Capsanthin	red	−	−	+	−	+	−	+
Capsanthin-5,6-epoxide	red	−	−	−	−	+	−	−
Capsorubin	red	−	−	+	−	+	−	+

Source: Govindarajan, 1986.

Notes
R – Red paprika; YB – Yellow mutant paprika; RB – Red bell; G – Mature green bell; CG – College gold; RY – California wonder yellow ripe fruit; * for green stages only; ** for green stage of all varieties. − compound not present, + compound present.

modifications in the cell. Chloroplast (rich in chlorophylls) changes to chromoplast, which contains only carotenoids. Red carotenoids are not reported in yellow pepper varieties in which hydroxy-α-carotene, mono- and diepoxides of cryptoxanthin, auroxanthin have been identified (Table 4.2).

The pungency and hotness principle of the *Capsicum* species is unanimously reported in the literature to be a group of compounds related to capsaicin (8-methyl 6-nonenoyl-vanillylamide) and are called capsaicinoids (Iwai *et al.*, 1979). The common structure of these compounds is composed of a vanillylamide moiety and C9–C11 fatty acids (Nakatani *et al.*, 1986; Hawer *et al.*, 1994; Esteves *et al.*, 1997; Peusch *et al.*, 1997; Laskaridou-Monnerville, 1999; Singh *et al.*, 2001; Sprague *et al.*, 2001; Reilly *et al.*, 2001a,b) (Figure 4.1). Levels of these capsaicinoids vary with genotype, maturity and are influenced by growing conditions and losses after processing (Zewdie and Bosland, 2001). The main pungency compounds of hot chilli peppers are capsaicin and dihydrocapsaicin (90%), while the rest is constituted by nordihydrocapsaicin, homocapsaicin, homodihydrocapsaicin and norcapsaicin (Rico Avila, 1983; Laskaridou-Morneville, 1999; Manirakiza *et al.*, 1999; Murakami *et al.*, 2001; Singh *et al.*, 2001; Reilly *et al.*, 2001a). Capsaicinoids concentrations can be converted to Scoville Heat Units (SHU) by multiplying the pepper dry weight concentrations in parts per million (ppm) by the coefficients of heat value for each compound; 9.3 for nordihydrocapsaicin, 16.1 for both capsaicin and dihydrocapsaicin

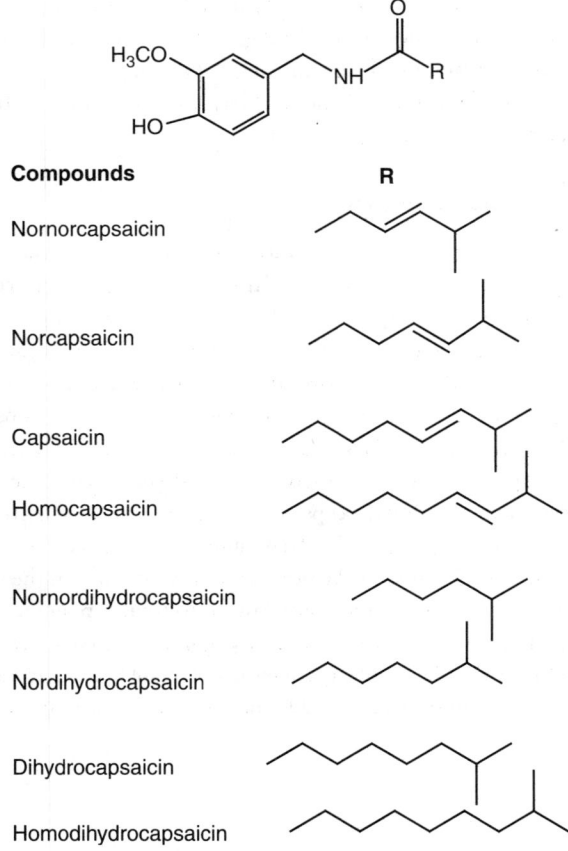

Figure 4.1 Chemical structures of capsaicinoid compounds.

(Govindarajan *et al.*, 1977; Zewdie and Bosland, 2001). A SHU is defined as the number of parts of sugar-water needed to neutralize the heat of one part sample extract (e.g. if the heat of a cayenne pepper is 30,000 SHU, it means that 30,000 parts of sugar-water are needed to dilute one part of cayenne pepper extract to the last point that hotness can be detected).

Besides the natural compounds occurring in the *Capsicum* species, anthropogenic activities have led to the discovery of various levels of organic and inorganic pollutants in *Capsicum* (PCBs, pesticides, heavy metals) and an increasing interest in spices has led to several methods for monitoring these pollutants (Manirakiza *et al.*, 2000).

As interest in spicy foods and in the pharmacological properties of capsaicinoids increases, there is a need to test and classify chemical principles responsible for the pungency and hotness of the *Capsicum* species. The existing methods, from spectrophotometric to chromatographic methods for the determination of these natural compounds, will be critically reviewed and compared, with regard to simplicity, precision and cost.

Biochemistry of the pungency principles in the *Capsicum* species

Although the contents of *Capsicum* depends on the particular variety, the capsaicinoid's profiles should not be a good chemotaxonomic indicator for *Capsicum* species (Zewdie and Bosland, 2001). According to Zewdie and Bosland, the statement that capsaicin and dihydrocapsaicin were always the major capsaicinoids was not true. The same authors confirmed that different genes control the synthesis of each capsaicinoid and thus explain the different capsaicinoid proportions within the same *Capsicum* species (Zewdie and Bosland, 2000). Besides the eight classic capsaicinoids reported in the literature (Figure 4.1), nonivamide (which presents pungent properties) [nonamide, N-(4-hydroxy-3-methoxyphenyl) methyl] has been confirmed as being naturally present in the *Capsicum* species.

Pattern occurrence of capsaicinoids in Capsicum fruits

Existing data from the analysis of *Capsicum* fruits of different varieties have shown considerable variations in the total capsaicinoid content and the impossibility of predicting proportions between capsaicinoids even within same species (Hall *et al.*, 1987; Zewdie and Bosland, 2001). With some exceptions, it is believed that capsaicin and dihydrocapsaicin are the major constituents (80%), while nordihydrocapsaicin and homodihydrocapsaicin vary widely from 1% to 12%. When detected, the remaining homologues are present at very low percentages (Constant and Cordell, 1996; Manirakiza *et al.*, 1999). Even if large variations in the total capsaicinoids content within and between *Capsicum* species following cultivation conditions exist, the ratio between the principal components capsaicin and dihydrocapsaicin is reported to vary narrowly; for instance, 1 : 1 for *C. annuum* L. to 2 : 1 for *C. frutescens* L. (Neumann, 1966). It has long been observed that pungency stimulants were concentrated in the pericarp tissues and not in the seeds or peduncles (Suzuki *et al.*, 1980). Individual capsaicinoid contribution to total capsaicinoids in relation to the age of flowering in the pericarp and placenta of *Capsicum* fruits is given in Figure 4.2. Several studies have concluded that levels of capsaicinoids increase with maturation (Hall *et al.*, 1987), remain constant (Sukrasno and Yeoman, 1993) or decrease slightly up to 60% after the maximum is reached.

Biosynthesis of capsaicinoids

Capsaicinoids, are synthesized from phenylpropanoid intermediates through the cinnamic acid pathway (Figure 4.3). One of the enzymes involved in the biosynthesis is the caffeic acid-3-0-methyltransferase which has been reported to be important in lignification

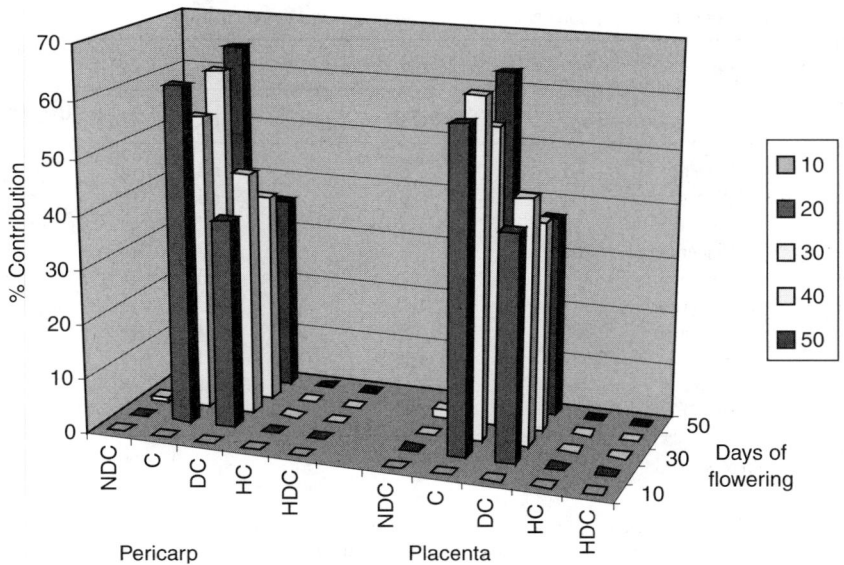

Figure 4.2 Individual capsaicinoid contribution to total capsaicinoids in relation with age of flowering in *Capsicum* fruit pericarp and placenta. (NDC – nordihydrocapsaicin, C – capsaicin, DC – dihydrocapsaicin, HC – homocapsaicin, HDC – homodihydrocapsaicin) (Suzuki *et al.*, 1980).

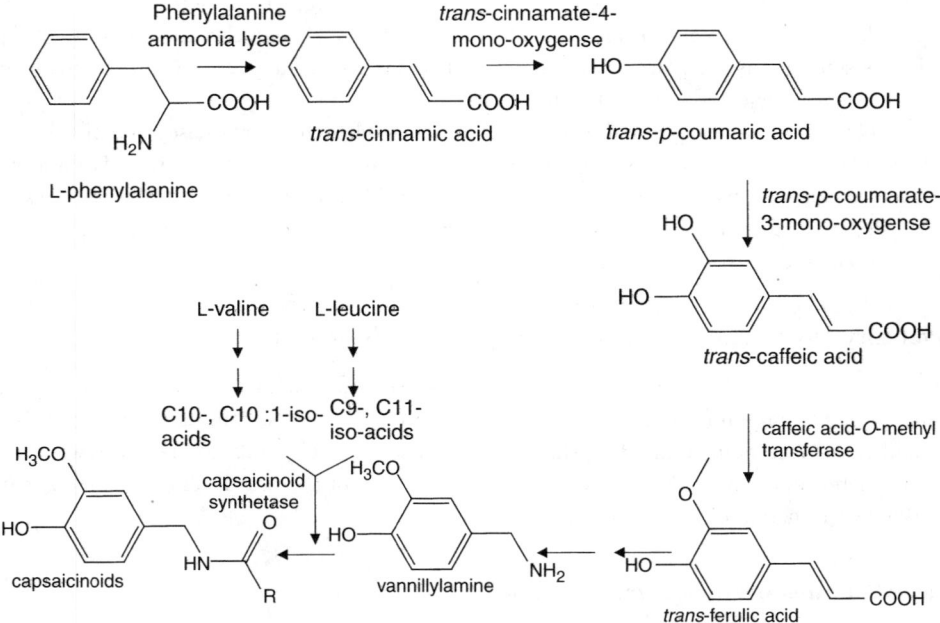

Figure 4.3 Capsaicinoids biosynthetic pathway (Fujiwake *et al.*, 1982a,b).

(Inoue *et al.*, 1998). In capsaicinoids synthesis, it catalyzes the conversion of caffeic acid to ferulic acid by methylating the hydroxyl group at the 3 position on the benzene ring. Ferulic acid is converted to vanillin. Vanillin is further converted to vanillylamine, which later condenses with branched chain fatty acid to form capsaicinoids. Extensive studies on the capsaicinoids

Table 4.3 Capsaicinoid composition in pericarp and placenta at different growth stages of fruits of *Capsicum annuum* var. *annuum* cv. Karayatsubusa

Flowering days	Part	Capsaicinoids (in μg/mg dry tissue)					
		Total	NDC	C	DC	HC	HDC
10	Pericarp	nd	nd	nd	nd	nd	nd
	Placenta	nd	nd	nd	nd	nd	nd
20	Pericarp	3.4	tr	2.1	1.3	nd	tr
	Placenta	30.4	tr	17.9	12.5	nd	tr
30	Pericarp	28.6	0.2	15.6	12.8	nd	tr
	Placenta	164.4	tr	101.9	62.5	nd	tr
40	Pericarp	40.9	tr	25.1	15.8	nd	tr
	Placenta	250.9	3.8	137.9	109.1	nd	tr
50	Pericarp	4.2	tr	2.7	1.5	nd	tr
	Placenta	52.9	tr	33.3	19.6	nd	tr

Source: Suzuki *et al.*, 1980.

Notes

Capsaicinoids were estimated with GC-MS. NDC – nordihydrocapsaicin; C – capsaicin; DC – dihydrocapsaicin; HC – homocapsaicin; HDC – homodihydrocapsaicin; nd – not detected; tr – trace.

biosynthesis with newly developed methods have led to the statement that, at all stages of growth, the total or individual capsaicinoids based on dry weight is far higher in the placenta than in the pericarp (Table 4.3). This statement allows the conclusion that the main site of capsaicinoids synthesis is the placental tissue of the fruits (Fujiwake *et al.*, 1982a,b; Bernal *et al.*, 1993a,b; Ochoa-Alejo and Gomez-Peralta, 1998).

Bernal *et al.* (1993a,b) suggested by means of *in vitro* studies that peroxidases were involved in the degradation of capsaicinoids. Pepper peroxidase, also mainly located in the placenta, was reported to oxidize capsaicin (Bernal *et al.*, 1993a) and dihydrocapsaicin (Bernal *et al.*, 1993b). Therefore, the turnover and degradation of capsaicinoids observed after *Capsicum* fruits reach maturation could be explained by the activity of peroxidases .

Inventory of analytical methods for capsaicinoids determination

The choice of an analytical method is generally guided by many factors, including the properties of the targeted compounds (polarity, stability, solubility), matrices, time assigned for analysis and additional factors introduced by the laboratory itself. For *Capsicum* compounds two objectives for determination can be followed: the determination of the heat level of a *Capsicum* fruit and the determination of its capsaicinoid composition.

Organoleptic and spectrophotometric methods

Procedures including organoleptic (Ramos, 1979) and spectrophotometric methods (Meilgaard *et al.*, 1987) have been developed. Organoleptic methods were preferred by the food industry for the direct heat level measure it offered, while it presented the disadvantage of requiring the extensive training of panelists and the monitoring of their sensitivity for the attainment of consistent results (Bajaj and Kaur, 1979). Spectrophotometric methods use vanadium oxytrichloride or phosphomolybdic acid to produce coloured solutions. Even if it is not specific, the result obtained is proportional to the heat level (Kosuge and Furuta, 1970; Rymal *et al.*, 1979) and therefore can be used as a tool for total capsaicinoid content.

Gas chromatographic methods

A variety of gas chromatographic (GC) methods have been developed for capsaicinoids analysis (Table 4.4) and most of them still present serious limitations. As shown in Table 4.4, sample preparation and extraction are not so different, proceeding by the grinding of the dried *Capsicum* or the chopping of the fresh fruits into small pieces and then extracting with different solvents. Most of the GC methods need a derivatization step to increase the volatility of the capsaicinoids, and, furthermore, an efficient clean up step is necessary (Todd *et al.*, 1977; Iwai *et al.*, 1979; Krajewska and Powers, 1987; Manirakiza *et al.*, 1999). Two types of derivatization procedures have been reported, trimethylsilylation of capsaicinoids (Lee *et al.*, 1976; Todd *et al.*, 1977; Iwai *et al.*, 1979; Fung *et al.*, 1982) and hydrolysis of capsaicinoids for yielding fatty acids and subsequent esterification (Jurenitsch *et al.*, 1979; Jurenitsch and Leinmuller, 1980). An on-column derivatization has also been reported (Krajewska and Powers, 1987). The use of structurally related codeine as an internal standard was reported (Manirakiza *et al.*, 1999), while tetracosane (Krajewska and Powers, 1987) and squalene (Hawer *et al.*, 1994) have also been used as internal standards. To overcome the problem of tailing peaks and to avoid the use of a derivatization step, Thomas *et al.* (1998) and Hawer *et al.* (1994) have recognized the use of a polar capillary analytical column (14% cyanopropylphenyl/86% methyl stationary phase, $30 \, \text{m} \times 0.25 \, \text{mm}$ i.d. $\times 0.25 \, \mu\text{m}$ film thickness) for interaction with polar functional group of the molecules. The resulting chromatogram is shown in Figure 4.4. Furthermore, the use of a thermionic selective detector (TSD) instead of flame ionization detection allows the elimination of sample clean-up (Thomas *et al.*, 1998). Unfortunately, GC/MS was not able to separate or differentiate nonivamide and nordihydrocapsaicin (Kosuge and Furuta, 1970; Saria *et al.*, 1981; Fung *et al.*, 1982; Reilly *et al.*, 2001b), leading to the misidentification of nonivamide as nordihydrocapsaicin. Although GC/MS is believed to offer good results, they still require specific sample preparation and a complex, costly instrument. Limits of quantification were superior to 50 ppm.

High Performance Liquid Chromatography methods

As for GC methods, HPLC methods are encumbered by significant difficulties (Table 4.5). For instance, normal phase HPLC was unable to separate all individual capsaicinoids (Dicecco, 1976; Cooper *et al.*, 1991; Constant *et al.*, 1995). Therefore, the results were given as total capsaicinoids even if three major capsaicinoids – capsaicin, dihydrocapsaicin and nordihydrocapsaicin – could be eventually separated. Furthermore, these HPLC methods were applied to relatively capsaicin rich matrices, while poor capsaicinoids matrices required sample clean-ups with a preconcentration step (Peusch *et al.*, 1997).

The reverse phase HPLC was limited by the non-resolution of nonivamide and capsaicin with regards to the similar chromatographic properties in HPLC (Saria *et al.*, 1981; Constant *et al.*, 1995). Constant and coworkers (1995) used complexation chromatography (AgNO$_3$) to separate the coeluting compounds. In most of these HPLC methods, the limit of quantification was superior to 150 ppm.

The use of the LC–MS (Reilly *et al.*, 2001a) has been reported to differentiate nonivamide and capsaicin by mass-to-charge ratio. The same authors have reported the use of LC–MS–MS with an electrospray ionization source operating in the selected reaction monitoring

Table 4.4 Gas chromatographic method for capsaicinoid determination in *Capsicum* products

Matrix	Sample preparation	Extraction	Analysis	Comments
Chilli peppers (*Capsicum* spp.)	Dry at 80°C overnight	With high speed blender/CHCl$_2$, sonication, activated charcoal, concentration, SPE clean-up on florisil, elution with EtAc	GC/MS, splitless, DB-5 analytical column, SIM mode, codeine as internal standard (IS) (Manirakiza et al., 1999)	Clean chromatogram, but only three capsaicinoids quantified and one detected
Capsicum fruits	Dry and grind to fine powder	Extraction with acetone/clean-up with HPLC	GC/MS on packed column with trimethylsilylation (TMS) as derivatization. α-cholestane used as internal standard (Iwai et al., 1979)	Rapid fouling of HPLC columns, increasing pressure in the system shortens the column life, old technique
Green *Capsicum* fruits	Freeze dry	Soxhlet extraction with acetone and liquid purification afterwards	GC/FID, packed analytical column, tetracosane (IS) and on column derivatization (Krajewska et al.,1987)	Old methodology
Red peppers	Dry, finely grinded (2–3 g taken)	Soxhlet extraction with acetone, purification by liquid liquid fractionation	GC/FID, (25 m fused silica capillary column bonded with crosslinked cyanopropyl-dimethylsiloxane (BP-10). Squalene as IS. Comparison of mass spectra (Hawer et al., 1994)	Good resolution of capsaicinoids and clean chromatogram, no derivatization step needed
Peppers samples	Pepper samples frozen, thawed and seeds removed, finely chopped to a uniform mixture of pepper tissue	Extraction with acetone	GC/TSD, more polar column (AT-1701, 14% cyanopropylphenyl, 86% methyl stationary phase, 30 m × 0.25 mm i.d. × 0.25 μm film thickness) (Thomas et al., 1998)	Very clean chromatogram, use of guard column, expensive procedure, no derivatization step needed, three compounds quantitated

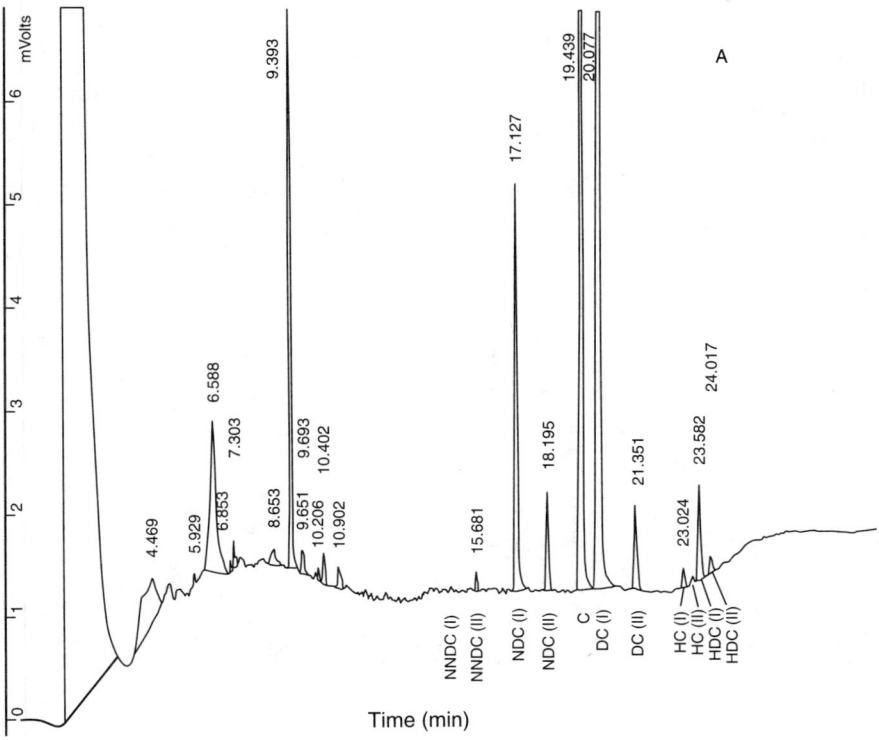

Figure 4.4 GC chromatogram of capsaicinoids obtained with a 14% cyanopropylphenyl/86% methyl stationary phase, 30 m × 9.25 mm i.d. × 0.25 μm film thickness (Thomas *et al.*, 1998).

(Reilly *et al.*, 2001b). The quantification of capsaicinoids using LC–MS–MS was more sensitive and exhibited greater accuracy, even at low analyte concentrations. A mass spectrum of product ions generated by collision-induced dissociation of capsaicin (*m/z* 306) is given in Figure 4.5, while a typical selected reaction monitoring profile from the analysis of capsaicinoids by LC–MS–MS is shown in Figure 4.6. The limit of quantification of LC–MS in *Capsicum* was 10 ng/ml, while it was ten times lower for LC–MS–MS.

Analytical recommendations

Although the choice of an analytical method for capsaicinoid determination usually depends upon the available instrumentation, the following considerations should be taken into account. The sensory, colorimetric, spectrophotometric or thin layer chromatography (TLC) procedures are not convenient for the determination of individual compounds and are, therefore, recommended only for the analysis of total capsaicinoids. Reversed-phase HPLC with a complexation step could be an alternative method. Although GC/MS is believed to offer good results, it still requires specific sample preparation and costly instrumentation. For its ability to identify capsaicinoids in dilute or very small samples and also for its high costs, LC–MS–MS should be employed for forensic analysis.

Table 4.5 Liquid chromatographic methods for capsaicinoid determination in *Capsicum* products

Matrix	Sample preparation	Extraction	Analysis	Comments
Capsicum oleoresins and pharmaceutical preparations	*Capsicum* oleoresins are sampled, container heated in waterbath, shaken for good homogeneity, SPE on silicagel	SPE on silica gel cartridge, wash with hexane and with methanol	HPLC separation, mobile phase 60% MeOH/40% water, Supelcosil C18, 25 cm × 0.46 mm, 5 μm particle size; elution in order of molecular weight; use of dimethoxyl-benzylmethyloctamide (DMMBO) as internal standard. UV detector (Cooper *et al.*, 1991)	Has been applied to material with higher concentrations of capsaicinoids; use of internal standard
Capsicum and extractives	*Capsicum* fruits are crushed or ground	Extraction in warm ethanol with reflux condenser or C18 SPE, elution with ethanol	LC with UV detector, fluorometer with excitation 280 nm and emission 325 nm; mobile phase used: acetonitrile 40%/60% water with 1% acetic acid v/v (AOAC, Official method 995.03, 1995)	No internal standard needed, based on empirical formula for the determination of the SHU
Chillies and paprika, *Capsicum frutescens* L. and *Capsicum annuum* L.	*Capsicum* powdered	Supercritical fluid extraction (SFE)/acetone, liquid liquid partition for further clean-up (chillies)	HPLC on 250 × 4 mm Lichrospher RP-18 column with a RP-18, 4 × 4 cm guard column, eluent mixture, acetonitrile/water/acetic acid (100:100:1); use of fluorescence detector (Peusch *et al.*, 1997)	SFE offers shorter extraction time, less sample preparation; the procedure is somewhat expensive, convenient for rich capsaicinoid rich compounds
Pepper sprays products	Equilibration at −20°C, evaporation of the volatile	Sample reestablished with methanol/n-butylchloride (1:1), if necessary 50 times diluted	LC–MS–MS, metaSil basic (100 × 3.0 mm), 3 μm, particle size, C2–C8 reverse phase, mobile phase: methanol/water (57.5:42.5) with formic acid 0.1% v/v. Electrospray ionization source operating in SIM mode (Reilly *et al.*, 2001)	Use of costly equipment, very high selectivity, can separate nonivamide with capsaicin with mass-to-charge ratio; well indicated for forensic analysis
Capsicum oleoresin	Chopped in fine tissue	Extracted with petroleum ether: acetone mixture and TLC separation (silica gel)	HPLC with conventional analytical columns (Perucka and Oleszek, 2000)	Evaluation of all capsaicinoids without any individual identification

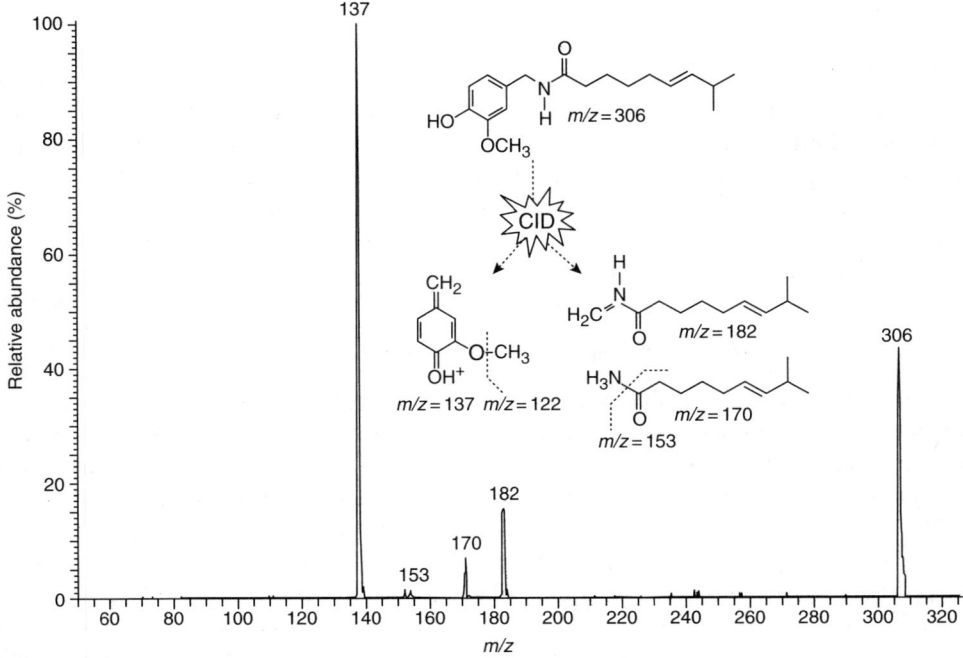

Figure 4.5 Mass spectrum of product ions generated by collision-induced dissociations of capsaicin at *m/z* 306 (Reilly *et al.*, 2001b).

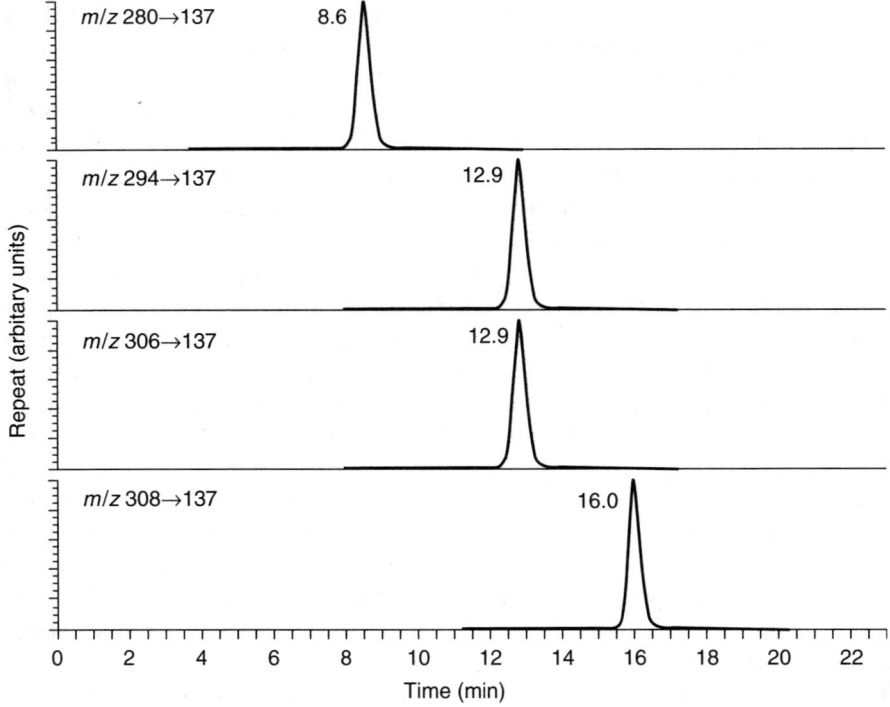

Figure 4.6 Typical selected reaction monitoring profile from the analysis of capsaicinoids by LC–MS–MS (Reilly *et al.*, 2001b). The identities of the peaks are octanoyl vanillamide (*m/z* 280–137), non-ivamide (*m/z* 294–137), capsaicin (*m/z* 306–137), and dihydrocapsaicin (*m/z* 308–137).

Capsaicinoids and toxicity

Many studies have been conducted on the most abundant capsaicinoids in the *Capsicum* pepper, capsaicin and dihydrocapsaicin. Capsaicin, the pungent phenolic compound of the *Capsicum* species, has shown a wide range of pharmacological properties, including antigenotoxic, antimutagenic and anticarcinogenic effects. Other studies, however, have shown it to be a tumour promoter and potential mutagen and carcinogen, resulting in capsaicin being termed as "a double edged sword" (Richeux *et al.*, 1999; Singh *et al.*, 2001).

Neurotoxicity, cytotoxicity and genotoxicity of capsaicinoids

The detection of painful stimuli occurs primarily at the peripheral terminals of specialized sensory neurons called nociceptors. The small diameter neurons transduce signals of a chemical, mechanical or thermal nature into action potentials and transmit their information to the central nervous system, ultimately eliciting a perception of pain and discomfort (Caterina and Julius, 2001). The localized desensitization of nociceptive afferents when capsaicin is applied in small doses has led to capsaicin being used as a therapeutic drug for pain relief (Minani *et al.*, 2001). For example, capsaicin has been evaluated as an analgesic in a variety of neuropathic pain conditions, including postherapeutic neuralgia, painful diabetic neuropathy, osteroarthritis, the post-surgical pain syndrome and Guillain-Barré syndrome (Markovits and Gilhar, 1997; Szallasi and Blumberg, 1999). However, at increased concentrations, a capsaicin-associated burning sensation can negate the beneficial and therapeutic effects (Minani *et al.*, 2001).

In order to determine the lethal toxic level of capsaicinoids and to extrapolate them to humans, many experiments were performed in mice, rats, guinea pig and rabbits. Pure capsaicin (16,000,000 SHU) was administered intravenously, subcutaneously in the stomach and applied topically until the death of the animals. The lethal toxic doses of capsaicin in mg per kg animal weight, were 0.56 for intravenous administration, 190 when consumed and 512 when applied topically. The probable cause of death in all cases was presumed to be respiratory paralysis. Guinea pig was the most sensitive to capsaicin, while rabbits were less sensitive. The acute toxicity of capsaicinoids as a food additive for humans was negligible. Moreover, if humans were as sensitive as mice, the acute fatal toxicity dose for a 70 kg person would be about 13 g of pure capsaicin which is very high regarding the *Capsicum* content and the quantity taken per one meal. Other studies have concluded that a person of 70 kg weight would have to consume nearly 2.5 L of Tabasco sauce to overdose and to become unconscious. People with few tastebuds in the mouth are not so bothered by the extreme heat. However, most of the people react very negatively to the super hot sauces, experiencing severe burning and sometimes blistering of the mouth and the tongue and immediate responses have included shortness of breath, fainting, nausea and spontaneous vomiting (Geppeti *et al.*, 1988; Palecek *et al.*, 1989; Midgren *et al.*, 1992).

Many other studies on capsaicinoid toxicity have shown that capsaicin directly inhibits protein synthesis by competition with tyrosine in a cell-free system, as well as in a cell culture system according to the cell type and the extracellular concentration (Cochereau *et al.*, 1996). Reports in the literature (Nagabhushan and Bhide, 1985; Lawso and Gannett, 1989; Morré *et al.*, 1995) indicate that capsaicin and its metabolites are able to induce DNA damage and mutagenicity observed in bacteria and *in vivo* micronucleus formation in mice. In the presence of Cu (II) and molecular oxygen, capsaicin was reported to cause strand scission in DNA through an oxidative mechanism (Morré *et al.*, 1995; Cochereau *et al.*, 1996; Cochereau *et al.*, 1997). For example, in Hela cells in culture, 100 μM capsaicin was reported to induce DNA fragmentation and chromatin condensation (Morré *et al.*, 1995).

Usage of capsaicinoids in self-defence weapons

An interesting application of capsaicinoids is their use in self-defence weapons. Defense sprays have become popular for both police use and personal protection. Most of them contain *o*-chlorobenzylidene, malononitrile, chloroacetophenone, oleoresin *Capsicum* or a combination of these ingredients. When applied topically, capsaicin produces a spontaneous inflammatory reaction in mucous membranes, and contact with eyes results in blepharospasm caused by irritation of the corneal nerves. Additional symptoms of eye contact include extreme burning heat, lacrimation, conjunctival edema and hyperemia (Gonzalez *et al.*, 1993). In the nasal mucosa, capsaicin produces burning pain, sneezing and a dose-dependent serious discharge (Lundblad *et al.*, 1984; Geppeti *et al.*, 1988). Contact with skin produces a burning sensation, erythema without vesiculation, while its inhalation results in transitory bronchoconstriction, cough and retrosternal discomfort (Collier and Fuller, 1984; Fuller *et al.*, 1985; Blanc *et al.*, 1991; Midgren *et al.*, 1992). Direct administration of extratracheal capsaicin aerosol in dogs resulted in apnea, brachycardia, hypotension, miosis and aqueous flare (Gonzalez *et al.*, 1993).

Conclusion

Capsaicinoids are a natural group of alkaloids attracting an increasing interest for their use as food additives and for their pharmacological properties. Human intoxication risk with capsaicinoids after ingestion was evaluated to be very low, while its use in self-defence weapons with possible contact with eyes could result in acute blepharospam. Their isolation from the *Capsicum* species and their complete quantification still requires relatively sophisticated apparatus. However, permanent research for low cost determination continues.

References

AOAC Official Method 995.03. Capsaicinoids in *capsicum* and their extractives. Liquid chromatographic method. First action 1995.

Bajaj, K. L. and Kaur, G. (1979) Colorimetric determination of capsaicin in *capsicum* fruits with the Folin-Ciocalteu reagent. *Mikrochim. Acta.*, 1, 81–86.

Bernal, A. M., Calderon, A. A., Pedreno, M. A., Munoz, R., Barcelo, A. R. and Merino, F. C. (1993a) Capsaicin oxidation by peroxidase from *Capsicum annuum* (var. *annuum*) fruits. *J. Agric. Food Chem.*, 41, 1041–1044.

Bernal, A. M., Calderon, A. A., Pedreno, A. M., Barcelo, A. R. and Merino, C. F. (1993b) Dihydrocapsaicin oxidation by *Capsicum annuum* (var. *annuum*) peroxidase. *J. Food Sci.*, 58(679), 611–613.

Blanc, P., Luc, D. and Juarez, C. (1991) Cough in hot pepper workers. *Chest*, 99(1), 27–32.

Caterina, M. J. and Julius, D. (2001) The vanilloid receptor: a molecular gateway to the pain pathway. *Ann. Rev. Neurosci.*, 24, 487–517.

Cochereau, C., Sanchez, D., Bourhaoui, A. and Creppy, E. E. (1996) Capsaicin, a structural analog of tyrosine inhibits the aminoacylation of tRNAtyr. *Toxicol. Appl. Pharmacol.*, 141, 133–137.

Cochereau, C., Sanchez, D. and Creppy, E. E. (1997) Tyrosine prevents capsaicin-induced protein synthesis inhibition in cultured cells. *Toxicology*, 117, 133–139.

Collier, J. and Fuller, R. (1984) Capsaicin inhalation in man and the effects of sodium cromoglycate. *Br. J. Pharmacol.*, 81, 113–117.

Constant, H. L., Cordell, G. A., West, D. P. and Johnson J. H. (1995) Separation and quantification of capsaicinoids using complexation chromatography. *J. Nat. Prod.*, 58(12), 1925–1928.

Constant, H. L. and Cordell, G. A. (1996) Nonivamide, a constituent of *capsicum* oleoresin. *J. Nat. Prod.*, 59(4), 425–426.

Cooper, T. H., Guzinski, J. A. and Fisher, C. (1991) Improved high-performance liquid chromatography method for the determination of major capsaicinoids in *capsicum* oleoresins. *J. Agric. Food Chem.*, 39, 2253–2256.

Dicecco, J. J. (1976) Gas-liquid chromatographic determination of capsaicin. *J. Assoc. Off. Anal. Chem.*, 59(1), 1–4.

Esteves, M. P., Girio, F. M., Amaral-Collaco, M. T., Andrade, M. E. and Empis, J. (1997) Characterization of starch from white and black pepper treated by ionizing radiation. *Sciences des Aliments*, 17, 289–298.

Fujiwake, H., Suzuki, T. and Iwai, K. (1982a) Intracellular distribution of enzymes and intermediates involved in the biosynthesis of capsaicin and its analogs in *capsicum* fruits. *Agric. Biol. Chem.*, 46, 2685–2689.

Fujiwake, H., Suzuki, T. and Iwai. K. (1982b) Capsaicinoid formation in the protoplast from the placenta of *capsicum* fruits. *Agric. Biol. Chem.*, 46, 2591–2595.

Fuller, R., Dixon, C. and Barnes, P. (1985) Bronchoconstrictor response to inhaled capsaicin in humans. *J. Appl. Physiol.*, 58 (4), 1080–1084.

Fung, T., Jeffery, W. and Beveridge, A. D. (1982) The identification of capsaicinoids in tear-gas spray. *J. Forensic Sci.*, 27, 812–821.

Geppeti, P., Fusco, B. and Marabine, S. (1988) Secretion, pain and sneezing induced the application of capsaicin to the nasal mucosa. *Br. J. Pharmacol.*, 93, 509–514.

Gonzalez, G., Rubia, P. and Callar, J. (1993) Reduction of capsaicin-induced ocular pain and neurogenic inflammation by calcium antagonists. *Invest. Ophthalmol. Vis. Sci.*, 34(12), 3329–3335.

Govindarajan, V. S. (1986) *Capsicum*-production, technology, and quality. Part III: chemistry of the color, aroma, and pungency stimuli. *CRC Crit. Rev. Food Sci. Nutr.*, 24(3), 245–355.

Govindarajan, V. S., Narasimhan, S. and Dhanara, S. J. (1977) Evaluation of spices and oleoresin II. Pungency of Scoville Heat Units. A standardized procedure. *J. Food Sci. Technol.*, 14, 28–34.

Hall, R. D., Holden, M. A. and Yeoman, M. M. (1987) The accumulation of phenylpropanoid and capsaicinoid compounds in cell cultures and whole fruit of the chilli pepper *Capsicum frutescens. Mill. Plant Cell, Tiss. Org. Cult.*, 8, 163–176.

Hawer, W. S., Ha, J., Hwang, J. and Nam, Y. (1994) Effective separation and quantitative analysis of major heat principles in red pepper by capillary gas chromatography. *Food Chem.*, 49, 99–103.

Howard, L. R., Talcott, S. T., Brenes, C. H. and Villalon, B. (2000) Changes in phytochemical and antioxidant activity of selected peppers cultivars (*Capsicum* species) as influenced by maturity. *J. Agric. Food Chem.*, 48, 1713–1720.

Inoue, K., Sewalt, V. J. H., Balance, G. M., Ni, W., Sturzen, C. and Dixon, R. A. (1998) Developmental expression and substrate specifities of α,α-caffeic acid-3-0-methyltransferase and caffeoylcoenzyme A 3-0-methyltransferase in relation to lignification. *Plant Physiol.*, 117, 761–770.

Iwai, K., Suzuki, T., Fujiwake, H. and Oha, S. (1979) Simultaneous microdetermination of capsaicin and its four analogues by using high-performance liquid chromatography and gas chromatography–mass spectrometry. *J. Chromatogr.*, 172, 303–311.

Jurenitsch, J., David, M., Heresch, F. and Kubelka, W. (1979) Detection and identification of new pungent compounds in fruits of *capsicum*. *Planta Med.*, 36, 61–67.

Jurenitsch, J. and Leinmuller, R. (1980) Quantification of nonylic acid vanillylamide and other capsaicinoids in the pungent principle of *capsicum* fruits and preparation by gas liquid chromatography on glass capillary columns (in German). *J. Chromatogr.*, 189, 389–393.

Kosuge, S. and Furuta, M. (1970) Studies on the pungent principle of *capsicum*. Part XIV: Chemical constitution of the pungent principle. *Agric. Biol. Chem.*, 34(2), 248–256.

Krajewska, A. M. and Powers, J. J. (1987) Gas chromatographic determination of capsaicinoids in green *capsicum* fruits. *J. Assoc. Off. Anal. Chem. Int.*, 70(5), 926–928.

Laskaridou-Monnerville, A.(1999) Determination of capsaicinoid dihydrocapsaicin by micellar electrokinetic capillary chromatography and its application to various species of *Capsicum*, Solanaceae. *J. Chromatogr. A*, 838, 293–302.

Lawso, T. and Gannett, P. (1989) The mutagenicity of capsaicin and dihydrocapsaicin in V79 cells. *Cancer Lett.*, 49, 109–114.

Lee, K. R., Suzuki, T., Kobashi, M., Hasegawa, K. and Iwai, K. (1976) Quantitative micro analysis of capsaicin, dihydrocapsaicin, nordihydrocapsaicin using mass fragmentography. *J. Chromatogr.*, 123, 119–128.

Lopez-Hernandez, J., Oruna-Concha, M. J., Simal-Lozano, J. M. E., Vazquez-Blanco, M. E. and Gonzalez-Castro, M. J. (1996) Chemical composition of padron peppers (*Capsicum annuum* L.) grown in Galicia (NW Spain). *J. Food Chem.*, 57(4), 557–559.

Lundblad, L., Xiao Y. H. and Lundberg, J. (1984) Mechanisms for reflexive hypertension induced by local application of capsaicin and nicotine to the nasal mucosa. *Acta. Physiol. Scand.*, 121, 277–282.

Machlin, L. J. (1995) Critical assessment of the epidemiologic data concerning the impact of antioxidant nutrients on cancer and cardiovascular disease. *Crit. Rev. Food Sci. Nutr.*, 35, 41–50.

Manirakiza, P., Covaci, A. and Schepens, P. (1999) Solid-phase extraction and gas chromatography with mass spectrometric determination of capsaicin and some of its analogues from chilli peppers (*Capsicum* spp.) *J. Assoc. Off. Anal. Chem. Int.*, 82(6), 1399–1405.

Manirakiza, P., Covaci, A. and Schepens, P. (2000) Single step clean up and GC-MS quantification of organochlorine pesticide residues in spice powder. *Chromatographia*, 52, 787–790.

Markovits, E. and Gilhar, A. (1997) Capsaicin: an effective topical treatment in pain. *Int. J. Dermatol.*, 36, 401–404.

Markus, F., Daood, H. G., Kapitany, J. and Biacs, P. A. (1999) Change in the carotenoid and antioxidant content of spice red pepper (paprika) as a function of ripening and some technological factors. *J. Agric. Food Chem.*, 47, 100–107.

Meilgaard, M., Civille, G. V. and Carr, B. T. (1987) *Sensory Evaluation Techniques*, CRC Press, Boca Rotan, FL, Vol.2.

Midgren, B., Hansson, L. and Karlsson, J. (1992) Capsaicin-induced cough in humans. *Am. Rev. Resp. Dis.*, 146(2), 347–351.

Minani, T., Bakoshi, S., Nakano, H., Mine, O., Muratami, T., Mori, H. and Ato, S. (2001) The effects of capsaicin cream on prostaglandin-induced allodynia. *Anesth. Analg.*, 93, 419–423.

Morré, D. J., Chueh, P. J. and Morré, D. M. (1995) Capsaicin inhibits preferentially the NADH oxidase and growth of transformed cells in culture. *Proc. Natl. Acad. Sci. U.S.A.*, 92, 1831–1835.

Muller, H. (1997) Determination of the carotenoid content in selected vegetables and fruit by HPLC and photodiode array detection. *Z. Lebensm. Unters. Forsch. A*, 204, 88–94.

Murakami, K., Ito, M., Htay, H. H., Tsubouchi, R. and Yoshimo, M. (2001) Antioxidant effect of capsaicinoids on the metal-catalyzed lipid peroxidation. *Biomed. Res. (Tokyo)*, 22(1), 15–17.

Nagabhushan, M. and Bhide, S. V. (1985) Mutagenicity of chilli extract and capsaicin in short-term tests. *Environ. Mutagen.*, 7, 881–888.

Nakatani, N., Inatani, R., Ohta, H. and Nishioka, A. (1986) Chemical constituents of peppers (*Piper* spp.) and application to food preservation: naturally occurring antioxidative compounds. *Environ. Health Perspect.*, 67, 135–142.

Neumann, D. (1966) On the biosynthesis of capsaicin. *Naturwissenchaften.*, 53, 131–136.

Ochoa-Alejo, N. and Gomez-Peralta, J. E. (1998) Activity of enzymes involved in capsaicin biosynthesis in callus tissue and fruits of chilli peppers (*Capsicum annuum* L.). *J. Plant Physiol.*, 141, 147–152.

Palecek, F., Sant'ambrogio, G. and Sant'ambrogio, F. (1989) Reflex responses to capsaicin intravenous, aerosol, and intratracheal administration. *J. Appl. Physiol.*, 67(4), 1428–1437.

Perucka, I. and Oleszek, W. (2000) Extraction and determination of capsaicinoids in fruit of hot pepper *Capsicum annum.* L. by spectrophotometry and high performance liquid chromatography. *Food Chem.*, 71(2), 287–291.

Peusch, M., Muller-Seitz, E., Muller, M. P. A. and Anklam, E. (1997) Extraction of capsaicinoids from chillies (*Capsicum frutescens* L.) and paprika (*Capsicum annuum* L.) using supercritical fluids and organic solvents. *Z. Lebensm. Unters. Forsch. A*, 204, 351–355.

Ramos, P. J. J. (1979) Further study on the spectrophotometric determination of capsaicin. *J. Assoc. Off. Anal. Chem.*, 62, 1168–1170.

Reilly, C. A., Crouch, D. J. and Yost, G. S. (2001a) Quantitative analysis of capsaicinoids in fresh peppers, oleoresin, capsicum, and pepper spray products. *J. Forensic Sci.*, 46(3), 502–509.

Reilly, C. A., Crouch, D. J., Yost, G. S. and Fatah, A. A. (2001b) Determination of capsaicin, dihydrocapsaicin, and nonivamide in self-defense weapons by liquid chromatography–mass spectrometry. *J. Chromatogr. A*, 912, 259–267.

Richeux, F., Cascante, M., Ennamany, R., Saboureau, D. and Creppy, E. E. (1999) Cytotoxicity and genotoxicity of capsaicin in human neuroblastoma cells SHSY-5Y. *Arch. Toxicol.*, 73, 403–409.

Rico Avila, J. (1983) Cultivo del pimiento de Carne gruessa en invernadero. Ministerio de Agricultura, Pesca y Alimentacion, Publicaciones de Extension Agraria, Madrid.

Rymal, K. S., Cosper, R. D. and Smith, D. A. (1979) Injection-extraction procedure for rapid determination of relative pungency in fresh Jalapeno peppers. *J. Assoc. Off. Anal. Chem.*, **67**, 658–659.

Saria, A., Lambeck, F. and Skofitsch, G. (1981) Determination of capsaicin in tissue and separation of capsaicin analogues by HPLC. *J. Chromatogr.*, **208**, 41–46.

Singh, S., Asad, S. F., Ahmad, A., Khan, N. U. and Hadi, S. M. (2001) Oxidative DNA damage by capsaicin and dihydrocapsaicin in the presence of Cu(II). *Cancer Lett.*, **169**, 139–146.

Sprague, J., Harrison, C., Rowbotham, D. J., Smart, D. and Lambert, D. G. (2001) Temperature-dependent activation of recombinant rat vanilloid VR1 receptors expressed in HEK293 cells by capsaicin and anandamide. *Eur. J. Pharmacol.*, **423**, 121–125.

Sukrasno, N. and Yeoman, M. M. (1993) Phenylpropanoid metabolism during growth and development of *Capsicum frutescens. Phytochem.*, **32**, 839–844.

Suzuki, T., Fujiwake, H. and Iwai, K. (1980) Intracellular localization of capsaicin and its analogues, capsaicinoids, in *capsicum* fruit. Microscopic investigation of the structure of the placenta of *Capsicum annuum* var.*annuum* cv. Karayatsubusa. *Plant Cell Physiol.*, **21**, 33–37.

Szallasi, A. and Blumberg, P. M. (1999) Vanilloid (*capsicum*) receptor and mechanisms. *Pharmacol. Rev.*, **31**, 139–212.

Thomas, B. V., Schreiber, A. A. and Weisskopf, C. P. (1998) Simple method for quantification of capsaicinoids in peppers using capillary gas chromatography. *J. Agric. Food Chem.*, **46**, 2655–2663.

Todd, P. H., Besinger, M. G. and Biftu, T. (1977) Determination of pungency due to *capsicum* by gas–liquid chromatography. *J. Food Sci.*, **42**, 660–665.

Zewdie, Y. and Bosland, P. W. (2000) Capsaicinoid inheritance in an interspecific hybridization of *Capsicum annuum* × *Capsicum chinense. J. Am. Soc. Hortic. Sci.*, **125**(4), 448–453.

Zewdie, Y. and Bosland, P. W. (2001) Capsaicinoids profiles are not good chemotaxonomic indicators for capsaicin species. *Biochem. Syst. Ecol.*, **29**(2), 161–169.

5 Biosynthesis of capsaicinoids in *Capsicum*

Keiko Ishikawa

In order to clarify the factors controlling capsaicinoid biosynthesis, the contents of capsaicinoids and their phenolic intermediates in: (1) the placentas of *Capsicum annuum* L. 'Jalapeño' and 'Shimofusa' during the fruit-developing period; (2) various organs of *C. annuum* L. 'Jalapeño' and; (3) the placentas and septa of tetraploid plants of *C. annuum* L. 'Shishitoh' were investigated. The content of free phenolics in placentas was less than one-fifteenth of the capsaicinoid content in placenta throughout the observation period. The contents of capsaicinoids and their phenolic intermediates in various organs such as peduncles, leaves, stems and roots were less than those in placentas. The contents of capsaicinoids and their phenolic intermediates in the placentas and septa of tetraploid plants were almost the same as those in diploid plants.

Introduction

Capsaicinoids are the pungency principles found in the fruits of *Capsicum*. The contents of capsaicinoids are quite important for the use of the fruits of *Capsicum*. So the regulation mechanism of the biosynthesis of capsaicinoids has been investigated intensively.

Capsaicinoid is a compound group that is an acid amide of vanillylamine and a C_9–C_{11} branched-chain fatty acid. At first, the pungency principle of *Capsicum* was thought to be a single compound, capsaicin. According to the description of Kosuge *et al.* (1957), Micko isolated capsaicin in 1898 and in 1923, Nelson determined the chemical constitution of capsaicin (8-methyl-N-vanillyl-6-nonenamide). In 1957, Kosuge *et al.*, identified another compound in capsaicin. On studying the chemical constitution, he named the second compound dihydrocapsaicin (8-methyl-N-vanillyl-nonaamide), and proposed to call the pungency principle 'capsaicinoid' (Kosuge *et al.*, 1961). To date, 22 compounds have been listed as capsaicinoids: capsaicin, dihydrocapsaicin, isomer of dihydrocapsaicin, homocapsaicin, homocapsaicin II, homodihydrocapsaicin, homodihydrocapsaicin II, norcapsaicin, nornorcapsaicin, etc. (Bosland and Votava, 1999). Krajewska and Powers (1988) reported sensory properties of capsaicin, dihydrocapsaicin, nordihydrocapsaicin and homodihydrocapsaicin. When solutions of low concentrations were examined, the differences in character, duration and location of pungency in the tongue and palate were observed. And although individual compounds had different values of Scoville Heat Units, the total pungency of the mixed compounds was highly correlated with the sum of pungency of individual compounds.

Ohta (1962) and then Fujiwake *et al.* (1980) reported that capsaicinoids were localized in the vacuole of the epidermal cells of placentas and septa, and the accumulation of capsaicinoids reached the maximum level when the size of the fruits reached the maximum. Experiments using radioactive substances revealed that capsaicinoids were biosynthesized from the two arms; the aromatic component from L-phenylalanine to vanillylamine, and C_9–C_{11} branched-chain

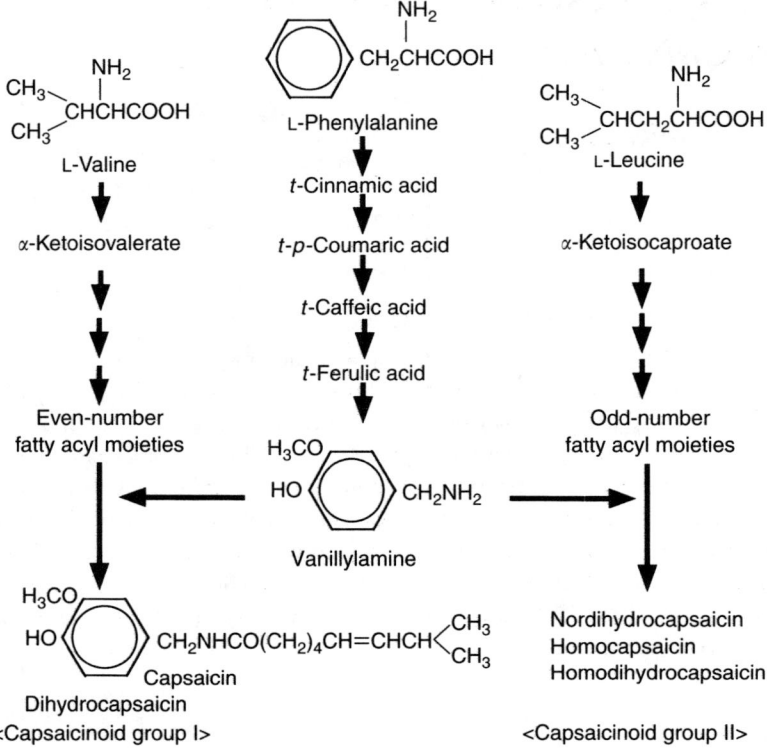

Figure 5.1 Proposed biosynthetic pathway of capsaicinoids.

fatty acid moieties from L-valine or L-leucine (Bennet and Kirby, 1968; Leete and Louden, 1968) (Figure 5.1). The intensive work of Iwai was particularly important for the establishment of the fundamental aspects of the biosynthetic pathway of capsaicinoids (Iwai, 1986). The pathway proposed for the biosynthesis of the vanillylamine moiety of capsaicinoids from L-phenylalanine is as follows: L-phenylalanine, *trans*-cinnamic acid, *trans-p*-coumaric acid, *trans*-caffeic acid, *trans*-ferulic acid, vanillin and vanillylamine. In the pathway of biosynthesis of capsaicinoids, the last step of condensation of vanillylamine and fatty acids is expected to be important, but the enzyme for that step has not so far been purified. Suppressive or stimulative factors are also unknown so far.

So we considered it to be important to characterize the biosynthetic status of capsaicinoids. At the beginning of this chapter, the time-course studies of the contents of capsaicinoids and their phenolic intermediates in the placenta of the two cultivars of *C. annuum* were described (Sakamoto *et al.*, 1994). One cultivar was the highly pungent 'Jalapeño', while the other was the non-pungent bell pepper 'Shimofusa'. In this study, the importance of phenolic intermediates was discussed. These results were slightly different from those of *C. frutescens* (Sukrasno and Yeoman, 1993).

The contents of the capsaicinoids and their phenolic intermediates in the various organs of *C. annuum* L. were described in a study by Ishikawa *et al.* (1998). In this study, the pungent 'Jalapeño' was used as a plant material. The contents of vanillylamine or phenolic intermediates in peduncles, leaves, stems and roots were less than that in the placenta.

The capsaicinoid contents in tetraploid plants of *C. annuum* L. 'Shishitoh' were described in a further study by Ishikawa *et al.* (2000). Tetraploid plants were established in order to investigate the biosynthesis of capsaicinoids in different physiological conditions. Tetraploid plants of *Capsicum* were easily obtained by the colchicine treatment of the seeds (Ishikawa *et al.*, 1997). The fruits of 'Shishitoh' are usually non-pungent, and consumed in the young green stage as vegetables in Japan. 'Shishitoh' produces a lot of elongated fruits with a thin pericarp, compared with a typical bell pepper like the 'California Wonder'. In a stress condition, such as dryness or high temperature, the fruits of 'Shishitoh' become easily pungent. So this cultivar is a suitable plant for the investigation of the factors affecting on the capsaicinoid contents. Our results showed that the seed number and the length of the fruits were reduced, and that the accumulation of capsaicinoids and their phenolic intermediates did not change.

Materials and methods

Capsaicinoid contents during the fruit-developing period

The seeds of *C. annuum* L. 'Jalapeño' and 'Shimofusa' (Japan Horticultural Production and Research Institute) were sown in May and the plants were grown for about ten months in the greenhouse. The fruits were removed two to twelve weeks after flowering and capsaicinoids were extracted from their placentas by the methods described next.

Capsaicinoid contents in the various organs

The placentas, peduncles, leaves, stems and roots of *C. annuum* L. 'Jalapeño' growing in the greenhouse for ten months were used for the analysis of capsaicinoid contents. Capsaicinoids were extracted from the placentas of the fruits two to twelve weeks after flowering by the methods described next.

Establishment of tetraploid plants of Capsicum

The seeds of *C. annuum* L. 'Shishitoh' (Nihon Horticultural Production Institute) were surface-sterilized with 1% sodium hypochlorite solution for 15 minutes and rinsed three times in sterile distilled water. About 40 seeds were then soaked in 10 ml of 1% colchicine solution in Petri dishes (9 cm in diameter), in which one sheet of filter paper was put. Four days after soaking, the seeds were transferred onto 'Murashige and Skoogs' medium (Murashige and Skoogs, 1962) with 3% sucrose and 0.2% Gellan Gum, and cultured at 25°C under continuous light condition. Several weeks after culture, plantlets were acclimatized and transferred to pots, and cultured at 25°C under continuous light condition. Ten weeks after the colchicine treatment, the DNA contents of the leaves of these plantlets were analysed by flow cytometry.

For the flow cytometric analysis, the leaf segment (about 0.5 cm^2) was chopped with a sharp razor blade in the buffer (Solution A of plant high resolution DNA kit type P, Partec, Germany). After chopping, the suspension was stained with staining solution (Solution B of plant high resolution DNA kit type P, Partec, Germany) and filtrated with 40 µm nylon mesh and immediately analysed with flow cytometer (Partec PA, Germany).

Morphology and capsaicinoids of the fruits of tetraploid 'Shishitoh'

Tetraploid plants of 'Shishitoh' were grown in a greenhouse for ten months with non-colchicine treated plants as the control. Four fruits were taken, and the seed number of the fruits was counted. The weight, length and diameter of the fruits, and the weight of placentas and septa were measured. Half of the placentas and septa were used for an analysis of capsaicinoid content.

Capsaicinoids were extracted from the placentas and septa of the fruits of tetraploid 'Shishitoh' by the methods described here.

Analysis of capsaicinoid contents by HPLC

The tissues were weighed and extracted twice with 80% ethanol. The extract was then evaporated to dryness and dissolved in 1 ml of dimethyl sulfoxide (DMSO). Free phenolic intermediates, including cinnamic acid, coumaric acid, caffeic acid, ferulic acid and vanillylamine, together with capsaicin and dihydrocapsaicin (CAPS), were quantified by the HPLC method. HPLC was carried out according to the method of Johnson *et al.* (1992) with a slight modification to the solvent gradient and detection wavelength (280 nm). The gradient conditions controlled by a Tosoh SC-8010 HPLC system were as follows: 0–50 minutes, linear gradient from 0% to 100% CH_3CN at 1 ml/min in 0.1% aqueous trifluoroacetic acid; 50–65 minutes, 100% CH_3CN. The analytical column used was an Inertsil ODS-2 type (4.6 mm i.d. × 15 cm, GL Sciences) with an injection volume of 20 μl. The effluent was monitored by a Tosoh UV-8010 detector and a Waters 991J photodiode-array detector. Authentic samples of vanillylamine, capsaicin and dihydrocapsaicin were presented by Professor H. Kamada, while authentic samples of cinnamic acid, coumaric acid, caffeic acid and ferulic acid were presented by Professor U. Sankawa.

Results and discussion

Capsaicinoid contents during fruit-developing period

As cinnamic acid, coumaric acid, caffeic acid and ferulic acid are ubiquitously distributed in the main stream of phenylpropanoid metabolism, their contents were summed and were presented as the content of phenylpropanoids. In addition, as CAPS are known to be the two major components that constitute more than 90% of total capsaicinoids, their contents were also summed and were presented as the CAPS content.

In the 'Jalapeño' placenta, CAPS started to accumulate after four weeks, and active accumulation continued until the eighth week (Figure 5.2). After reaching the maximum value, the CAPS contents remained practically constant for as long as observation was continued. In the case of 'Shimofusa', little CAPS accumulated, even at ten weeks, the content being about one-thousandth of that in 'Jalapeño'.

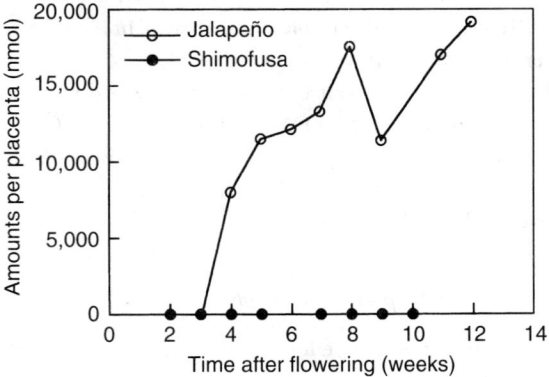

Figure 5.2 Accumulation of CAPS in the placentas of *C. annuum* L. 'Jalapeño' and *C. annuum* L. 'Shimofusa'.

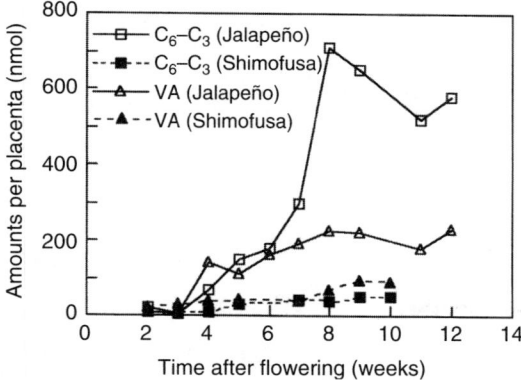

Figure 5.3 Accumulation of phenylpropanoids (C_6–C_3) and vanillylamine (VA) in the placentas of *C. annuum* L. 'Jalapeño' and *C. annuum* L. 'Shimofusa'. C_6–C_3: cinnamic acid, coumaric acid, caffeic acid and ferulic acid.

Phenylpropanoids were also actively accumulated in the 'Jalapeño' placentas, together with CAPS accumulation (Figure 5.3). However, their content was less than one-fifteenth of the CAPS content throughout the observation period. On the other hand, vanillylamine started to accumulate after four weeks, although the amount was less than that of phenylpropanoids.

In the case of 'Shimofusa', no remarkable changes were observed in the contents of phenyl-propanoids and vanillylamine during the fruit-developing period. The maximum contents were not markedly different from those in 'Jalapeño', in comparison with the differences between the CAPS contents in 'Jalapeño' and 'Shimofusa'. It is of interest to note that a substantial amount of phenolics was also produced in the placenta of 'Shimofusa' which is not pungent.

Sukrasno and Yeoman (1993) have reported that no free phenolic intermediates were detected in the fruit of *C. frutescens*. However, we detected free phenolics in the placentas of two cultivars of *C. annuum*. The maximum total content was almost the same as that of glycosylated phenolics observed in the fruits of *C. frutescens* (about 3,000 nmol per fruit). We cannot account for the importance of glycosylation of the phenolic intermediates in the capsaicinoid biosynthesis; however, the difference between the free and glycosylated forms might be attributable to the subgenus of the *Capsicum* fruits.

Sukrasno and Yeoman (1993) also reported that the capsaicinoid accumulation in the fruit concurred with the disappearance of cinnamoyl glycosides, and suggested that cinnamoyl glyco-sides were accumulated mainly in the seeds, rather than in the placenta, on the basis of tissue compartmentation analyses of the phenolics in well-matured fruit. In our study of the placentas of 'Jalapeño', however, no accumulation of phenolics occurred prior to that of CAPS.

Connecting these observations with the fact that capsaicinoids were accumulated in the placenta, a translocation system for the phenolic intermediates from the seeds to placenta seems to have been involved. More detailed kinetic studies on the phenolics in various tissues of *Capsicum* fruits will be required to elucidate the translocation system for the phenolic intermediates and its role in the biosynthesis of capsaicinoids.

Capsaicinoid contents in the various organs

The CAPS were specially accumulated in the placenta (Figure 5.4). The amount of CAPS in the peduncles, leaves, stems and roots was as small as *c*.0.5 μmol/g, which was only 0.7% of the amount in placentas.

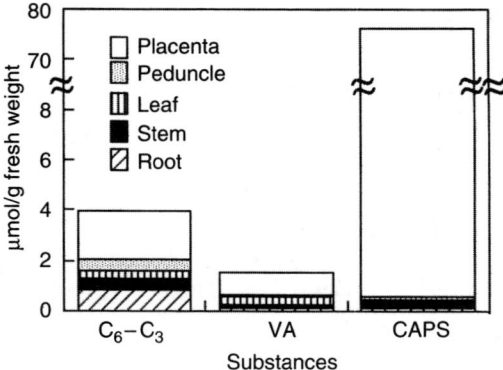

Figure 5.4 The contents of capsaicinoids in various organs. CAPS: CAPS; C_6–C_3, cinnamic acid, coumaric acid, caffeic acid and ferulic acid; VA: vanillylamine.

The average content of phenylpropanoids (C_6–C_3) in the placenta and other organs was 1.97 and 0.54 μmol/g, respectively. The average content of vanillylamine in the placenta and other organs was 0.90 and 0.14 μmol/g, respectively. The amounts of both phenylpropanoids and vanillylamine in placenta were higher than those in other organs.

Our result showed that the amounts of these phenolic intermediates and vanillylamine in placenta and in other organs were not correlated to the production of CAPS.

Morphology and capsaicinoids of the fruits of tetraploid 'Shishitoh'

Germination rates were 100% in both control and colchicine treated conditions. Polyploidy was determined by the flow cytometric analysis of the leaves. Main peaks of 2C and 4C were observed in the leaves of diploid and tetraploid plants, respectively (Figure 5.5). By the flow cytometric analysis, about 20% of the plants were identified as tetraploid. These plantlets had short, thick roots and hypocotyls.

After acclimatization, these tetraploid plantlets were grown in a greenhouse together with their diploid control plants for about ten months, and the fruits were collected (Figure 5.6).

The average seed number of tetraploid 'Shishitoh' was about 40% of that of diploid, although the average fresh weight of placenta and septum of tetraploid was about 90% of that of diploid (Table 5.1). Because a lot of tiny undeveloped seeds were observed on the placentas of tetraploid fruits, it was expected that the development of the embryos might be suppressed in tetraploid plants.

Although the diameter of the diploid and tetraploid fruits was about 2 cm, the average length of the tetraploid fruits was only 4.5 cm, which was 60% of the average length of the diploid fruits. It is expected that the elongation of the pericarp cells of tetraploid fruit was suppressed. It is reported that reduction of the seed number induced the suppression of the fruit development (Polowick and Sawhney, 1985).

The contents of CAPS and phenolic intermediates in the fruits of tetraploid 'Shishitoh' were almost the same as those in the fruits of diploid 'Shishitoh' (Table 5.2). The capsaicin contents of the placentas and septa of diploid and tetraploid fruits were both about 2 μg/g fresh weight. Our results agreed with the report by Yazawa *et al.* (1989b), in which the capsaicin content of

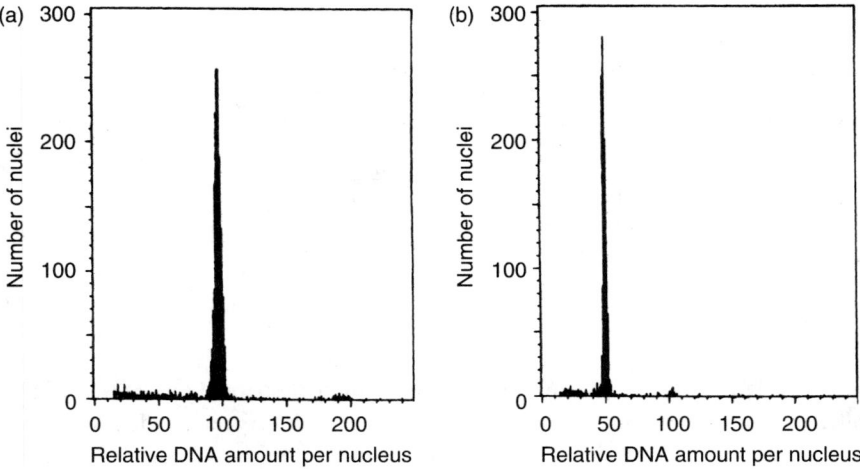

Figure 5.5 Typical flow cytometric histograms of the nuclei isolated from (a) a young diploid leaf and (b) a young tetraploid leaf.

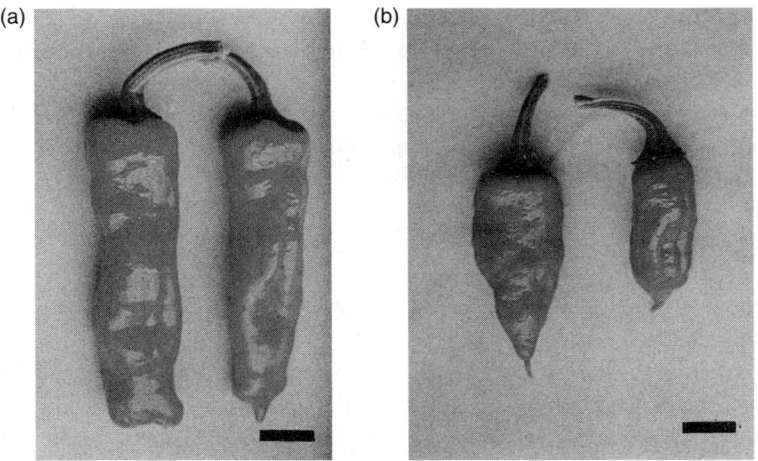

Figure 5.6 Typical fruits of (a) diploid and (b) tetraploid 'Shishitoh' Bar shows 1 cm.

Table 5.1 Morphology of the mature fruits of diploid and tetraploid 'Shishitoh'

Characters	Diploid	Tetraploid
No. of seeds	73.0 ± 15.8	27.3 ± 5.2
The length of the fruits (mm)	69.9 ± 4.8	44.7 ± 1.2
The diameter of the fruits (mm)	21.5 ± 0.4	20.0 ± 1.2
The weight of the fruits (g)	8.11 ± 1.13	5.47 ± 0.35
The weight of the placenta and interlocular septum (mg)	621.0 ± 75.2	546.2 ± 50.2

Note
Mean $\pm$ SE, $n = 4$.

Table 5.2 The contents of capsaicinoids of the mature fruits of
diploid and tetraploid 'Shishitoh'

Capsaicinoids	Diploid	Tetraploid
Vanillylamine	1.43 ± 0.30	0.97 ± 0.56
Caffeic acid	77.38 ± 21.68	74.20 ± 13.17
Ferulic acid	4.18 ± 1.58	7.33 ± 2.39
Capsaicin	2.06 ± 1.56	1.60 ± 0.54
Dihydrocapsaicin	2.16 ± 1.50	1.70 ± 0.57

Mean ± SE (μ g/g), $n = 4$.

'Shishitoh' was 18 μg/g dry weight. We also detected 14 mg/g fresh weight in the fruit of 'Shishitoh' in a different condition, the same level with the pungency cultivars such as 'Jalapeño' (Sato *et al.*, 1999). It is concluded that the pungency of tetraploid fruits of 'Shishitoh' was not affected.

The ratio of CAPS content was almost 1 : 1, which agreed with the report by Yazawa *et al.* (1989b), and also with the result of 'Jalapeño' (Sakamoto *et al.*, 1994). One of the famous pungent Japanese peppers 'Takanotsume' contains about 1800 μg/g dry weight of capsaicin and 300 μg/g dry weight of dihydrocapsaicin (Yazawa *et al.*, 1989a). It is expected that the ratio of capsaicin to dihydrocapsaicin is different among the cultivars.

Our results showed that although the seed number and the fruit length of tetraploid 'Shishitoh' were reduced, the content of capsaicinoids and their phenolic intermediates were the same as that of diploid 'Shishitoh'.

In order to clarify the factors controlling the biosynthesis of capsaicinoids, we are now investigating the other conditions that may change the pungency of *Capsicum*.

Conclusions

In the breeding and production of *Capsicum*, the regulation of capsaicinoid biosynthesis is one of the most important subjects.

In order to clarify the biosynthetic status, the contents of the capsaicinoids and their phenolic intermediates in the placentas of the pungent fruits of *C. annuum* L. 'Jalapeño', and the non-pungent fruits of *C. annuum* L. 'Shimofusa', during fruit-developing periods were investigated. The content of phenolic intermediates in the placentas of 'Jalapeño' was less than one-fifteenth of the capsaicinoid content throughout the observation period. Also in peduncles, leaves, stems and roots, the contents of capsaicinoids and phenolic intermeditates were less than those in placentas.

In order to investigate the biosynthesis of capsaicinoids in different physiological conditions, tetraploid plants of *C. annuum* L. 'Shishitoh', were obtained by colchicine treatment of the seeds. Although the seed number and the fruit length of the tetraploid plants were reduced to 40% and 60%, respectively, of diploid plants, the contents of capsaicinoids and their phenolic intermediates were almost the same.

References

Bennett, D. J. and Kirby, G. W. (1968) Constitution and biosynthesis of capsaicin. *J. Chem. Soc.*, C, 442–446.

Bosland, P. W. and Votava, E. J. (1999) *Pepper: Vegetable and Spice Capsicums.* pp. 91–95. Wallingford: CABI publishing.

Fujiwake, H., Suzuki, T. and Iwai, K. (1980) Intracellular localization of capsaicin and its analogues in *Capsicum* fruit II. The vacuole as the intracellular accumulation site of capsaicinoid in the protoplast of *Capsicum* fruit. *Plant Cell Physiol.*, 21(6), 1023–1030.

Kosuge, S., Inagaki, Y. and Uehara, K. (1957) Studies on the pungent principles of *Capsicum*. Part I. On the chemical constitution of the pungent principles. (1) On isolation of the pungent principles. *J. Agric. Chem. Soc. Japan*, 32(8), 578–581.

Kosuge, S., Inagaki, Y. and Okamura, H. (1961) Studies on the pungent principles of red pepper. Part VIII. On the chemical constitutions of the pungent principles. (5) On the chemical constitution of the pungent principle II. *J. Agric. Chem. Soc. Japan*, 35(10), 923–927.

Krajewska, A. M. and Powers, J. J. (1988) Sensory properties of naturally occurring capsaicinoids. *J. Food Science*, 53(3), 902–905.

Leete, E. and Louden, M. C. L. (1968) Biosynthesis of capsaicin and dihydrocapsaicin in *Capsicum frutescens*. *J. Am. Chem. Soc.*, 90(24), 6837–6841.

Ishikawa, K., Mishiba, K., Yochida, H. and Nunomura, O. (1997) Establishment of tetraploid plants of *Capsicum annuum* L. by colchicine treatment with the analysis of flow cytometry. *Capsicum and Eggplant Newslet.*, 16, 22–25.

Ishikawa, K., Janos, T., Sakamoto, S. and Nunomura, O. (1998) The contents of capsaicinoids and their phenolic intermediates in the various tissues of the plants of *Capsicum annuum* L. *Capsicum and Eggplant Newslet.*, 17, 44–47.

Ishikawa, K., Kuboki, H., Sato, K., Maitani, T. and Nunomura, O. (2000) Morphology and the contents of capsaicinoids of mature fruits of tetraploid plants of *Capsicum annuum* L. cv. 'Shishitoh'. *Jpn. J. Food Chem.*, 7(2), 74–77.

Iwai, K. (1986) Biosynthesis of some useful food constituents and their applications (Review article). *Nippon Nogeikagaku Kaishi.*, 60(3), 219–226.

Johnson, T. S., Ravishankar, G. A. and Venkataraman, L. V. (1992) Separation of capsaicin from phenylpropanoid compounds by high-performance liquid chromatography to determine the biosynthetic status of cells and tissues of *Capsicum frutescens* Mill. *in vivo* and *in vitro*. *J. Agric. Food. Chem.*, 40, 2461–2463.

Murashige, T. and Skoog, F. (1962) A revised medium for rapid growth and bioassays with tobacco tissue cultures. *Physiol. Plant.*, 15, 473–497.

Ohta, Y. (1962) Physiological and genetical studies on the pungency of pepper IV capsaicin-secreting organs and receptacles in the fruits of *Capsicum annuum* L. *Jpn. J. Breeding*, 12, 43–47.

Polowick, P. L. and Sawhney, V. K. (1985) Temperature effects on male fertility and flower and fruit development in *Capsicum annuum* L. *Sci. Hort.*, 25, 117–127.

Sakamoto, S., Goda, Y., Maitani, T., Nunomura, O. and Ishikawa, K. (1994) High-performance liquid chromatographic analyses of capsaicinoids and their phenolic intermediates in *Capsicum annuum* to characterize their biosynthetic status. *Biosci. Biotech. Biochem.*, 58(6), 1141–1142.

Sato, K., Sasaki, S. S., Goda, Y., Yamada, T., Nunomura, O., Ishikawa, K. and Maitani, T. (1999) Direct connection of supercritical fluid extraction and supercritical fluid chromatography as a rapid quantitative method for capsaicinoids in placentas of Capsicum. *J. Agr. Food Chem.*, 47(11), 4665–4668.

Sukrasno, N. and Yeoman, M. M. (1993) Phenylpropanoid metabolism during growth and development of *Capsicum frutescens* fruits. *Phytochem.*, 32(4), 839–844.

Yazawa, S., Ueda, M., Suetome, N. and Namiki, T. (1989a) Capsaicinoids contents in the fruit of interspecific hybrids in *Capsicum*. *J. Japan. Soc. Hort. Sci.*, 58(2), 353–360.

Yazawa, S., Suetome, N., Okamoto, K. and Namiki, T. (1989b) Content of capsaicinoids and capsaicinoid-like substances in fruit of pepper (*Capsicum annuum* L.) hybrids made with 'CH-19 Sweet' as a parent. *J. Japan. Soc. Hort. Sci.,* 58(3), 601–607.

6 Biotechnological studies on *Capsicum* for metabolite production and plant improvement

G. A. Ravishankar, B. Suresh, P. Giridhar, S. Ramachandra Rao and T. Sudhakar Johnson

Capsicum is a versatile plant used as a vegetable, a pungent food additive, a colourant and a pharmaceutical. Capsaicinoids and carotenoids are the major chemical constituents of importance, which add high commercial value to this plant. The studies on the production pathway of these secondary metabolites and their regulation is a subject which needs attention in order to develop a biochemical understanding of their formation. Genetic engineering and biotechnology work need to be carried out to achieve the targets of pre-harvest improvement and post-harvest characterization for value addition of this plant. Nutraceutical, pharmaceutical and colour principles of *Capsicum* will become useful in years to come. This review focuses on the chemistry, biosynthetic pathway of capsaicinoid and carotenoid, genetic engineering, downstream processing, industrial prospects and the relevance of biotechnology.

Introduction

The genus *Capsicum*, which is commonly known as chilli, "red chile", "chilli peppers", "hot red pepper", "tabasco", "paprika", "cayenne", etc., is a member of the family Solanaceae. This genus has approximately 22 wild species and five domesticated species (Table 6.1) namely *Capsicum annuum*, *Capsicum frutescens*, *Capsicum baccatum*, *Capsicum chinense* and *Capsicum pubescens* (Bosland, 1994). It is a small perennial shrub and the fruit is a berry. The *Capsicum* plant grows optimally in conditions of 7.27°C with annual precipitation of 0.3–4.6 m and a soil pH of 4.3–8.7 (Simon *et al.*, 1984). They grow best in well drained, sandy or silt-loam soil and are cold sensitive. Plantings are established by seedling or transplanting. Hot and dry weather is desirable for the fruit ripening.

Chilli bred for generating better yield, should have superior genetic potential, protection against insects and pests and improved quality. Hybridization in chillies was tried by both inter and intraspecific crosses and backcrosses. Haploid breeding and single seed descent have also been tried. As it is a self-pollinating crop, commercial production of hybrid chillies has been successful by hand emasculation. Molecular marker-assisted selection techniques provide new tools for the breeding of chillies. In this regard, isozyme and molecular markers have been applied to chillies. Prince *et al.* (1993) have reported a saturated isozyme and RFLP map of chilli which contains 192 chilli and tomato genomic cDNA clones with 19 linkage groups with a total coverage of 720 cM. Specific map positions of 26 RFLP markers in seven linkage groups were not determined and vast regions of the chilli genome remain unmapped.

The total capsaicinoid content, and its composition in various chilli varieties, is given in Table 6.2 (Jurenitsch *et al.*, 1979). Capsaicin contents of various chilli varieties have been reported to range from 0.2% to 1.0% (Tewari, 1990). Very high values of above 1.0% have been reported in some highly pungent Indian varieties of *C. annuum* fruits (Ananthasamy *et al.*, 1960;

Table 6.1 List of major *Capsicum* species

Species	New world distribution
C. annuum L.	Colombia north to southern USA
C. baccatum L.	Argentina, Bolivia, Brazil, Paraguay, Peru
C. buforum A.T. Hunz.	Brazil
C. campylopodium Sendt.	Brazil (southern)
C. cardenasii Heiser & Smith	Bolivia
*C. chacoense A.T. Hunz.	Argentina, Bolivia, Paraguay
*C. chinense Jacq.	Latin and South America
*C. coccineum (Rusby) A.T. Hunz.	Bolivia, Peru
C. cornutum (Hiern) A.T. Hunz.	Brazil (southern)
C. dimorphum (Miers) O.K.	Colombia
C. dusenii Bitter	Brazil (southeast)
*C. eximium A.T. Hunz.	Argentina, Bolivia
*C. galapagoensis A.T. Hunz.	Galapagos Islands
*C. frutescens	
C. geminifolium (Dammer) A.T. Hunz.	Colombia, Ecuador
C. hookerianum (Miers) O.K.	Ecuador
C. lanceolatum (Greenm.) Morton & Standley	Guatemala, Honduras, Mexico
C. leptopodum (Dunal) O.K.	Brazil
C. minutiflorum (Rusby) Hunz.	Argentina, Bolivia, Paraguay
C. mirabile Mart. ex Sendt.	Brazil (southern)
C. parvifolium Sendt.	Brazil (northeast), Colombia, Venezuela
*C. praetermissum Heiser & Smith	Brazil (southern)
C. pubescens R. & P.	Latin and South America
C. schottianum Sendt.	Argentina, Brazil (S), Paraguay (SE)
C. scolnikianum A.T. Hunz.	Peru
C. tovarii nom. nud.	Peru
C. villosum Sendt.	Brazil (southern)

Note
* Widely used species.

Table 6.2 Capsaicinoids content and oleoresin yield of some world varieties of chillies

Variety/source	Capsaicinoids in chilli (%)	Oleoresin yield (%)	Capsaicinoids in oleoresin (%)
Mombasa, Uganda	0.80–0.85	12–12.5	6.8–6.9
Mombasa, African	0.42	13.1	3.2
Small chillies, African	0.82	13.3	6.2
Bahamian	0.51	12.5	4.1
Bird chillies, India	0.36	8.7	4.1
Santaka, Japan	0.30	11.5	2.6
Sannam, India	0.33	16.5	2.0
Mundu, India	0.23	16	1.4
Jwala, India	0.63	9	7
Green chillies, India	0.69	8	8.6

Source: Govindarajan, 1985.

Deb *et al.*, 1963; Thirumalachar, 1967). Mathew *et al.* (1971) reported that the capsaicin contents of Indian chillies varied between 0.2% and 0.5%.

Most of the bigger red-coloured fruits cultivated and marketed the world over, including chillies, paprika and *Capsicum*, belong to the species *C. annuum*, while the highly pungent ones belong to *C. frutescens*.

Capsicum oleoresin

The term "oleoresin" has been used for desolventized total extracts by a specified solvent. In its use as a food additive, the best oleoresin of *Capsicum* is that which contains those components which constitute colour and flavour (pungency, aroma and sensory factors). Oleoresins are essentially divided into three types. *Oleoresin paprika* is used as a food colouring agent in processed meats, dairy products, soups, sauces and snacks. *Oleoresin red pepper* is a source of both colour and pungency, essentially used in canned meats, sausages, in some snacks and in a dispersed form in some drinks, such as gingerale. *Oleoresin Capsicum* (African) is the most pungent and used for the counter-irritant property in plasters and pharmaceutical preparations.

Despite the advantages in the use of oleoresin over the ground spices, the world production and use of spice oleoresins is only around 1,600–1,800 tons and represents only 10% of total international spice trade. Oleoresin chillies with a pungency range of $0.25–1.0 \times 10^6$ Scoville units (1.66–6.66% capsaicinoids) and colour up to 20,000 units are used by larger food industries in the USA and the UK (Johnson, 1993).

Pungency and its measurement

Kobayashi (1925) revealed that the vanillylamide moiety and acyl residues with appropriate chain lengths were required for exerting pungency. Nelson (1919) had reported that pungency depends on chain length. The alkyl residue with $C_8 H_{17}$ showed the strongest pungency. Decrease in pungency was found to be associated with both longer and shorter alkyl chain lengths (Nelson, 1919). Pungency seems to depend not only on the presence of the methoxy group but also on the position of the hydroxy group.

The basic principle of pungency evaluation using an organoleptic method was established by Scoville (1912). Scoville quantified the pungency of *Capsicum* fruits by examining diluted ethanolic sweetened solutions of *Capsicum* extract to determine the greatest dilution at which definite pungency could be recognized. Scoville (1912) expressed the greatest dilution as the reciprocal of the dilution in Scoville Heat Units. Scoville Heat Units for pure capsaicin are reported as $15–17 \times 10^6$ (Suzuki *et al.*, 1957; Todd, 1958). The pungency of various chilli varieties varies from 0 to 300,000 SHU (Table 6.3).

Suzuki *et al.* (1957) compared the pungency determined by the organoleptic method to that of chemical methods. Hartman (1970) correlated SHU to the proportion of capsaicinoid content to GLC. Johnson *et al.* (1993) developed the HPLC method for the separation of capsaicin and other phenyl propanoides. Govindarajan *et al.* (1977) proposed a standardized procedure for the evaluation of the pungency of SHU by which a linear regression was obtained between SHU and the capsaicin content of samples. The pungency, total capsaicinoid content and oleoresin yield of various chilli pepper varieties of the world are presented in Table 6.4.

Capsaicin and four related compounds – dihydrocapsaicin, nordihydrocapsaicin, homocapsaicin and homodihydrocapsaicin – are all responsible for pungency. Of these, the first two compounds are equally pungent and constitute 80–90% of the total capsaicinoids; the latter two are present in minor amounts and possess only half the pungency (Maga, 1975). The pungency

Table 6.3 Pungency of various chilli pepper varieties

Pepper type	Heat rating (in Scoville Heat Units)
Habanero	200,000–300,000
Red Amazon	75,000
Pequin	75,000
Chiltecpin	70,000–75,000
Tabasco	30,00–50,000
Cayenne	35,000
Arbol	25,000
Japone	25,000
Smoked Jalepeno (Chipotle)	10,000
Serrano	7,000–25,000
Puya	5,000
Guajillo	5,000
Jalepeno	3,500–4,500
Poblano	2,500–3,000
Pasilla	2,500
TAM Mild Jalepeno-1	1,000–1,500
Anaheim	1,000–1,400
New Mexican	1,000
Ancho	1,000
Bell & Pimento	0

Note
0–5,000: mild; 5,000–20,000: medium; 20,000–70,000: hot;
70,000–300,000: extremely hot.

Table 6.4 Trade types of chillies worldwide

Trade name	Source	Pungency Scoville units (% capsaicinoids)	Total capsaicinoids	Principal use	
Fukien rice chillies	China	120,000 (0.8)	0.5–0.95	↑	
Pusa Jwala	India	High	0.6–0.7		A
Chillies	Sierra Leone		0.5–0.7		
Usimilagai	India		0.6–0.7	↑	
Bahamian	USA	75,000 (0.5)	0.5		B
Tabasco	USA		ca. 0.5		
Chiltecpin	Mexico	Medium	ca. 0.5		
Hantaka	Japan		0.3–0.5		
Funtua	Nigeria	60,000 (0.4)	0.4		
Jalepeno	Mexico	Low to medium	—		
Sannam	India		0.3	↓	C
Chillies	Kenya		0.2		
Dandicut cherry	Pakistan and Bangladesh	30,000 (0.2)	0.2	↓	

Source: Govindarajan, 1985.

Note
A: Oleoresin; B: Chilli powder; C: Pickle.

Table 6.5 Pungency threshold of major
components of capsaicinoid

Name of the compound	Pungency (SHU $\times 10^6$)
Capsaicin	16.0
Dihydrocapsaicin	16.0
Nordihydrocapsaicin	9.1
Homodihydrocapsaicin	8.6
Homocapsaicin	8.6

threshold of major compounds of capsaicinoid is given in Table 6.5. The pungency of capsaicin is regulated by many factors. Capsaicin has three structural features: (a) the 4-hydroxy-3-methoxy benzyl (vanillyl) group (I); (b) the optimal length of linear alkyl chain (II); and (c) an acid amide linkage (III) (Govindarajan and Satyanarayana, 1991).

Any change in one or more of these structural features decreases or even abolishes the pungency stimulation. Changes in 4-hydroxy-3-methoxy substitution on the aromatic moiety (I), such as dimethoxy, dioxy methylene groups, or a shift to the 2,3 position, almost abolishes the pungency. The carboxyl group present appears essential for pungency. The length of alkyl chain influences pungency, e.g. shorter and longer chain or branched chain acids stimulate little pungency.

Uses and biological significance of capsaicinoids

Effect of capsaicin on human physiology

Chillies, when taken with food, stimulate our taste buds and thereby increase the flow of saliva which contains the enzyme amylase, which in turn helps in the digestion of starchy or cereal foods. The stimulating effects involving receptors in the mucous membranes of the mouth, nose and throat, the gastrointestinal tract, and the vascular receptors after adsorption and circulation through the blood could evoke central excitation and consequent responses in the internal organs. Chillies and other spices, have been considered psychological stimulants in animals (Molnar, 1965) and in man (Glatzel and Gewarz, 1968; Meyer-Bahlburg, 1972). Long-term capsaicin inhalation desensitizes the respiratory tract and protects it form various gaseous irritant-induced pulmonary damage from cigarette smoke, formalin, nitrogen dioxide and ether. *Capsicum* preparations are used as counter-irritants to lumbago, neuralgia and rheumatoid disorders. Taken internally, *Capsicum* has a tonic and carminative action and is specially useful in atoric dyspepsia. It is sometimes added to tanin or rose gargles for pharyngitis and to relax sore throat. It is administered in the form of powder, tincture, liniment, plaster, ointment and medicated wool. Pharmacopoeial requirements are chiefly met by highly pungent varieties of *C. frutescens*.

As a medicinal plant, the *Capsicum* species has been used as a carminative, digestive irritant, stomachic, stimulant, rubefacient and tonic. The *Capsicum* plants have also been used as folk remedies for dropsy, colic diarrhoea, asthma, arthritis, muscle cramps and toothache. *C. frutescens* L. has been reported to have hypoglycemic properties. Prolonged contact with the skin may cause dermatitis and blisters, while excessive consumption can cause gastroenteritis and kidney damage. Paprika and cayenne pepper may be cytotoxic to mammalian cells *in vitro*. Consumption of red pepper may aggravate symptoms of duodenal ulcers. High levels of ground hot pepper have induced stomach ulcers, and cirrhosis of the liver in laboratory animals.

The discovery of the medicinal applications of capsaicinoids have sparked innovative ideas for their use. The medicinal use of *Capsicums* has a long history, dating back to the Mayas who used them to treat asthma, cough and sore throats. The Aztecs used chilli pungency to relieve toothaches. The pharmaceutical industry uses capsaicin as a counter-irritant balm for external application (Carmichael, 1991). It is the active ingredient in Heet and Sloan's liniment, two liniments used for sore muscles. The capsaicin is used to alleviate pain. Its mode of action is thought to be the stimulation of nerve endings which release a neurotransmitter called substance P, which informs the brain that something painful is occurring. Capsaicin causes an increase in the amount of substance P released. Eventually the substance P is depleted and further releases from the nerve endings are reduced. Moreover, reducing substance P also helps in reducing long-term inflammation, which can cause cartilage break down.

Creams containing capsaicin have reduced post-operative pain for mastectomy patients and for amputees suffering from phantom limb pain. Prolonged use of the cream has also been found to help reduce the itching in dialysis patients, the pain from shingles (Zoster) and cluster headaches. Creams made of capsaicin, namely Zostrix and Axsain, have been found to bring relief without any side-effects in the case of patients suffering from arthritis and pain in the feet. Capsaicin has also shown bacteriostatic activity against several bacterial species, such as *Bacillus cereus* and *B. subtilis*, even at 1/10,000 dilution (Gal, 1965). In contrast, *Aspergillus niger* and *Penicillium chrysogenum* decomposed capsaicin in a medium within ten days of incubation. Gutsu *et al.* (1982) proposed that capsaicinoid may function as an immunity factor. They realized that resistance to diseases in *Capsicum* plants seems to be related to capsaicinoid content.

Formation and accumulation of capsaicinoids

Though the capsaicinoids are simple phenolic amides, they are remarkable for their very high pungency. In spite of large variations in the total capsaicinoids content within and between *Capsicum* spp. and due to cultivation conditions, the ratio of principal components, capsaicin and dihydrocapsaicin are about 1:1 for *C. annuum* and 2:1 for *C. frutescens* (Jurenitsch and Leinmueller, 1980). It was found that, upon maturation the epidermal cells of dissepiment (the surface layers on the inner side of the pod) undergo segmentation and proliferation perpendicular to the surface. Capsaicinoids first formed in an oily state between the outer wall and cuticular layer of these epidermal cells in the chilli and precipitated out as colourless tetragonal and hexagonal crystals on the drying of the fruits (Furuya and Hashimoto, 1955). With the distribution of capsaicinoids in the pericarp, the dissepiment and placenta of the fruit changed the ratio from 1:1 to 2:1 (Lee, 1977). The capsaicinoids were first detected around 20 days after flowering in both placenta and pericarp and rose sharply up to 40 days. Then, there was a sharp decrease after 50 days (Iwai *et al.*, 1979).

Iwai *et al.* (1979) found that at all stages of growth, the total and individual capsaicinoids were many times higher in the placenta than in the pericarp. They concluded that the placenta is the site of synthesis of the capsaicinoids. The capsaicinoids were reported to be formed under continuous light during post-harvest ripening. The initial formation was observed after four days of ripening and increased by 2.5-fold after seven days of ripening. No pungency principles were detected during ripening in dark. In the placenta, the formation of dihydrocapsaicin and nordihydrocapsaicin, which are vanillylamides of saturated branched-fatty acids, was higher than that of capsaicin, which is a vanillylamide of unsaturated fatty acid. The major compounds in the placenta after ten days of ripening were dihydrocapsaicin, nordihydrocapsaicin and capsaicin in the proportion of 44%, 31% and 25%, respectively. Iwai *et al.* (1977a,b) reported that the proportion of nordihydrocapsaicin to total capsaicinoids was 31% in fruits after ten days of post-harvest ripening under continuous light, while it was 7–15% in naturally ripened hot peppers. They also noticed increased dihydrocapsaicin content when vanillylamine and isocapric acid were externally supplied under continuous light during post-harvest ripening. No capsaicinoids were found in the dark, even in the presence of vanillylamine and isocapric acid. In fruits ripened with vanillylamine and isocapric acid under continuous light, there was a 9-fold increase in capsaicinoids. The portion of dihydrocapsaicin was extraordinarily high; as much as 92% of the total capsaicinoids in fruits ripened in the presence of vanillylamine and isocapric acid. Fujimoto *et al*. (1980) reported the intracellular localization of capsaicin and its analogues in *Capsicum* fruit. They also mentioned that the vacoule is the primary intracellular accummulation site in the protoplast of *Capsicum* fruit.

It is well-known that capsaicinoid content is also influenced by environmental factors such as temperature, light exposure, fertilization, etc. The formation of capsaicinoids is influenced by temperature, and it is highly likely that higher night-time temperature may be responsible for higher capsaicinoids. The influence of light on high capsaicinoid formation during post-harvest ripening was attributed to the induction/activation of some enzyme systems, such as phenylalanine ammonia lyase (PAL), that are involved in the biosynthesis of capsaicinoids, thereby resulting in the accumulation of capsaicin. The levels of valine and leucine in the placenta of *Capsicum* fruit are closely related to the content and composition of capsaicinoids (Suzuki *et al*., 1981).

The pungency is controlled by a single dominant allele. With respect to any correlation of fruit size and pungency, the capsaicin content in *Capsicum* fruit had once been considered to be inversely proportional to the size of the fruit (Trenov and Khristov, 1966).

Biosynthesis of capsaicin

Capsaicin is the major metabolite in *Capsicum* and is produced mainly in the placenta of the fruits. A degree of variability of capsaicin in the varieties of the same species has been recorded by several researchers (Quagliotti and Ottaviana, 1971). The various intermediate steps of capsaicinoid and biosynthesis through phenyl propanoid metabolism have been well studied.

The biosynthetic pathway of capsaicinoids has been thoroughly evaluated (Figure 6.1). Earlier studies of the biosynthetic pathways of capsaicin and its analogues were reported by Bennet and Kirby (1968) and Leete and Louden (1968). Bennet and Kirby (1968) showed that L-phenylalanine, *p*-coumaric acid, ferulic acid, caffeic acid and vanillylamine are involved in the biosynthesis of capsaicin. Capsaicinoids are synthesized from phenyl propanoid intermediates. The capsaicinoid structure is the acid amides of vanillylamine and C_9 to C_{11} isotype fatty acids (Kosuge and Furata, 1970; Bennet and Kirby, 1968). In nature, capsaicin and dihydrocapsaicin are the major analogues occupying more than 90% of the total capsaicinoids, whereas homocapsaicin, homodihydrocapsaicin and nordihydrocapsaicin are the minor analogues (Iwai *et al*., 1979). All these analogues are biosynthesized from L-phenylalanine and L-valine or L-phenylalanine and L-leucine in the placenta of *Capsicum* fruits by phenyl propanoid metabolism (Iwai *et al*., 1979).

Trans-cinnamic acid, *trans-p*-coumaric acid, *trans*-caffeic acid and *trans*-ferulic acid were also reported to be involved in the biosynthesis of capsaicin and its analogues (Bennet and Kirby, 1968). Fujiwake *et al*. (1980b) described an assay method for the capsaicinoid synthesizing enzyme activity with labelled (^{14}C-methoxy) vanillylamine and also with some properties of the enzyme prepared from *C. annuum* which catalyzes the formation of capsaicinoid from vanillylamine and C_9 and C_{11} isotype fatty acids, such as 7-methyloctanoyl CoA, 8-methylnonenoyl CoA, 8-methyl-6-nonenoyl CoA or 9-methyldecanoyl CoA. Isotype fatty acid reaction requires

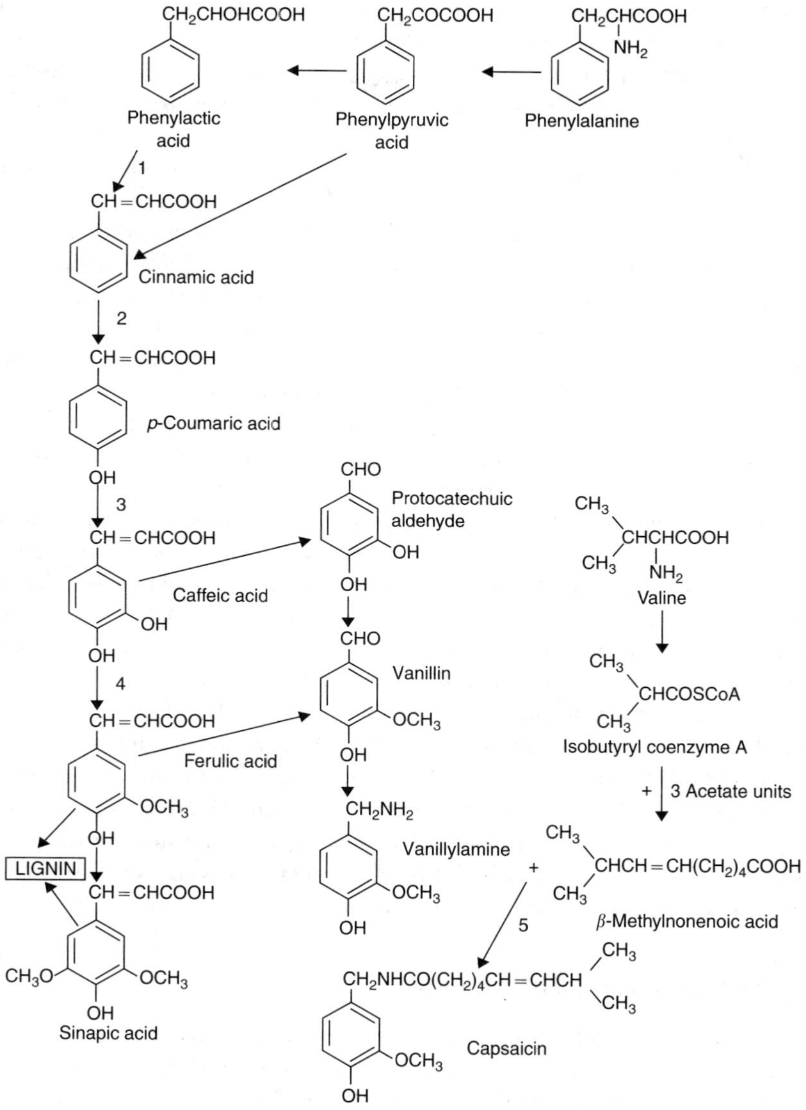

Figure 6.1 Proposed biosynthetic pathway of capsaicin and vanillin (from Yeoman *et al.*, 1980). The enzymes involved are: 1, Phenylalanine ammonia lyase (PAL); 2, *trans*-Cinnamic acid 4-hydroxylase (Ca$_4$H); 3, *trans-p*-Coumaric acid 3-hydroxylase (Ca$_3$H); 4, Caffeic acid *O*-methyltransferase (COMT); 5, Capsaicinoid synthase (CS).

CoA, ATP and Mg^{++} as cofactors. Among acyl-CoAs examined, 7-methyloctanoyl (iso-C$_{9:0}$-CoA) was used most effectively as an acyl donor for capsaicinoid formation, while dihydrocapsaicin (60%) was the major product when equivalent amounts of iso C$_{9:0}$-CoA, iso-C$_{10:0}$ CoA, iso-C$_{10:1}$-CoA and iso-C$_{11:0}$-CoA were added together as acyl donors.

The enzymatic formation of capsaicinoid from vanillylamine and iso-C$_{10:0}$ is composed of two reaction steps. Namely the first step is the enzymatic conversion of iso-C$_{10:0}$ to iso-C$_{10:0}$-CoA using ATP and Mg^{++} as the cofactors, and the second step is the enzymatic condensation of iso-$C_{10:0}$-CoA and vanillylamine to yield capsaicinoid. It is difficult to reveal the mechanism of capsaicinoid biosynthesis at subcellular organelles level using a conventional technique for cell fractionation on the intact fruits or tissue slices, because the subcellular organelles (especially the vacuole) are disrupted during cell fractionation. Using protoplast as the experimental tool, on the other hand, can be replaced with intact fruits and tissue slices to overcome the problem of disruption of the subcellular organelles during the course of the study of biosynthetic pathway of capsaicinoid.

Fujiwake *et al.* (1982a) investigated the phenylalanine ammonia lyase, *trans*-cinnamate 4-monooxygenase and capsaicinoid synthetase activities in the subcellular fractions from protoplasts of the placenta of *Capsicum* fruits, along with the subcellular distribution of intermediates of capsaicinoid biosynthesis, *trans*-cinnamic acid, *trans-p*-coumaric acid and capsaicinoid. The activity of *trans* cinnamate 4-monooxygenase and capsaicinoid synthetase was mainly in the vacuolar fractions, while the activity of PAL was in the cytosol fraction.

Non-pungency is a recessive trait; pungency is inherited as a single major gene at locus C (Andrews, 1995). However, virtually nothing is known about the genes that control the synthesis of individual capsaicinoids, or that control the abundance of total capsaicinoids. (Suzuki *et al.*, 1980; Sukrasno and Yeoman, 1993). The site of synthesis and accumulation of the capsaicinoids is the epidermal cells of the placenta (Suzuki *et al.*, 1980). Within the cells, capsaicin-synthesizing activity has been demonstrated in the vacuolar fraction and capsaicinoids have been demonstrated to accumulate in vacuoles (Suzuki *et al.*, 1980; Fujiwake *et al.*, 1982b). Ultimately, capsaicinoids are secreted extracellularly into receptacles between the cuticle layer and the epidermal layer of the placenta (Fujiwake *et al.*, 1982b). These filled receptacles of capsaicinoids often appear as pale yellow to orange droplets on the placenta of the most pungent chilli fruits. Padilla and Yahia (1998) described the evolution of capsaicinoids during the development, maturation and senescence of the fruit in three varieties of hot chilli peppers widely used in Mexico and the relation with the activity of peroxidases in these fruits, namely Habenero, De arbol, piquin. They conjectured that peroxidase activity increased at the time when capsaicinoids started to decrease. There was an inverse relationship between the evolution of the capsaicinoids and peroxidase activity that might indicate that this enzyme was involved in capsaicinoid degradation.

Molecular biology of capsaicin synthesis

The capsaicin biosynthetic pathway has two distinct branches, one of which utilizes phenylalanine and gives rise to the aromatic component vanillylamine via the phenylpropanoid pathway. The second branch forms the branched-chain fatty acids by elongation of deaminated valine. The early steps in the phenylpropanoid pathway are expressed in many plants and in many plant cell types as the products of these reactions are intermediates for a wide range of plant secondary products. cDNA and, in some cases, genomic clones have been characterized from many plants for three of these enzymes; phenylalanine ammonia lyase (PAL) (Estabrook and Sengupta-Gopalan, 1991; Joos and Hahlbrock, 1992; Lee *et al.*, 1992; Nagai *et al.*, 1994; Pellegrini *et al.*, 1994), Cinnamate 4-Hydroxylase (Ca_4H) (Fahrendrorf and Dixon, 1993; Hotze *et al.*, 1995; Kawai *et al.*, 1996; Akashi *et al.*, 1997; Schopfer and Ebel, 1998), and Caffeic Acid 0-Methyl Transferase (COMT) (Gowri *et al.*, 1991; Jaeck *et al.*, 1996; Lee *et al.*, 1998). Capsaicin is the result of the condensation of 8-methyl-6-nonenoic acid with vanillylamine by capsaicinoid synthetase. In plants, the condensation of the acetyl and malonyl groups during fatty acid synthesis requires at least three separate classes of 3-ketoacyl-ACP synthases, E.C. 2.3.1.41 (Kauppinen, 1992). The first condensation to form a 4-carbon product is carried out by 3-ketoacyl-ACP synthase III. Intermediate (4–16 carbon) chain lengths are produced by 3-ketoacyl-ACP synthase I. Finally, the 16-carbon palmitoyl-ACP is elongated by 3-ketoacyl-ACP synthase II to form stearoyl-ACP. Elongation of the branched-chain fatty acids for the synthesis of capsaicinoids is predicted to require 3-ketoacyl-ACP synthase I activity. Aluru *et al.* (1998a,b) and Curry *et al.* (1999) have isolated a 3-ketoacyl-ACP synthase gene from the Habanero chilli, *C. chinense*, by screening cDNA libraries of transcripts from placental tissues. A cDNA library of the habanero placental transcripts was constructed in the phage vector system, lambda ZapII (Stratagene). The cDNA was synthesized from mRNA isolated from the placental tissue of an immature habanero fruit at approximately 70% of the maximal capsaicinoid accumulation. Following amplification, the library was screened using a differential screening strategy with a probe for COMT from alfalfa (M63853, Gowri *et al.*, 1991). The hybridizing clones were characterized and the clone with the largest insert was sequenced. cDNA clones containing transcripts abundant in pungent habanero but not detectable in non-pungent habanero were selected for further analysis and the clones with the largest insert were sequenced.

The cDNA clone for chilli 3-ketoacyl-ACP synthase contains a 1911 nucleotide transcript; 7 nucleotide 5'UTR, 1467 nucleotide coding region, 437 nucleotide 3'UTR and a 36 nucleotide poly A tail. The predicted translation product of the coding region is a 488 amino acid protein with a sequence identity/similarity greater than 75–88% (Altschul *et al.*, 1997) to 3-ketoacyl-ACP synthase I from the plant; AF026148, *Perilla frutescens*; L13242, *Ricinus communis* U24177, *Arabidopsis thaliana* M60410, *Hordeum vulgare* (Siggaard-Anderson *et al.*, 1991). Transcripts for this gene are abundant and developmentally regulated in the placenta of immature fruit. Transcripts are not detected in other tissues of the plant such as the root, stem, leaf, flower, seed and fruit wall. Assignment of the name 3-ketoacyl-ACP synthase to the chilli cDNA clone is based on sequence similarity.

Similarly, by using heterologous probes Curry *et al.* (1998) have isolated the cDNA forms of *Pal*, *Ca4h* and *Comt* from a library of cloned placental transcripts. These genes encode the first, second and fourth step of the phenylpropanoid branch of the capsaicinoid pathway. Based on the pattern of expression of these three genes during fruit development and across fruit of different pungency levels, Curry *et al.* (1998) have developed a hypothesis about the regulation of transcription for capsaicinoid biosynthetic enzymes. Transcripts of biosynthetic genes accumulate in the placenta early in fruit development and then decline in abundance; transcript levels of biosynthetic genes are proportional to the degree of pungency, with the hottest chilli having the greatest accumulation of transcripts. Curry *et al.* (1998, 1999) have employed these transcript levels as a screening tool of a cDNA library of habanero placental tissue. Using this differential approach they have isolated a number of cDNA clones and confirmed their differential patterns of expression; two of the clones putatively encode enzyme activities predicted for capsaicinoid biosynthesis, β-ketoacyl synthase and a transaminase. Lee *et al.* (1998) isolated and characterized 0-diphenol-0-methyltransferase cDNA clone in hot pepper. Matsui *et al.* (1997) have performed purification and molecular cloning of bell pepper fruit fatty acid hydroperoxide lyase.

Carotenoids and their biosynthesis

Carotenoids are a class of hydrocarbons (carotenes) and their oxygenated derivatives (xanthophylls). Carotenoids play an important role in the transfer of energy to chlorophylls, in addition to the photoprotection of chlorophylls in photosynthetic organisms. Paprika and paparika oleoresins are used as natural colour additives in a wide variety of foods, drugs and cosmetics. Capsaicinoids and carotenoids are found in the oily fraction of several varieties of chilli peppers. The carotenoid colourants are normally commercialized as a dried paprika powder or as a main

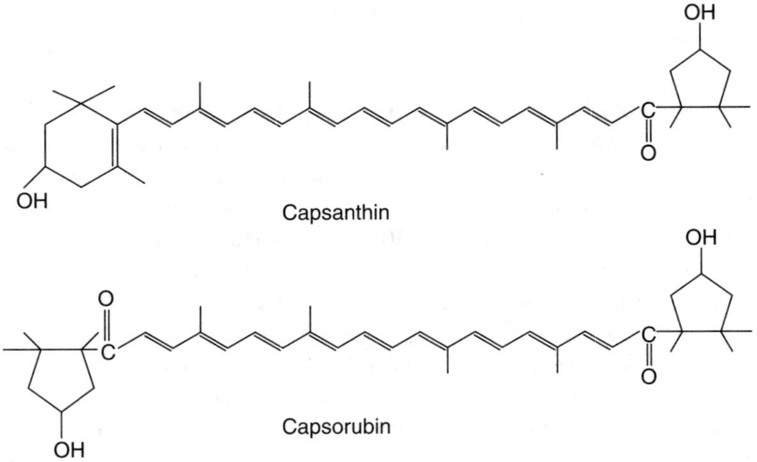

Figure 6.2 Structure of major carotenoids of *Capsicum*.

component of oleoresins obtained after extraction with hexane. The colour and pungency are dependent on the *Capsicum* variety. The Guajillo variety is the most frequently used in the Mexican colourant industries due to its low capsaicin content. About 20 carotenoids contribute to the *Capsicum* pod colour and to the colour value of paprika powder and oleoresin (Harkay, 1974). The ketocarotenoids, capsanthin and capsorubin (Figure 6.2) are unique *Capsicum* carotenoids. The major red colour in paprika comes from the carotenoids capsanthin, capsanthin 5,6-epoxide and capsorubin, while the yellow colour is from β-carotene, zeaxanthin, violoxanthin, antheroxanthin, β-cryptoxanthin and cucurbitaxanthin A (Reeves, 1987). Yellow colour constituents act as a precursor for red colour.

Capsanthin, the major carotenoid in ripe fruits, contributes up to 60% of the total carotenoids. Capsanthin and capsorubin increase proportionally with advanced stages of ripeness, with capsanthin being the more stable of the two. The amount of carotenoids in fruit tissue depends on factors such as cultivar, maturity stage and growing conditions (Kanner *et al.*, 1977). The majority of carotenoids are tetrapenes, formed from the joining of eight molecules of isoprene via the divalent unit. Difference in carotenoid structure is based on the structural alterations in one or both halves of the molecule, which includes biochemical reactions such as hydrogenation, dehydrogenation, cyclization, addition of oxygen in various forms, hydroxylation, epoxidation, double bond migration, methyl migration, chain elongation and chain shortening. All the carotenoids present in the *Capsicum* sp. are C_{40} isoprenoids containing nine conjugated double-bonds in the central polyenic chain. The changes in the end groups (β,κ,ε,3-hydroxy,5,6-epoxide) will change the chromosphere properties of each pigment. Recently, Maoka *et al.* (2001) reported the isolation of a series of apocarotenoids from the fruits of red paprika *C. annuum* by spectroanalysis, namely apo-14′zeaxanthinal (4), apo-13-zeaxanthinone (6), apo-12′-capsorubinal (9), apo-8′-capsorubinal (10) and 9,9′-diapo-10,9′-retro-carotene-9,9′-dione (11). The other six known apocarotenoids were identified to be apo-8′-zeaxanthinal (1), apo-10′-zeaxanthinal (2), apo-12′-zeaxanthinal (3), apo-15-zeaxanthinal (5), apo-11-zeaxanthinal (7) and apo-9-zeaxanthinone (8) which have not been previously found in paprika. These apocarotenoids were assumed to be oxidative cleavage products of C_{40} carotenoid, such as capsanthin in paprika.

Molecular biological studies of carotenoids

Developmentally regulated transcription was found to be the major mechanism that controls carotenogenesis in fruits and flowers. Merav *et al.* (2000) have cloned plant genes for carotenoid biosynthesis enzymes and analysed the regulation of their expression. Developmentally regulated transcription was found to be the major mechanism that controls carotenogenesis in fruits and flowers. To alter the accumulation of carotenoids they have genetically manipulated the pathway in tobacco and tomato. To this end they have over-expressed the following genes: *Ipi* (isopentyl pyrophosphate isomerase), *Psy* (phytoene synthase), *CrtB* (phytoene synthase from cyanobacteria), *CrtI* (phytoene desaturase from *Erwinia herbicola*), *Pds* (phytoene desaturase from tomato) and *CrtO* (β-carotene ketolase from *H. pluvialis*). These genes were fused to various flower-specific promoters and a transit peptide from the tomato *PDS* was used in the case of the bacterial and algal genes. Petals of transgenic tobacco plants expressing the *crtB* gene accumulated phytoene, the first intermediate in the carotenoid pathway. When co-transformed with constructs containing *CrtI* and *Ipi* genes, these plants did not show any significant change. To obtain white flowers, a desired trait in certain ornamentals, they have produced transgenic tomato plants over-expressing sequences of the native *Pds* or *Psy* in a flower specific mode. The petals of these plants were white due to the silencing of the relevant genes. Over-expressing the algal gene *CrtO* in tobacco resulted in the accumulation of astaxanthin in the nectaries, changing their colour from yellow to red. These results demonstrate the feasibility for genetic engineering of carotenoid biosynthesis in plants.

Uses of carotenoids

Carotenoids are important in human nutrition as a source of vitamin A (eg. from β-carotene) and as a preventive agent for cancer and heart disease. In addition, carotenoids add colour to foods and beverages (eg. orange juice). And in addition, carotenoids are the precursors of many important chemicals responsible for the flavour of foods and the fragrance of flowers. Carotenoids have been considered important only as precursors of vitamin A. However, there has been significant interest in the evaluation of carotenoids for roles that are unrelated to their conversion to vitamin A. Recent studies have emphasized the role of carotenoids in disease prevention. When ingested they have shown important biological functions, such as antioxidant activity and free radical scavenging (Hornero-Mendez *et al.*, 2000).

Health aspects of carotenoids

In human nutrition, carotenoids play an important role as a source of provitamin A. In the gastrointestinal tract β-carotene gets converted to vitamin A, which plays an important role in the regulation of vision, growth and reproduction (Ong and Choo, 1997). More recently, however, the protective effects of carotenoids against serious disorders such as cancer (Peto *et al.*, 1981; Shekelle *et al.*, 1981), heart disease, squamous cell carcinoma of the lung, oral tumours and degenerative eye disease have been recognized, and have stimulated intensive research into the role of carotenoids as antioxidants and as regulators of the immune response system. The antioxidant property of β-carotene by its effective radical trapping was studied by Burton and Ingold (1984). Kunert and Tappel (1983) reported the efficient reduction of lipid peroxidation by prior treatment with β-carotene in guinea pigs.

Extraction and downstream processing of pungency principles and pigments of *Capsicum*

Post-harvest features

Provitamin-A carotenoids of *Capsicum*, namely β-carotene and β-cryptoxanthin, are reduced during the processing of paprika. Esterified carotenoids, which are found as ketocarotenoids, capsanthin and capsorubin, are more stable than zeaxanthin (free, monoesterified and diesterified forms), β-cyrptoxanthin (free and mono-esterified form) and β-carotene (free form) (Howard, 2001). Loss of provitamin-A activity of 67% and 81% has been reported for paprika varieties Agridulce and Bola, respectively. (Minguez-Mosquera and Horneo-Mindez, 1994). The natural antioxidants present in the fruit prevent the degradation of colour (Kanner *et al.*, 1979). Tocoferol acts as an oxidation barrier, while ascorbic acid is useful for tocoferol regeneration and carotenoids prevent lipid oxidation (Esterbauer, 1991). Though there are reports of a decrease of capsaicinoid upon food processing, the pungency factor is generally not affected under the conditions of oleoresin preparation. However, prolonged storage of pasteurized yellow wax pepper for a period of four months has caused a reduction in capsaicin and dihydrocapsaicin by 30% and 10%, respectively (Lee and Howard, 1999). Thus, characteristic features of raw materials need to be considered for the processing and extraction of the constituents for pungency or colour components.

Extraction and down stream processing

The oleoresin is oil-soluble, but when emulsified becomes water-dispersible. Oleoresins are extracted by percolation with hexane and hexane/acetone/alcohol isopropyl (3:2:1) at ambient temperature, with the paprika/solvent ratio being 1:4.

Light shows a strong degradative effect on the colour of all oleoresins. The oleoresin extracted with hexane/acetone/isopropyl alcohol (3:2:1) has the least stability, especially in the presence of air. In darkness and in the presence of air, an induction period became evident for all oleoresins and the oleoresin extracted with hexane/acetone/ isopropyl alcohol (3:2:1) showed a colour degradation rate constant 1.8 times higher than the others. The commercial oleoresin needs to be made more stable by the addition of antioxidant to facilitate storage.

The stability of paprika oleoresin is strongly dependent on light and on the type of solvent used for its extraction. Balakrishnan and Verghese (1997) reported that after hexane extraction, it was possible to reduce pungency in oleoresin by fractionation with 70% alcohol v/v methanol solutions, recovering 87% of carotenoids and 83% of capsaicinoids after separation. Amaya Guerra *et al.* (1997) reported the extraction of oleoresin from dried Guajillo pepper with the four solvents ethanol, acetone, ethyl acetate and hexane. Selective extraction of capsaicinoids and carotenoids from chilli guajillo "puya" flour was studied by Santamaria *et al.* (2000) (Figure 6.3). They reported that when ethanol was used as a solvent 80% of capsaicinoids and 73% of carotenoids were extracted, representing an interesting alternative for the substitution of hexane in industrial processes. Additionally, when the flour was pretreated with cellulases and pectinases and extracted in ethanol, the yield increased to 11% and 7% for carotenoid and capsaicinoid, respectively. They have also proposed selective two-stage extraction process after the treatment with enzymes. The first step uses 230% (v/v) ethanol and releases up to 60% of the initial capsaicinoids, and the extraction step with industrial ethanol permits the recovery of 83% of carotenoids present in the flour.

Manuel Jaren-Galan *et al.* (1999) reported the extraction of oleoresins from paprika (*C. annuum*) with supercritical carbon dioxide. They have reported that higher extraction volumes,

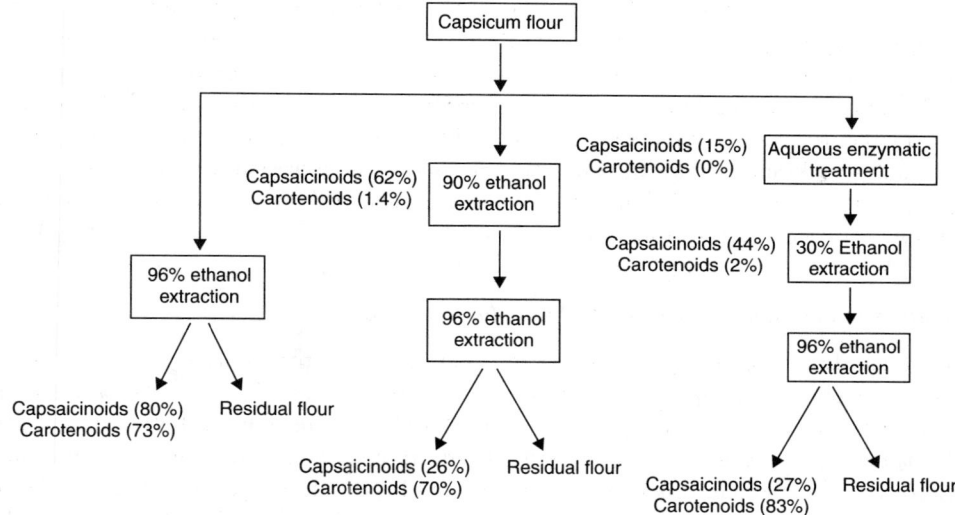

Figure 6.3 A protocol for extraction of capsaicinoids and carotenoids by using three extraction options: water, industrial alcohol, ethanol (30% v/v).

Source: Santamaria *et al.* (2000).

increasing extraction pressures and the use of co-solvents, such as 1% ethanol or acetone, resulted in higher pigment yields. Pigments isolated at lower pressures consisted of β-carotene exclusively and at higher pressures resulted in more proportions of red carotenoids and smaller amounts of β-carotene. Kiss *et al.* (2000) have used the microwave-assisted extraction of pigments from paprika powders.

Immobilized cell cultures of *Capsicum* and biotransformation to produce secondary metabolites of importance

Biotransformation is a growing field of biotechnology and encompasses both enzymatic and plant/microbial biocatalysis. It is a process through which the functional groups of organic compounds, for example, substrates or precursors or intermediates, are modified either stereo- or regio- specifically by living cultures, entrapped cells or enzymes or permeabilized cells to a chemically different product. Biotransformations can produce novel compounds, and it is possible to enhance the productivity of a desired compound. It also overcomes the problems associated with chemical synthesis.

The production of high value food metabolites, fine chemicals and pharmaceuticals can be achieved by biotransformations using biological catalysts in the form of enzymes and whole cells (Armstrong *et al.*, 1993; Rhodes *et al.*, 1994; Berger, 1995; Cheetham, 1995; Dornenburg and Knorr, 1996 a,b; Meyer *et al.*, 1997; Scragg, 1997; Krings and Berger, 1998; Ramachandra Rao and Ravishankar, 2000a; Ravishankar and Ramachandra Rao, 2000). Consumer sensitivity about synthetic food additives has stimulated interest in the manufacture of natural and nature identical food ingredients using novel biotechnological methods. From an industrial point of view, biotransformations performed by plant cell culture systems can be desirable when a given reaction is unique to plant cells and the product of the reaction has a high value.

The production of capsaicin in cell cultures of the *Capsicum* species, *C. frutescens* and *C. annuum* was proposed by Yeoman *et al.* (1980). They described the formation of *trans*-cinnamic acid from phenlyalanine, an aromatic amino acid, by PAL, which is further converted into *p*-coumaric acid by cinnamic acid 4-hydroxylase. The conversion of *p*-coumaric acid into caffeic acid and ferulic acids is initiated by two enzymes, *p*-coumaric acid 3-hydroxylase and caffeic acid *O*-methyltransferase, respectively. Ferulic acid can divert into lignin pathway or can be converted into vanillin and vanillylamine by oxidation and oxidative deamination reactions, respectively. Vanillylamine condenses with β-methylnonenoic acid, obtained from valine, and the pungent principle, capsaicin is synthesized by capsaicin synthetase.

Enhancement of capsaicin production by media manipulation

A careful analysis of secondary metabolite profiles during the culture of plant cells indicate that most of the secondary compounds are produced during the post-exponential or stationary phase of growth (Lindsey and Yeoman, 1985). The biochemical factors underlying this phenomena are the channeling of precursors from growth related processes to secondary metabolism. This observation for capsaicin production in immobilized cell culture system has already been reported (Ravishankar *et al.*, 1988). The above situation could be induced by subjecting the cells to nutrient stress. We found that nitrates and phosphates stress enhances the capsaicin production in immobilized cells by 13- and 5-fold, respectively, in comparison with free cell culture systems. Mathematical modelling of capsaicin production in immobilized cells of *Capsicum* was studied by Suvarnalatha *et al.* (1993) to optimize physical parameters, such as the bead strength of calcium alginate used for immobilization and the medium constituents for enhanced yield.

Influence of elicitors on secondary metabolite production

Secondary metabolite production in plant cell cultures can be elicited using a range of elicitors (Di Cosmo and Talleri, 1985). Elicitation is envisaged to overcome the problem of low productivity of plant cells for industrial applications (Knorr *et al.*, 1993). Treatment of immobilized cells and placental tissues with various elicitors, such as fungal extracts (*A. niger* and *Rhizopus oligosporus*) and bacterial polysaccharides, curdlan and xanthan were performed. It was found that curdlan was most effective in eliciting capsaicin synthesis (Johnson *et al.*, 1991). Immobilized cells responded more effectively than placental tissues for curdlan treatment. Curdlan and xanthan in combination enhanced capsaicin production by nearly 8-fold for curdlan treatment. Suvarnalatha *et al.* (1993) showed the optimization for capsaicinoid formation of immobilized *C. frutescens* cells using Response Surface Methodology.

Attempts have been made to increase the production of capsaicin in both freely suspended and immobilized cell cultures of *C. frutescens* by the feeding of phenylpropanoid intermediates. A 6- to 7-fold increase was found in the capsaicin accumulation upon precursor biotransformation. The feeding of intermediate precursors to *Capsicum* cell cultures not only increased the capsaicin accumulation (Figure 6.4) but also shortened the time required to produce high amounts of capsaicin (Johnson *et al.*, 1990, 1991). Immobilized placenta, which is the site of synthesis of capsaicin in the fruit, was administered with intermediates of capsaicin pathway which resulted in large amounts of capsaicin accumulation, which is similar to the content of pungent variety of *Capsicum* fruit (Johnson and Ravishankar, 1996). During biotransformation studies to increase capsaicin yields, it was found that low capsaicin producing *Capsicum* cell cultures formed vanillin when fed with phenylpropanoid compounds – protocatechuic acid, caffeic acid, ferulic acid, vanillylamine, coniferyl aldehyde and veratraldehyde.

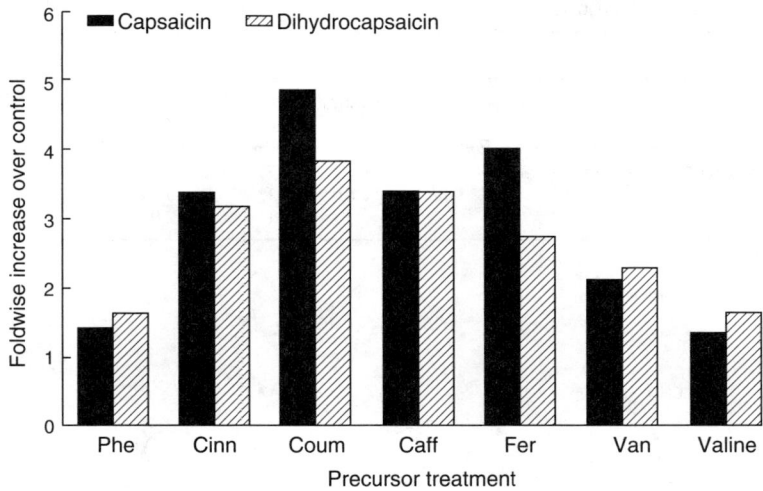

Figure 6.4 Influence of feeding intermediate metabolites (added individually at 2.5 mM final conc.) on capsaicin and dihydrocapsaicin formation in immobilized placenta of *C. frutescens*. Abbreviations for the compounds used, Phe: phenylalanine; Cinn: cinnamic acid; Coum: *p*-coumaric acid; Caff: caffeic acid; Fer: ferulic acid; Van: vanillylamine; Cultures were treated with precursors at the 2.5 mM level on day zero and analysis was done after five days of culture.

Protocatechuic aldehyde and caffeic acid biotransformation leading to vanillin and capsaicin production

Freely suspended cells and immobilized cell cultures of *C. frutescens* biotransform externally-fed protocatechuic aldehyde to vanillin, whereas caffeic acid treated cultures accumulate more capsaicin than vanillin. The addition of S-adenosyl-L-methionine (SAM), a methyl donor, to protocatechuic aldehyde-treated immobilized *C. frutescens* cell cultures results in 14.08 mg/L vanillin accumulation, which is 2.5-fold higher than that in cultures with protocatechuic aldehyde alone. This result suggests that the influence of SAM on O-methylation of protocatechuic aldehyde results in a higher accumulation of vanillin. The increase in vanillin accumulation correlates well with an increase in the specific activity of caffeic acid O-methyltransferase in protocatechuic aldehyde and SAM-treated immobilized *C. frutescens* cell cultures. Capsaicin accumulation is also increased in protocatechuic acid and caffeic acid-fed immobilized *C. frutescens* cell cultures (Table 6.6). The formation of vanillin from protocatechuic aldehyde involves O-methylation at the *meta* position; further oxidation of vanillin leads to the formation of vanillic acid, which upon further demethylation or oxidation of protocatechuic aldehyde yields protocatechuic acid (Figure 6.5). (Ramachandra Rao and Ravishankar, 2000b).

Ferulic acid, vanillylamine and coniferyl aldehyde biotransformation leading to vanillin and capsaicin production

Other phenylpropanoid compounds, ferulic acid, vanillylamine and coniferyl aldehyde were also tested for the biotransformation to capsaicin and vanillin in free cell and immobilized cell cultures of *C. frutescens* (Figure 6.6).

Table 6.6 Influence of intermittent feeding of protocatechuic aldehyde and caffeic acid together (1.25 mM each) to immobilized cell cultures of *C. frutescens*

Addition of precursor (day)	3rd day		6th day		9th day		12th day		15th day	
	Van	Cap	Van	Cap	Van	Cap	Van	Cap	Van	Cap
3	—	—	4.65 ± 0.43	1.4 ± 0.23	3.98 ± 0.35	1.73 ± 0.18	3.73 ± 0.43	3.78 ± 0.48	3.18 ± 0.4	2.9 ± 0.28
6	—	—	—	—	3.7 ± 0.4	2.28 ± 0.3	3.13 ± 0.33	3.15 ± 0.23	2.73 ± 0.3	3.65 ± 0.4
9	—	—	—	—	—	—	3.1 ± 0.28	2.9 ± 0.23	2.08 ± 0.25	1.9 ± 0.2

Note
Van, vanillin; Cap, capsaicin; values are mg/L $\pm$ SD: $n = 3$.

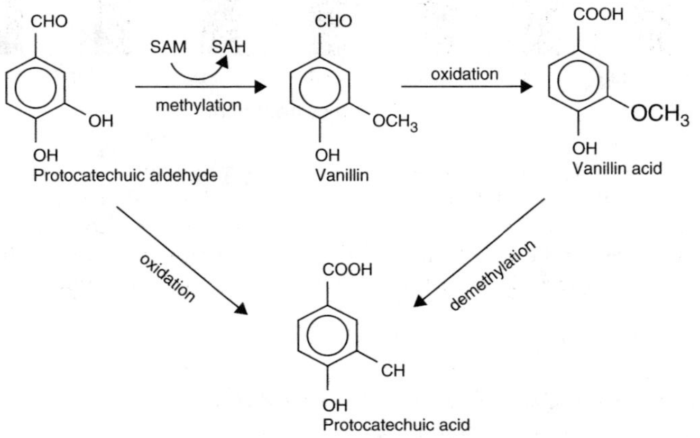

Figure 6.5 Probable biosynthetic pathway of vanillin from protocatechuic aldehyde in *C. frutescens* cell cultures.

Ferulic acid administered *Capsicum* cultures are biotransformed to vanilla flavour metabolites – vanillin, vanillic acid, vanillyl alcohol, *p*-hydroxybenzoic acid, *p*-coumaric acid and protocatechuic acid. Additionally, an increase in the capsaicin accumulation is observed in free cells and immobilized cell cultures. Immobilized *Capsicum* cells show a 4.0 and 5.23 times increase in vanillin and vanillyl alcohol accumulation on the 10th day over its freely suspended cultures in 2.5 mM ferulic acid supplementation whereas vanillic acid, protocatechuic acid and *p*-hydroxybenzoic acid accumulation were 3.14-, 7.4-, and 1.27-fold over freely suspended cultures (Ramachandra Rao, 1998; Ramachandra Rao and Ravishankar, 2000b).

β-cyclodextrin (Figure 6.7) was used to effectively biotransform the phenylpropanoid precursor to capsaicin and vanillin. The precursor moiety fills the lumen of β-cyclodextrin molecule and renders it soluble. This results in higher biotransformation efficiency. β-cyclodextrin (BCD) used as a BCD–ferulic acid mixture treatment reveals that there is a 1.8-fold increase in vanillin accumulation (18.0 mg/L) in four days of incubation in immobilized *Capsicum* cell cultures. The addition of BCD favours increase in the vanillin accumulation in a shorter duration. The effect of reducing agent, dithiothreitol (DTT) at a 1:1 ratio (2.5 mM each) on ferulic acid biotransformation shows that a 4.35-fold higher vanillin accumulation (23.0 mg/L) over only ferulic acid-fed freely suspended *Capsicum* cell cultures. At a higher concentration of DTT (5.0 mM) treatment shows significant reduction in vanillin accumulation (Ramachandra Rao, 1998).

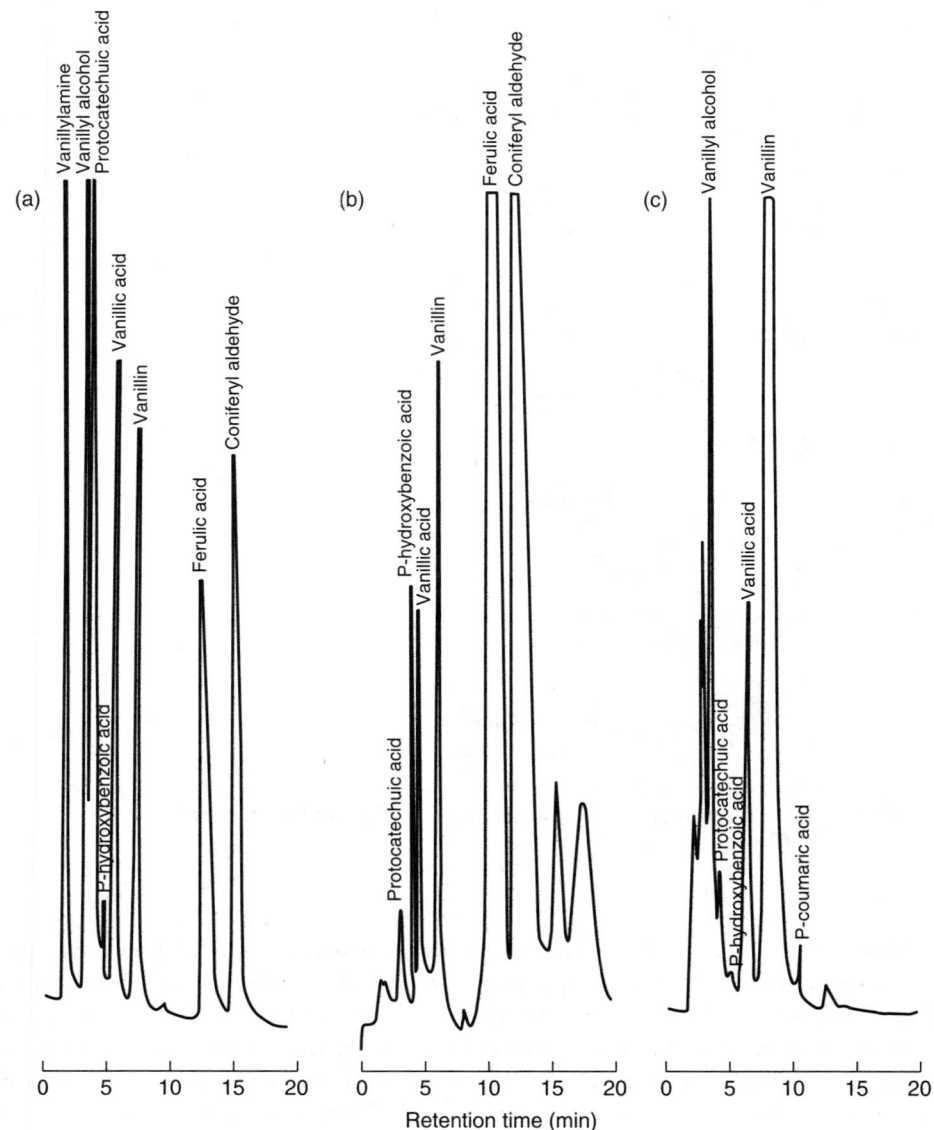

Figure 6.6 HPLC profile of authentic vanilla flavoured metabolites (a), coniferyl aldehyde-fed *Capsicum* cultures (b) and natural vanilla extract (c).

The effect of adsorbents Amberlite XAD-7 and XAD-4 at 2% (w/v) on ferulic acid biotransformation results in an increase in the vanilla flavour metabolite accumulation, as well as capsaicin in freely suspended cultures treated with ferulic acid (2.5 mM). The results show that all the metabolites are adsorbed by adsorbents with different affinities. Adsorbents also stimulate the excretion of metabolites from *Capsicum* cell cultures. The addition of XAD-7 (hydrophilic resin) stimulates 2.2- and 3.2-fold in vanillin and vanillic acid accumulation on the 8th day and the excretion of these metabolites from cells into medium is 95% and 75%, respectively.

(a)

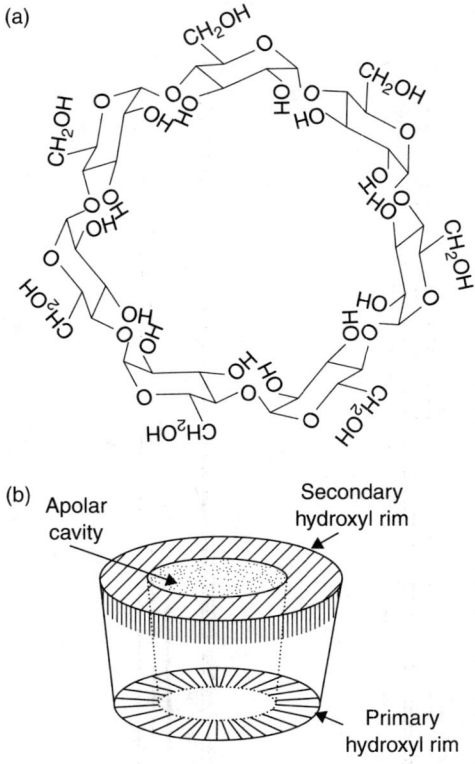

(b)

Apolar cavity

Secondary hydroxyl rim

Primary hydroxyl rim

Figure 6.7 (a) Structure of β-cyclodextrin. (b) Molecular shape of β-cyclodextrin.

The adsorption is 64% and 89% for vanillin and vanillic acid on the 8th day. The production of *p*-hydroxybenzoic acid is significantly increased with XAD-7 treatment and the adsorption is 96.3%. XAD-7 stimulates the capsaicin production by 1.34 times on 6th day, corresponding to 60% adsorption. XAD-4 favours more vanillin accumulation, and *p*-hydroxybenzoic acid is 23% adsorption, suggesting that XAD-4 is less selective. The addition of adsorbents exert positive influences on the accumulation of vanilla flavour metabolites and capsaicin. The addition of XAD-4 favours more vanillin accumulation, while XAD-7 favours vanillic acid and *p*-hydroxybenzoic acid accumulation in ferulic acid-fed *C. frutescens* cultures (Ramachandra Rao, 1998).

Vanillylamine-fed cultures show formation of vanillin and its metabolites, which suggests that *Capsicum* cultures can perform oxidative deamination by a reversible reaction (Johnson *et al.*, 1996). Coniferyl aldehyde-fed *Capsicum* cultures show that a range of vanilla flavour metabolites form and accumulated ferulic acid is the major metabolite. The flavour profile obtained in the biotransformation compared well with the profile of vanilla extract obtained from natural vanilla cured bean, which suggests that biotransformed vanillin has a similar note of vanilla to that found in the vanilla bean, and can be used for an alternative source of vanilla flavoured metabolites. Veratraldehyde is also bioconverted to vanilla metabolites by *Capsicum* cell cultures (Ramachandra Rao, 1998).

Isoeugenol biotransformation leading to the formation of vanilla compounds

Isoeugenol, a clove principle, is biotransformed to vanillin and vanilla flavoured metabolites in both freely suspended and immobilized *C. frutescens* cell cultures (Figure 6.8). The addition of isoeugenol also stimulates its biotransformation to capsaicin. The addition of β-cyclodextrin together with isoeugenol, each at 2.5 mM, results in 23 mg/L of vanillin accumulation on 4th day, which is 1.62 times more than in cultures treated with isoeugenol alone. Isoeugenol biotransformation is more effective in immobilized cells. Based on metabolites identified, the pathway of vanillin formation from isoeugenol has been proposed (Figure 6.9) (Ramachandra Rao and Ravishankar, 1999). The formation of ferulic acid from isoeugenol may take place by oxidation of the aromatic side-chain. In the second step, formation of vanillin from ferulic acid may occur via feruloyl CoA through a reaction similar to β-oxidation of fatty acids (Zenk, 1965).

Biotransformation studies using *Capsicum* cell cultures have revealed that *Capsicum* contains a repository of enzymes that can perform a broad range of reactions including oxidation, reduction, methylation, oxidative deamination and demethylation to produce a range of vanilla flavoured metabolites. These findings suggest that these cultures can be adopted to generate

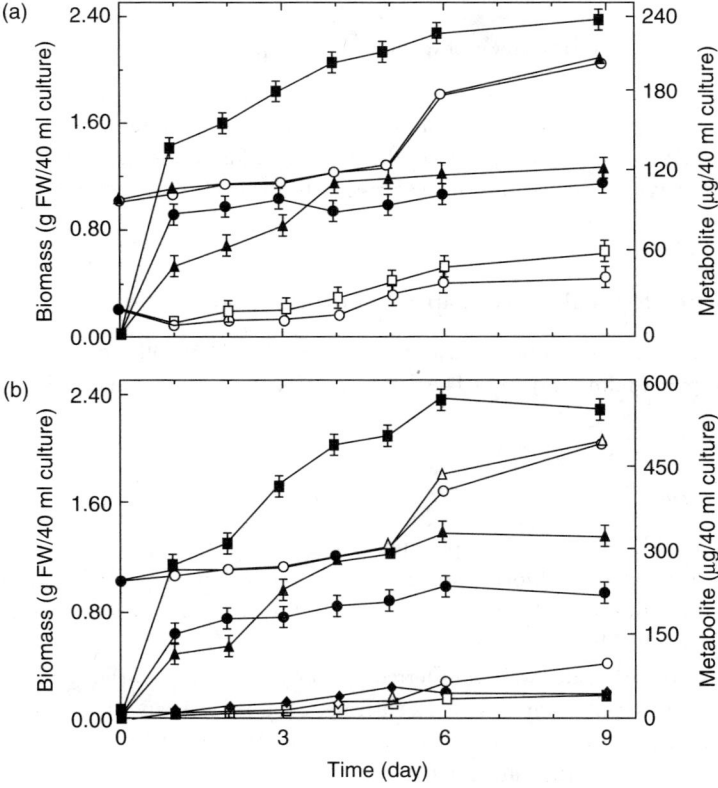

Figure 6.8 Growth pattern, accumulation of vanilla flavoured metabolites and capsaicin in 1.25 mM (a) and 2.5 mM (b) isoeugenol-fed immobilized *C. frutescens* cell cultures (mean ± SD, *n* = 3). Δ, Biomass in control cultures; O, isoeugenol-fed cultures; ▲, vanillin content in beads; ●, vanillin in medium; ■, total vanillin content; ◆, ferulic acid; ◇, total capsaicin in control; □, total capaicin in isoeugenol-fed cultures.

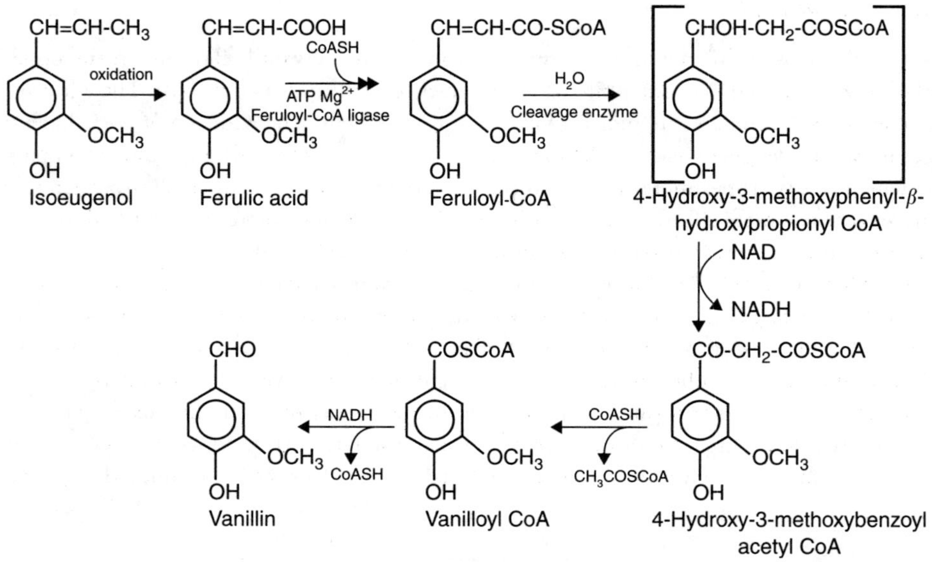

Figure 6.9 Possible biosynthetic pathway of vanillin formation from isoeugenol.

novel compounds. Though the amounts of vanilla flavoured metabolites is relatively low in view of the commercialization of these processes directly, the information generated on biosynthetic pathways and enzymes can be used to express at high levels using metabolic engineering approaches.

Column reactor for scale-up of capsaicin production

The scale-up of capsaicin production using immobilized cells in column reactor (Figure 6.10) has been attempted (Johnson, 1993). The details of inputs of operation and yield of capsaicin are presented.

Inputs of operation

1   Twenty-day-old *Capsicum* cells or placenta (100 g fresh weight);
2   Alginate and calcium chloride solutions required to envelope 100 g of cells/placenta in beads. One liter of sodium alginate 2.5% (w/v) with cells extruded into two litres of calcium chloride dihydrate 0.9% (w/v);
3   Beads washed with water were transferred by compressed air into the vessel containing 1 L of MS medium supplemented with 3% sucrose, 2 mg/L 2,4-D and 0.5 mg/L kinetin;
4   Airflow (mixture of CO_2 + air 2:1 for initial seven days of culture 4:1 for the latter seven days of production) at the rate of 0.4 Wm;
5   Incubation at $25 \pm 2°C$ in continuous light of 1000 lux;
6   pH adjustment to 5.8 during culture;
7   Replenishment of entire medium after seven days with fresh medium;
8   Capsaicin recovery and analysis in two weeks culture with two harvests at a seven-day interval.

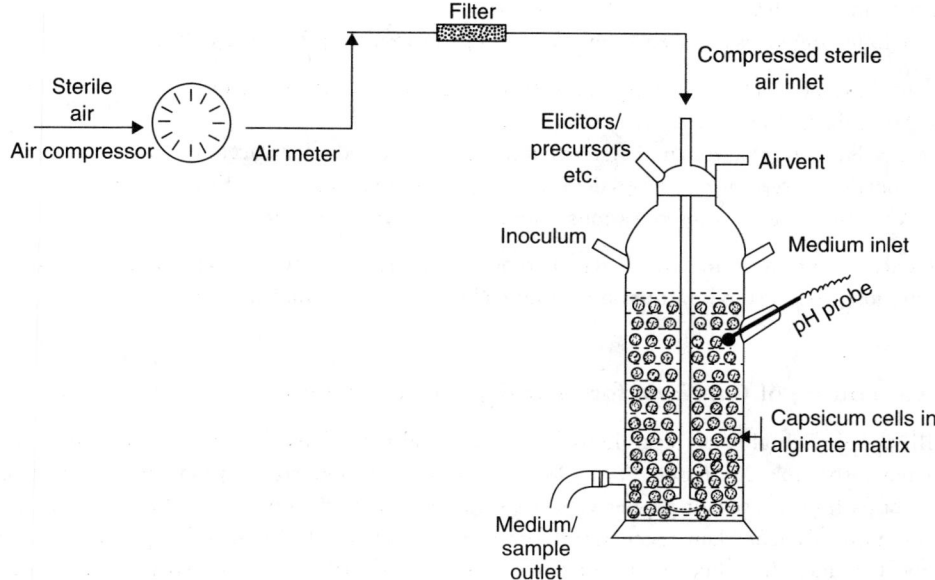

Figure 6.10 Lab scale column reactor for immobilized cells of *C. frutescens* for capsaicin production.

Yield components

The airlift culture vessel was run with immobilized cells or placenta for two weeks under three different conditions: (I) using standard medium (SM); (II) SM + elicitor; and (III) SM + precursor (coumaric acid). The results were as follows:

Yield in standard medium

1   Immobilized cells: 35 mg/100 g fresh cells/15 days or 0.4% on dry wt. basis;
2   Immobilized placenta: 240 mg/100 g fresh placenta/15 days or 1.0% on dry wt basis.

Yield in standard medium + elicitor

The elicitors used were curdlan at 8 mg/L concentration for cells and *Rhizopus oligosporus* mycelial extract equivalent to 2.5 g mycelium per litre medium administrated for placental immobilized cultures.

1   Immobilized cells: 76 mg/ 100 g/15 days fresh cells or 0.95% on dry wt basis;
2   Immobilized placenta: 300 mg/100 g fresh placenta/15 days or 1.25% on dry wt. basis.

Yield in standard medium + precursor (Coumaric acid 2.5 mM)

1   Immobilized cells: Not done;
2   Immobilized placenta: 1148 mg/100 g fresh placenta/15 days or 7.6% on dry wt. basis.

By the methods described above we were able to get a maximum production of 1.15 g/L. However increases can be obtained using continuous cultivation with the adsorption of capsaicin from the medium and by recycling the medium to minimize the cost. However, the plant may not be viable if it has less than 10,000 litre capacity.

The following steps are needed for the scale up of capsaicin producing culture:

- Improvement in productivity, stability and viability of cells in culture;
- Minimization of microbial contamination;
- Development of automated systems of immobilizing the cells/placenta;
- Continuous separation of capsaicin from the medium and the recycling of nutrients;
- Analysis and evaluation of various immobilization methods.

Hence an intensive industrial R&D trend is required to solve the problems of capsaicin production process through immobilization technology for commercial exploitation.

Tissue culture of *Capsicum* for *in vitro* plant regeneration

Chilli pepper is highly susceptible to fungal and viral pathogens and these cause considerable damage to the crop (Morrisson *et al.*, 1986). One of the solutions to this problem is the development of pathogen-resistant pepper varieties using genetic transformation techniques, which in turn require efficient plant regeneration protocols (Liu *et al.*, 1990). Most pepper regeneration protocols induce shoot buds on cotyledon (Sripichitt *et al.*, 1987) and hypocotyl explants using growth hormones, such as BA and IAA, then transferring buds to rooting medium (Gunay and Rao, 1978; Phillips and Hubstenberger, 1985; Agarwal *et al.*, 1989; Arroyo and Revilla, 1991; Valera-Montero and Ochoa-Alejo, 1991; Christopher and Rajam, 1994, 1996). Silver nitrate has been used for promoting shoot development and plant regeneration from cotyledons of chilli pepper (Hyde and Phillips, 1996). Shoot induction and subsequent plant regeneration have been obtained from decapitated, rooted hypocotyl explants or half-seed explants using hormone-free MS medium or medium supplemented with BA and IAA (Valera-Montero and Ochoa-Alezo, 1991; Ezura *et al.*, 1993; Ramirez-Malagon and Ochoa-Alezo, 1996). Pepper cultivars also differ markedly in their regeneration requirements (Ochoa-Alezo and Ireta-Moreno, 1990; Christopher and Rajam, 1994, 1996; Szasz *et al.*, 1995; Hyde and Phillips, 1996). Hussain *et al.* (1999) reported a highly efficient three-stage protocol for the regeneration of chilli pepper (*C. annuum* L.) from cotyledonary explants. This protocol used phenyl acetic acid (PAA) in both the shoot-bud induction medium and the medium for the induction of buds from the cotyledons was MS medium, supplemented with BA + PAA. Somatic embryos could be a method of choice because of the difficulties encountered in the elongation of *in vitro* organogenic shoot buds. Harini and Lakshmi Sita (1993) reported somatic embryogenesis in peppers. Later, Binzel *et al.* (1996) and Marla *et al.* (1996) reported the induction of direct somatic embryogenesis and plant regeneration in *C. annuum*.

Hairy root cultures of *Capsicum*

Hairy roots of chilli pepper, *C. frutescens* cv. Cayenne have been obtained using *Agrobacterium rhizogenes* mediated transformation by Sekiguchi *et al.* (1996), where they have reported that capsaicin is not produced in hairy root cultures. Later Yamakawa *et al.* (1998) successfully transformed *C. frutescens* with a phenylalanine ammonia lyase gene, where the PAL cDNA from parsley (*Petroselium*) was linked to the CaMV 35 S promoter and subsequently transferred into hairy roots. They found that hairy roots harbouring the PAL transgene showed different PAL activity with slow growth and altered morphology. Increased PAL activity was studied by Sekiguchi *et al.* (1999) where the PAL activity was increased by the expression of parsley PAL 2 cDNA. The increase in PAL activity leads to increased lignin biosynthesis. They have also observed alteration in metabolism of aromatic compounds and also the content of aminoacid. These studies show that there is a scope for the increase of phenylpropanoid biosynthesis by transformation. The fast growing hairy roots with stable metabolic production capabilities could be a model system for further studies on phenylpropanoid pathway.

Genetic transformation and transgenic *Capsicum*

Recent years have witnessed a rapid development of genetic engineering approaches to several crop improvements (Manoharan *et al.*, 1998). This is mainly due to the vast advancement in the understanding of growth and physiological processes of the plant at molecular level. However, development of an efficient and reproducible tissue culture regeneration protocol is the first step in utilizing the power and potential of this technology. Although excellent progress has been made in obtaining transgenic plants from many species of the solanaceae, pepper has lagged behind due to the unavailability of an efficient regeneration protocol (Liu *et al.*, 1990; Ebida and Hu, 1993). Sweet pepper is an important vegetable crop around the world, yet it suffers great losses due to infection by various viruses, including CMV (Cucumber mosaic virus). Genetic engineering of sweet pepper for useful traits, such as virus-resistance, is dependent on an efficient and reliable transformation and regeneration protocol. Despite several recent articles describing systems for the regeneration of both chilli and sweet peppers cultivars, elongation of shoot buds is a formidable job (Ochoa-Alezo and Ireta-Moreno, 1990). Wang *et al.* (1991) reported, for the first time, the recovery of plantlets from cultured sweet pepper hypocotyls and cotyledons with GUS gene expression. Later, Zhu *et al.* (1996) reported transgenic sweet pepper plants from *Agrobacterium*-mediated transformation. They reported the regeneration of fertile transgenic sweet pepper (*C. annuum* var.*grossum*) plants at a relatively high rate from various explants that were co-cultivated with *A. tumefaceins* strain GV3111-SE harbouring a plasmid that contained the cucumber mosaic virus coat protein (CMV-CP) gene. Similarly, *Agrobacterium*-mediated transformation in *Capsicum* has been reported by Professor Lakshmi Sita's group at the Indian Institute of Science, Bangalore, who were first successful in getting transgenic *Capsicum* of a high-pungent variety with GUS and NPT II gene insertion (Manoharan *et al.*, 1998).

As *Capsicum* is a self-pollinated crop there is a lot of scope in using the barnase gene expression for conferring male sterility and for using the barstar gene for male restorer lines similar to the application realized in Canola (*Brassica napus*). This would go a long way in getting hybrids of the required type by effective cross-pollination.

The genus *Capsicum* is also susceptible to the bacterial pathogen *Xanthomonas campestris*. Studies have shown that *avr*Bs2, an avirulence gene of this pathogen triggers disease resistance in pepper plants containing the Bs2 resistance gene and contribute to bacterial virulence on susceptible host plants. (Gassman *et al.*, 2000). The *avr*Bs3 gene of the same organism has been found to be specific for avirulence in pepper (Bonas *et al.*, 1993). Analysis of disease resistance loci for diseases caused by *X. campestris* has been reported by Pflieger *et al.* (1998). A PCR-based approach was used to isolate resistance gene analogs (RGAs) in pepper with primers corresponding to the Nucleotide Binding Site of the *RPS2* (*Arabidopsis*), *N* (tobacco) and L6 (flax) genes and to the kinase domains of the *Fen* and *Pto* (tomato) genes. Pepper PCR products were cloned, sequenced and localized on the interspecific pepper maps. Cloned resistance genes (*Pto*, *Cf2*, *N*) and Pathogenesis-Related protein genes were used as a probes for heterologous RFLP mapping (Pflieger *et al.*, 1998).

It has been observed that hot pepper (*C. annuum*) exhibits a hypersensitive response (HR) against infection by many tobacco viruses. A clone (CaPR-4) encoding a putative pathogenesis related protein 4 was isolated by Park *et al*. (2001) by differential screening of the cDNA library prepared for resistant pepper plant leaves inoculated with tobacco mosaic virus (TMV) pathotype PO. This study was extended to demonstrate capR-4 gene expression in pepper plants by various signal molecule, such as jasmonic acid and other abiotic elicitors. Such information has been useful in developing systemic acquired resistance (SAR) approaches for disease management. Similarly, the expression of SAR study would shed light on the defence system of *Capsicum* for future use in the genetic engineering of plant to confer resistance.

There are relatively few studies on the tissue-specific expression in *Capsicum* fruits. A recent report by Sung *et al.*, (2001) has shown that a regulatory role for flower and fruit development in *C.annuum* may be developed by an interaction of protein products through MADS-box genes, canMADS1 (isolated from floral bud) and CanMADS6. This was found by using the OSMADS 1 rice MADS-box gene as a probe.

Metabolic engineering of *Capsicum*, with regard to enhanced capsaicin or carotenoid production, will be an area of commercial importance. Already cloning of the carotenoid pathway gene has been demonstrated in several systems, including rice (Ye *et al.*, 2000). However, there is a need to get a high pigment and low pungency product from *Capsicum* which will be of value for pharmaceuticals, consumer purposes and as a food colourant. Attempts to get both a placenta-specific expression of capsaicinoids and a fruit wall specific expression of carotenoids is also needed. Such developments need to be completed for disease resistance, which would result in the overall improvement of the *Capsicum* plant for pre- and post-harvest applications for augmenting qualitative and quantitative output of the products from *Capsicum*.

Industrial prospects and relevance of biotechnology

Capsaicin is fast becoming a number one plant based pharmaceutical in the world due to its benefits as a pain reliever for arthritis and as a nutraceutical owing to its natural antioxidant properties. Similarly, carotenoids of paprika have been useful as natural colourants and as potent antioxidants for use in designer foods. It is estimated that the world market for food colourants is US$7000 million annually, of which US$2000 million is for natural colours for food applications. The use of paprika carotenoid is for both direct use as a carotenoid colourant in processed foods, but also as colourant for meat, especially chicken by way of feeding it to the poultry birds. Since the use of chemical methods of colouring processed foods is diminishing, scope for the application of a natural colourant is enormous.

Already several companies are manufacturing capsaicin-based products for a number of diversified applications. The details are given in Table 6.7.

The need for the technological improvement of *Capsicum* oleoresin for pungency and for colourant is growing as industry demands high pungency oleoresin for capsaicin-related applications. Both these can be met by biotechnological intervention, firstly by increasing capsaicinoid yield by pathway engineering and secondly, by suppressing capsaicin production in paprika *Capsicum* with high colour value. The use of antisense gene technology for low pungency fruit and the over-expression of carotenoid pathway is a great prospect. Since the technology of oleoresin preparation for colour also involves compensation of the pungency factor it is necessary to get zero pungency in colour-rich fractions. This is achievable through genetic engineering, as described earlier.

Overall focus in the near future should be on the improvement of *Capsicum* plants for preharvest management practice such as disease resistance, nematode resistance and high yields of oleoresin for specific uses.

Table 6.7 Some industrial applications of capsaicinoids and carotenoids of *Capsicum* – a representative listing

Company	Brand/Product	Uses	Remarks
1. Burlington-Biomedical and Scientific Co.	Denatonium Capsaicinate	Aversive agent, biocide, antifoulant and flavorant	US patent # 5,891,919
2. Duke University	Aerosol formulation with capsaicin and essential oils	To reduce cigarette smoking	US patent # 893,371
3. New Mexico Tech Research station	Protecting objects/ items for grazing animals and chewing rodents	Protectant containing habareno pepper	US patent # 5,674,476
4. Sabina International Ltd	Alleviation of premenstrual symptoms	Herbal formulation with *Capsicum*	US patent # 5,707,630
5. Department of Health and Human Services	Bird feed	Formulation preventing consumption of bird feed by other animals	US patent # 5,879,696
6. The Procter & Gamble Co.	Dermal	Prevents irritation for thioglycolate depilatories	US patent # 4,546,112
7. Kalamazoo, Holdings Inc	Insecticidal	Crop protection from worms and beetles	US patent # 5,599,803
8. Bioquimex, Spain	BIO-RED L	Carotenoid for poultry feed	Also potent antioxidant

Conclusion

Capsicum, which is traditionally consumed as a vegetable or food additive, is now gaining importance for its capsaicinoids and carotenoids, owing to their nutraceutical and therapeutic value. Production of these constituents can be achieved by genetic engineering and biotechnology, however, adequate efforts have not been made in this direction. The challenge of obtaining a zero capsaicin and high colour value crop will be of great economic value, which is achievable through genetic engineering. The scientific challenge lies in obtaining the tissue-specific expression of the metabolites in the desired loci of the plant. Production of capsaicinoids through immobilized cell cultures will be an area of biotechnological utility. However, the scale-up of the process has not yet been addressed.

In future years the demand for capsaicinoids and carotenoids of *Capsicum* is bound to increase due to the demand for nutraceuticals in processed foods and for use as a herbal medicine.

Acknowledgements

This work was supported by the Council of Scientific and Industrial Research (CSIR), New Delhi, India. The authors would like to thank Dr V. Prakash, Director, CFTRI for his encouragement and support.

References

Agarwal, S., Chandra, N. and Kothari, S. L. (1989) Plant regeneration in tissue culture of pepper (*Capsicum annuum* L. cv. *mathania*) *Pl. Cell Tissue Org. Cult.* 16, 47–55.

Akashi, T., Akoi, T., Takahashi, T., Kameya, N., Nakamura, I. and Ayabe, S. (1997) Cloning of cytochrome P450 cDNAs from cultured *Glycyrrhiza echinata* L. cells and their transcriptional activation by elicitor treatment. *Plant Sci.* 93, 39–47.

Altshcul, S. F., Madden, T. L. and Schaffer, A. A. (1997) Gapped BLAST and PSI BLAST a new generation of protein data base search programs. *Nucleic Acids Res.* 25, 3389–3402.

Aluru, M., Curry, J. and O'Connel, M. (1998a) Nucleotide sequence of a 3-oxacyl-[Acyl-carrier-Protein] synthase (beta-keto-acyl-ACP synthase) gene (Accession No. AF085148) from habanero chile PGR98-181. *Plant Physiol.* 118, 1102.

Aluru, M., Curry, J. and O'Connel, M. (1998b) Nucleotide sequence of a probable aminotransferase gene (Accession No. AF085149) from habanero chile PGR98-182. *Plant Physiol.* 118, 1102.

Amaya Guerra, C. A., Sernma Sldivar, S. R. O., Cadenas, E. and Nevera Munoz, A. (1997) Evaluation of different solvent systems for the extraction and fractionation of oleoresins from guajillo peppers. *Arch. Latinoam. Nutr.* 47, 127–130.

Ananthasamy, T. S., Kamat, U. N. and Pandya, H. G. (1960) Capsicum contents of chilli varieties. *Curr. Sci.* 29, 271.

Andrews, J. (1995) *Peppers, the Domesticated* Capsicums, University of Texas Press, Austin, USA., pp. 25–50.

Armstrong, D. W., Brown, L. A., Porter, S. and Rutten, R. (1993) Biotechnological derivation of aromatic flavour compounds and precursors. In: P. Schreier and Winterhalter, P. (eds) *Progress in Flavour Precursor Studies*. Allured Publ. Corp., Carol Stream, pp. 425–438.

Arroyo, R. and Revilla, M. A. (1991) *In vitro* plant regeneration from cotyledon and hypocotyl segments in two bell pepper cultivars. *Plant Cell Reports* 10, 414–416.

Balakrishnan, K. V. and Verghese, J. (1997) Studies on the recovery of pungency-free colour matter from Indian *Capsicum* extracts. *Acta Alimentar.* 26, 9–21.

Bennet, D. J. and Kirby, G. W. (1968) Constitution and biosynthesis of capsaicin. *J. Chem. Soc.* C 442–445.

Berger, R. G. (1995) *Aroma Biotechnology.* Springer-Verlag, Berlin, Heidelberg, pp. 79–90.

Binzel, M. L., Sankla, N., Joshi, S. and Sankhla, D. (1996) Induction of direct somatic embryogenesis and plant regeneration in pepper (*Capsicum annuum* L.). *Plant Cell Rep.* 15, 536–540.

Bonas, U., Conrads-strauch, J. and Balbo (1993) Resistance in tomato to *Xanthomonas campestris* pv *vesticatoria* is determined by alleles of the pepper-specific virulence gene avr Bs3. *Mol. Gen. Gent.* Apr. 238(1–2), 261–269.

Bosland, P. W. (1994) Chillies: History, cultivation and uses, In: G. Charalambous (ed.) *Spices, Herbs and Edible Fungi.* Elsevier Publ., New York. pp. 347–366.

Burton, G. W. and Ingold, K. U. (1984) β-carotene: an unusual type of lipid antioxidant. *Science* 224, 569–573.

Carmichael, J. K. (1991) Treatment of Herpes Zoster and post therapeutic neuralgia. *Am. Family Physician* 44, 203–210.

Cheetham, P. S. J. (1995) Biotransformations: new routes to food ingredients. *Chem. Ind.* April, 265–268.

Christopher, T. and Rajam, M. V. (1994) *In vitro* clonal propagation of *Capsicum* spp. *Plant Cell Tiss. Org. Cult.* 38, 25–27.

Christopher, T. and Rajam, M. V. (1996) Effect of genotype explant and medium on *in vitro* regeneration of red pepper. *Plant Cell Tiss. Org. Cult.* 46, 245–250.

Curry, J., Mendoza, M. and O'Connel, M. (1998) Nucleotide sequence of a caffeic acid 3-O-methyltransferase gene (Accession No. AF081214) from habanero chile PGR 98-170. *Plant Physiol.* 118, 711.

Curry, J., Aluru, M., Mendoza, M., Nivarez, J., Melendrez, M. and O'Connel, M. (1999) Transcripts for possible capsaicinoid biosynthetic genes are differentially accumulated in pungent and non pungent *Capsicum* spp. *Plant Sci.* 148, 47–57.

Deb, A. R., Ramanujam, S., Krishnamurthy, G. S. R. and Thirumalachar, D. K. (1963) Capsaicin content of some new pusa varieties of chillies. *Ind. J. Technol.* 1, 59–60.

Di Cosmo, F. and Tallevi, S. G. (1985) Plant cell cultures and microbial insult: interactions with biotechnological potential. *Trends Biotechnol.* 3, 110–111.

Dornenburg, H. and Knorr, D. (1996a) Generation of colours and flavours in plant cell and tissue cultures. *Crit. Rev. Plant Sci.*, 15, 141–168.

Dornenburg, H. and Knorr, D. (1996b) Production of the phenolic flavour compounds with cultured cells and tissues of *Vanilla planifolia* species. *Food Biotechnol.* 10, 75–92.

Ebida, I. A. A. and Ching-yh Hu (1993) *In vitro* morphogenic responses and plant regeneration from pepper (*Capsicum annuum* L. cv. Early California Wonder) seedling explants. *Plant Cell Rep.* 13, 107–110.

Estabrook, E. M. and Sengupta-Gopalan, C. (1991) Differential expression of phenylalanine ammonia-lyase and chalcone synthase during soybean nodule development. *Plant Cell* 3, 299–308.

Esterbauer, H. (1991) Role of vitamin E in preventing the oxidation of low-density lipoprotein. *Am. J. Clin. Nutr.*, 53, 314–321.

Ezura, H., Nishiyama, S. and Kasumi, M. (1993) Efficient regeneration of plants independent of exogenous growth regulators in bell pepper (*Capsicum annuum* L.) *Plant Cell Rep.* 112, 676–680.

Fahrendorf, T. and Dixon, R. A. (1993) Stress responses in alfalfa (*Medicago sativa* L.) XVII: molecular cloning and expression of the elicitor inducible cinnamic acid 4-hydroxy-lyase cytochrome P450. *Arch. Biochem. Biophys.* 305, 509–515.

Fujimoto, H., Suzuki, T. and Iwai, K. (1980) Intracellular localization of capsaicin and its analogues in *Capsicum* fruit II. The vacuole as the intracellular accumulation site of capsaicinoid in the protoplasts of *Capsicum* fruit. *Plant Cell Physiol.* 21, 1023–1030.

Fujiwake, H., Suzuki, T. and Iwai, K. (1980) Intracellular localization of capsaicin and its analogues in *Capsicum* fruits II. The vacuole as the intracellular accumulation site of capsaicinoid in the protoplast of *Capsicum* fruit. *Pl. Cell Physiol.* 21, 1023–1030.

Fujiwake, H., Suzuki, T. and Iwai, K. (1982a) Capsaicinoid formation in the protoplast from the placenta of *Capsicum* fruits. *Agric. Biol. Chem.* 46, 2591–2592.

Fujiwake, H., Suzuki, T. and Iwai, K. (1982b) Intracellular distribution of enzymes and intermediates involved in biosynthesis of capsaicin and its analogues in *Capsicum* fruits. *Agric. Biol. Chem.* 46, 2685–2689.

Furuya, T. and Hashimoto, K. (1955) Distribution of capsaicin secretory cells in *Capsicum*. *Yakugaku Zasshi.* 74, 771.

Gal, I. E. (1965) Neurene angaben uber capsicidin. *Experimentia* 21, 383.

Gassmann, W., Dahlbeck, D., Chesnokova, O., Minsavage, G. V., Jones, J. B. and Staskawicz, B. J. (2000) Molecular evolution of virulence in natural field strains of *Xanthomonas campestris* pv. *Vesicatoria*. *J. Bacteriol.* 182(24), 7053–7059.

Glatzel, A. and Gewarz, D. (1968) Ihre wirkungen auf dengesunder und kraneken mensschen, Nicolissche verlagsbuch handling. *Herford.*

Govindarajan, V. S. (1985) Capsicum production, technology, chemistry and quality. Part-II. Processed products, standards, world production and trade. *CRC Critical Rev. Nutr.* 23(3), 207–288.

Govindarajan, V. S., Narasimhan, S. and Dhanraj, S. (1977) Evaluation of spices and oleoresins II Pungency of *Capsicum* by Scoville heat units – a standardized procedure. *J. Food Sci. Technol.* 14, 28–34.

Govindarajan, V. S. and Satyanarayana, M. N. (1991) *Capsicum*-production, technology, chemistry and quality. Part V. Impact on physiology, pharmacology, nutrition and metabolism; structure, pungency, pain and desensitization sequences. *CRC Crit. Rev. Food Sci. Nutr.* 29, 435–474.

Gowri, G., Bugos, R. C., Campbell, W. H., Maxwell, C. A. and Dixon, R. A. (1991) Molecular cloning and expression of alfalfa S-adenosyl L-methionine; caffeic acid 3-O-methyl transferase, a key enzyme of lignin biosynthesis. *Plant Physiol.* 97, 7–14.

Gunay, S. and Rao, P. S. (1978) *In vitro* plant regeneration from hypocotyl and cotyledon explants of red pepper (*Capsicum annuum*). *Plant Sci. Lett.* 11, 365–372.

Gutsu, E. V., Baleshova, N. N., Lazur'evskii, G. V. and Timina, O. O. (1982) *Izv. Akad. Nauk, Mold. SSR. Ser. Biol. Khim. Nauk.*, 24, 81–86.

Harini, I. and Lakshmi Sita, G. (1993) Direct somatic embryogenesis and plant regeneration from immature embryos of chilli (*Capsicum annuum* L.). Plant Sci. 89, 107–112.

Harkay, V. (1974) Storage experiments with raw material of seasoning paprika with particular reference to the red colour pigment components. *Acta. Alim. Acad. Sci. Hung.* 3, 239–249.

Hartman, K. T. (1970) A rapid gas-liquid chromatographic determination for capsaicin in *capsicum* species. *J. Food. Sci.* 35, 543–574.

Hornero-Mendez, D., Ricardo Gomez-Ladron de Guevara., Isabel, M. Minguez-Mosquera (2000) Carotenoid biosynthesis changes in five red pepper (*Capsicum annuum* L.) cultivars during ripening. Cultivar selection for breeding. *J. Agric. Food Chem.* 48, 3857–3864.

Hotze, M., Schroder, G. and Schroder, J. (1995) Cinnamate 4-hydroxylase from *Catharanthus roseus* and a strategy for the functional expression of plant cytochromes P450 proteins as transnational fusions with P450 reductase in *Escherichia coli*. *FEBS Letters* 374, 345–350.

Howard, L. R. (2001) Antioxidant vitamin and phytochemical content of fresh and processed pepper fruit (*Capsicum annuum*). In: Robert, E. and Wildman, C. (eds) *Hand book of Nutraceuticals and Functional Foods*. CRC press, New York, 209–229.

Hussain, S., Jain, A. and Kopthari, S. L. (1999) Phenylacetic acid improves bud elongation and *in vitro* plant regeneration efficiency in *Capsicum annuum* L. *Plant Cell Rep.* 19, 64–68.

Hyde, C. L. and Phillips, G. C. (1996) Silver nitrate promotes shoot development and plant regeneration of chilli pepper (*Capsicum annuum* L.) via organogenesis. *In Vitro Cell Dev. Biol. – Plant* 32, 72–80.

Iwai, K., Lee, K. R., Kobashi, M. and Suzuki, T. (1977a) Formation of pungent principles in fruits of sweet pepper, *Capsicum annuum* L. var. *grsossum* during post harvest ripening under continuous light. *Agri. Biol. Chem.* 41, 1873–1876.

Iwai, K., Suzuki, T., Lee, K. R., Kobashu, M. and Oka, S. (1977b) *In vivo* and *in vitro* formation of dihydrocapsaicin in sweet pepper fruits, *C. annuum* L. var. *grossum. Agric. Biol. Chem.* 41, 1877–1882.

Iwai, K., Suzuki, T. and Fujiwake, H. (1979) Formation and accumulation of pungent principle of hot pepper fruits, capsaicin and its analogues, in *Capsicum annuum* var. *annuum*, CV. Karayatasubusaat different growth stages after flowering. *Agric. Biol. Chem.* 43, 2493–2498.

Jaeck, E., Martz, F., Stiefel, V., Fritig, F. and Legrand, M. (1996) Expression of class I O-methyl transferase in healthy and TMV infected tobacco. *Mol. Plant Microbe Interact.* 9, 681–688.

Johnson, T. S. (1993) Studies on production of capsaicin in immobilised cells and placental tissues of *Capsicum annuum* and *Capsicum frutescens*. PhD thesis, University of Mysore, Mysore, India.

Johnson, T. S. and Ravishankar, G. A. (1996) Precursor biotransformation in immobilized placental tissue of *Capsicum frutescens* Mill 1. Influence of feeding intermediate metabolites of capsaicinoid metabolites of capsaicinoid pathway on capsaicin and dihydrocapsaicin accumulation. *J. Plant Physiol.* 147, 481–485.

Johnson, T. S., Ravishankar, G. A. and Venkataraman, L. V. (1990) *In vitro* capsaicin production by immobilized cells and placental tissues of *Capsicum annuum* L. grown in liquid medium. *Plant Sci.* 70, 223–229.

Johnson, T. S., Ravishankar, G. A. and Venkatraman, L. V. (1991) Elicitation of capsaicin production in freely suspended and immobilised cell cultures of *Capsicum frutescens* Mill. *Food Biotechnol.* 5, 197–205.

Johnson, T. S., Ravishankar, G. A. and Venkatraman, L. V. (1993) Separation of capsaicin from the phenyl propanoid compounds by High Performance Liquid Chromatography to determine the biosynthetic status of cells and tissues of *Capsicum frutescens*. Mill *in vivo* and *in vitro*. *J. Agric. Food Chem.* 40, 2461–2463.

Johnson, T. S., Ravishankar, G. A., Venkataraman, L. V. (1996) Biotransformation of ferulic acid to vanillin and capsaicin in immobilized cell cultures of *Capsicum frutescens. Plant Cell Tiss. Org. Cult.* 44, 117–121.

Joos, H. J. and Hahlbrock, K. (1992) Phenylalanine ammonia lyase in Potato (*Solanum tuberosum* L.) Genomic complexity, structural comparison of two selected genes and modes of expression. *Eur. J. Biochem.* 204, 621–629.

Jurenitsch, J., Kubelka, W. and Jentsch, K. (1979) Identification of cultivated taxa of *Capsicum* taxanomy, anatomy and comparison of pungent principle. *Planta Med.*, 35, 174–183.

Jurenitsch, J. and Leinmueller, R. (1980) Quantification of nanonoic acid vanillylamide and other capsaicinoids in the pungent principle of *Capsicum* fruits and preparations by gas–liquid chromatography on glass capillary columns. *J. Chromatogr.* 189, 389–397.

Kanner, J., Harel, S., Palevitch, D. and Ben-gera, I. (1977) Colour retention in sweet paprika powder as affected by moisture content and ripening stage. *J. Food Technol.* 12, 59–64.

Kanner, J., Mendal, H. and Budowski, P. (1979) Prooxidant and antioxidant effects of ascorbic acid and metal salts in a β-carotene–linoleate model system. *J. Food Sci.*, 42, 60–64.

Kauppinen, S. (1992) Structure and expression of the KAS12 gene encoding a beta-ketoacyl carrier protein synthase I isozyme from barley. *J. Biol. chem.* 267, 23999–24006.

Kawai, S., Mori, A., Shiokawa, T., Kajita, S., Katayama, Y. and Morohoshi, N. (1996) Isolation and analysis of cinnamic acid 4-hydroxylase homologous genes from a hybrid aspin, *Populus Kitakamiensis. Biosci. Biotechnol. Biochem.* 60, 1586–1597.

Kiss, G. A., Forgacs, E., Cserhati, T., Mota, T., Morai, S. H. and Ramos, A. (2000) Optimization of the microwave assisted extraction of pigments from paprika (*Capsicum annuum* L.) powders. *J. Chromatogr.* 889 (1–2), 41–49.

Knorr, D., Caster, C., Dorenberg, H., Dorn, R., Graf, S., Havkin-Frenkel, D., Podstolski, A. and Werrmann, U. (1993) Biosynthesis and yields improvement of food ingredients from plant cell and tissue cultures. *Food Technol.* 39, 139–142.

Kobayashi, S. (1925) *Rikagaku Kenkyusho Hokoku,* 4, 527.

Kosuge, S. and Furata, M. (1970) Studies on the pungent principle of *Capsicum* part XIV. Chemical constituents of pungent principle. *Agric. Biol. Chem.* 34, 248–256.

Krings, U. and Berger, R. G. (1998) Biotechnological production of flavours and fragrances. *Appl. Microbiol. Biotechnol.* 49, 1–8.

Kunert, K. J. and Tappel, A. L. (1983) The effect of vitamin C on *in vivo* lipid peroxidation in Guinea pigs as measured by pentane and ethane production. *Lipids* 18, 271.

Lee, S. W. (1977) Physico-chemical studies on the after ripening of hot pepper fruits. VI.Hot taste components in different parts and capsaicin analogues. *J. Korean Agric. Chem. Soc.* 14, 157.

Lee, S. W., Robb, J. and Nazar, R. (1992) Truncated phenylalanine ammonia-lyase expression in tomato (*Lycopersicum esculentum*). *J. Biol. Chem.* 267, 11824–11830.

Lee, B., Choi, D. and Lee, K. W. (1998) Isolation and characterization of O-diphenol, O-methyltransferase cDNA clone in hot pepper (*Capsicum annuum* L.). *J. Plant Biol.* 41, 9–15.

Lee, Y. and Howard, L. R. (1999) Firmness and phytochemical losses in pasteurized yellow banana peppers (*Capsicum annuum*) as affected by calcium chloride and storage. *J. Agric. Food Chem.* 47, 700–703.

Leete, E. and Louden, M. C. L. (1968) Biosynthesis of capsaicin and dihydrocapsaicin in *Capsicum frutescens. J. Am. Chem. Soc.* 90, 6837.

Lindsey, K. and Yeoman, M. M. (1985) Immobilized plant cells. In: Yeoman, M. (ed.) *Plant Cell Culture Technology.* Blackwell, Oxford, pp. 229–267.

Liu, W., Parrot, W. A., Hildrbrand, D. F., Collins, G. B. and Williams, E. G. (1990) *Agrobacterium* induced gall formation in bell pepper (*Capsicum annuum* L.) and formation of shoot like structures expressing introduced genes. *Plant Cell. Rep.* 9, 360–364.

Maga, J. A. (1975) *Capsicum. CRC Crit. Rev. Food Sci. Nutr.*, 6, 177–199.

Manoharan, M., Sree Vidya, C. S. and Lakshmi Sita, G. (1998) *Agrobacterium*-mediated genetic transformation in hot chilli (*Capsicum annuum* L. var. Pusa jwala). *Plant Science* 131, 77–83.

Manuel Jaren-Galen, Uwe Nienaber and Steven J. Schwartz. (1999) Paprika (*Capsicum annuum*) oleoresin extraction with super critical carbon dioxide. *J. Agric. Food Chem.*, 47, 3558–3564.

Maoka, T., Fujiwara, Y., Hashimoto, K. and Akimoto, N. (2001) Isolation of a series of apocarotenoids from the fruits of the red paprika *Capsicum annuum* L. *J. Agric. Food Chem.* 49(3), 1601–1606.

Marla, L., Benzel, N., Sageeta Joshi, and Sankhla, D. (1996) Induction of direct somatic embryogenesis and plant regeneration in pepper *Capsicum annuum. Plant Cell Rep.* 15, 536–540.

Mathew, A. G., Lewis, Y. S., Jagadishan, R., Nambudri, E. S. and Krishnamurthy, N. (1971) Oleoresin *Capsicum. Flavour Ind.* 2, 23–26.

Matsui, K., Shibutani, M., Shibata, Y. and Kajiwara, T. (1997) In: Dordrecht (ed.) *Physiology, Biochemistry and Molecular Biology of Plant Lipids.* Kluwer Academic, Boston, pp. 348–350.

Merav, H., Varda, M. and Joseph, H. (2000) Genetic engineering of the carotenoid biosynthesis pathway. Abs# 1015. *Plant Biology Abstracts,* July 15–19, San Diego, USA.

Meyer, H. P., Kiner, A., Imwinkelried, R., Shaw, N. and Lonza, A. G. (1997) Biotransformations for fine chemical production. *Chimia* 51, 287.

Meyer-Bahlburg, H. F. L. (1972) Pilot studies on stimulant effects of *Capsicum* species. *Nutr. Metab.* 14, 245.

Minguez-Mosquera, M. I. and Horneo-Mendez, D. (1994) Comparative study of the effect of paprika processing on the carotenoids in peppers (*Capsicum annuum*) of the Bola and Agridulce varieties. *J. Agric. Food Chem.* 42, 1555–1560.

Molnar, J. (1965) Die pharmakologischen wirkungen des capsaicins, des schemeckenden wirkstoffes in paprika. *Arzneim. Forsch.* 15, 178.

Morrisson, R. A., Koning, R. E. and Evans, D. A. (1986) Pepper. In: Evans, D. A., Sharp, W. R. and Ammirato, P. V. (eds) *Handbook of Plant Cell Culture*, Vol. 4. McGraw Hill, NY, pp. 554–573.

Nagai, N., Kitauchi, G., Shimosaka, M. and Okazaki, M. (1994) Cloning and sequencing of a full length cDNA coding for phenylalanine ammonia lyase from tobacco cell culture. *Plant Physiol.* 104, 1091–1092.

Nelson, E. K. (1919) The constitution of capsaicin, the pungent principle of *Capsicum. J. Am. Chem. Soc.* 41, 1114–1123.

Ochoa-Alezo, N. and Ireta-Moreno, I. (1990) Cultivar differences in shoot-forming capacity of hypocotyl tissues of chilli pepper (*Capsicum annuum* L.) cultures *in vitro. Sci. Hort.* 42, 21028.

Ong, A. S. H. and Choo, Y. M. (1997) Carotenoids and tocols from palm oil. In: Fereidoon shahidi (ed.) *Natural Antioxidants, Chemistry, Health Effects and Applications.* Chapter 8. AOCS Press. Champaign, Illinois, pp. 133–149.

Padilla, M. C. and Yahia, E. M. (1998) Changes in capsaicinoids during development, maturation and senescence of chilli peppers and relation with peroxidase activity. *J. Agric. Food. Chem.* 46(6), 2075–2079.

Park, C. J., Shin, R., Park, J. M., Lee, G. J., Yoo, T. H. and Paek, K. H. (2001) A hot pepper cDNA encoding a pathogenesis-related protein 4 is induced during the resistance response to tobacco mosaic virus. *Mol. Cell.* 11(1) 122–127.

Pellegrini, L., Rohfritsch, O., Fritig, B. and Legrand, M. (1994) Phenyl alanine ammonia lyase in tobacco, molecular cloning and gene expression during the hyper sensitive reaction to Tobacco mosaic virus and the response to a fungal elicitor. *Plant Physiol.* 106, 877–886.

Peto, R., Doll, R., Buckley, J. D. and Sporn, M. B. (1981) Can dietary beta-carotene materially reduce human cancer rates? *Nature* (London) 290, 201.

Pflieger, S., Veronique Lefebvre., Carol, C., Radwanski, E., Livingston, K., Jahn, M. K. and Palloix, A. (1998) Disease resistance loci in pepper: genome distribution and characterization. *Intl. Plant and Animal Genome VI conference.* Jan 18–22, San Diego, CA, USA.

Phillips, G. and Hubstenberger, J. F. (1985) Organogenesis in pepper tissue cultures. *Plant Cell Tiss. Org. Cult.* 4, 261–269.

Prince, J. P., Ochard, E. and Tanksley, S. D. (1993) Conservation of a molecular linkage map of pepper and comparison of synteny with tomato. *Genome* 36, 404–417.

Quagliotti, L. and Ottaviana (1971) Genetic analysis of the variability in capsaicin content in two pepper varieties. *Genet. Agric.* 25, 56–60.

Ramachandra Rao, S. (1998) Studies on biotransformation to produce phytochemicals of importance using plant cell cultures. PhD Thesis, University of Mysore, Mysore, India.

Ramachandra Rao, S. and Ravishankar, G. A. (1999) Biotransformation of isoeugenol to vanillin and capsaicin in freely suspended and immobilised cell cultures of *Capsicum frutescens*: study of the influence of β-cyclodextrin and fungal elicitor. *Proc. Biochem.* 35, 341–348.

Ramachandra Rao, S. and Ravishankar, G. A. (2000a) Vanilla flavour production by conventional and biotechnological routes. *J. Sci. Food Agric.* 80, 289–304.

Ramachandra Rao, S. and Ravishankar, G. A. (2000b) Biotransformation of protocatechuic aldehyde and caffeic acid to vanillin and capsaicin in freely suspended and immobilized cell cultures of *Capsicum frutescens, J. Biotechol.* 76, 137–146.

Ramirez-Malagaon, R. and Ochoa-Alezo, N. (1996) An improved and reliable chilli pepper (*Capsicum annuum* L.) plant regeneration method. *Plant Cell. Rep.* 16, 226–231.

Ravishankar, G. A. and Ramachandra Rao, S. (2000) Biotechnological production of phytopharmaceuticals. *J. Biochem. Mol. Biol. Biophys.* 4, 73–102.

Ravishankar, G. A., Sarma, K. S. and Venkataraman, L. V. (1988) Effect of nutritional stress on capsaicin production in immobilized cell cultures of *Capsicum annuum. Curr. Sci.*, 57, 381–383.

Reeves, M. J. (1987) Re-evaluation of *Capsicum* colour data. *J. Food. Sci. Technol.* 52, 1047–1049.

Rhodes, M. J. C., Spencer, A. and Hamil, J. D. (1994) Plant tissue culture in the production of flavour compounds. *Biochem. Soc. Trans.* 19, 702–706.

Santamaria, R. I., Reyes-Durate, M. D., Barzana, E., Fernando, D., Gama, F. M., Mota, M. and Lopez-Munguia, A. (2000) Selective enzyme-mediated extraction of capsaicinoids and carotenoids from chili Guajillo Puya (*Capsicum annuum* L.) using ethanol as solvent. *J. Agric. Food. Chem.* 48, 3063–3067.

Schopffer, C. R. and Ebel, J. (1998) Identification of elicitor induced cytochrome p450s of soybean (*Glycine max*) using differential display of mRNA. *Mol. Gen. Genet.* 258, 315–322.

Scoville, W. C. (1912) Note on *Capsicum. J. Am. Pharm. Assoc.* 1, 453.

Scragg, A. G. (1997) The production of aromas by plant cell cultures. In: Schepier, T. (ed.) *Adv. Biochem. Engg. Biotechnol.* Vol. 55, Springer-Verlag, Berlin, pp. 239–263.

Sekiguchi, S., Yamakawa, T., Kodama, T., Smith, S. M. and Yeoman, M. M. (1996) Establishment of hairy root culture of chilli pepper (*Capsicum frutescens*). *Plant Tiss. Cult. Lett.* 13, 219–221.

Sekiguchi, S., Yamakawa, T., Kodama, T., Smith, S. M. and Yeoman, M. M. (1999) Characterization of chilli pepper hairy roots expressing the Parsley PAL2 cDNA. *Plant Biotechnol.* 16(2), 153–158.

Shekelle, R. B., Liu, S., Raynor. Jr, W. J., Lepper, M., Liza, C., Rosoff, A. H., Paul, O., Shryock, A. M. and Stamler, J. (1981) Dietary vitamin A and risk of cancer in the western electric study. *Lancet.* **1981-II** 1185.

Siggaard-Anderson, M., Kauppinen, S. and Von Wettstein-Knowles, P. (1991) Primary structure of a cerlenin-binding beta-ketoacyl-[acyl carrier protein] synthase from barley chloroplasts. *Proc. Nat. Acad. Sci.* USA 88, 4114–4118.

Simon, J. E., Chadwick, A. F. and Craker, L. E. (1984) *An Indexed Bibiliography. 1971–1980. The Scientific Literature on Selected Herbs and Aromatic and Medicinal Plants of the Temeperate Zone.* Archon Books, Hamden, CT. p. 770.

Sripichitt, P., Nawata, E. and Shigenaga (1987) *In vitro* shoot forming capacity of cotyledon explant in red pepper (*Capsicum annuum*.L.cv.yatsufusa) *Jap. J. Breed.* 37, 133–142.

Sukrasno, N. and Yeoman, M. M. (1993) Phenyl propanoid metabolism during growth and development of *Capsicum frutescens* fruits. *Phytochemistry.* 32, 839–844.

Sung, S. K., Moon, Y. H., Chung, J. E., Lee, S. Y., Park, H. G. and An, G. (2001) Characterization of MADS box genes from hot pepper. *Mol. Cells.* 11(3), 352–356.

Suvarnalatha, G., Chand, R. N., Ravishankar, G. A. and Venkataraman, L. V. (1993) Computer aided modelling or optimization for capsaicinoid production of immobilized *Capsicum frutescens* cells. *Enz. Microbial. Technol.* 15, 710–715.

Suzuki, J. I., Tausig, F. and Morse, R. E. (1957) Some observations on red pepper. A new method for the determination of pungency in red pepper. *Food Technol.* 11, 100–104.

Suzuki, T., Fuziwake, H. and Iwai, K. (1980) Intercellular localization of capsaicin and its analogues, capsaicinoid in *Capsicum* fruit, 1.Microscopic investigation of the structure of the placenta of *Capsicum annuum* var. *annuum* cv. *Karayatsubusa, Pl. Cell Physiol.* 21, 839–853.

Suzuki, T., Kawada, K. and Iwai, R. (1981) The precursors effecting the composition of capsaicin and its analogues in the fruits of *Capsicm annuum* var. *annum* cv. *Karayatsubusa. Agric. Biol. Chem.* 45, 535–537.

Szasz, A., Nervo, G. and Fari, M. (1995) Screening for *in vitro* shoot forming capacity of seedling explants in bell pepper (*Capsicum annuum* L.) genotypes and efficient plant regeneration using thidizuron. *Plant. Cell Rep.* 14, 666–669.

Tewari, V. P. (1990) Development of high capsaicin chillies (*Capsicum annuum* L.) and their implications for the manufacture of export products. *J. Plant. Crops* 18(1), 1–13.

Thirumalachar, D. K. (1967) Variability for capsaicin content in chilli. *Curr. Sci.* 36, 269–270.

Todd, Jr. P. H. (1958) Detection of foreign pungent compounds. Oleoresin *Capsicum* found *Capsicum* and chilli spices. *Food Technol.* 12, 468–469.

Trenov, R. and Khristov, S. (1966) Nauchni Tr.Nauchnoizsled.Inst. Kon-servana Pran-St., *Plovdiv*, 4, 101.

Valero-Montero, L. L. and Ochoa-Alezo, N. (1991) A novel approach for chilli pepper (*Capsicum annuum* L.) Plant regeneration, shoot induction in rooted hypocotyls. *Plant. Sci.* 84, 215–219.

Wang, Y. W., Yang, M. Z., Pan, N. S. and Chen, Z. L. (1991) Gus expression in hypocotyls and cotyledons of transformed sweet paper. *Acta Bot. Sin.* 33, 780–786.

Yamakawa, T., Sekiguchi, S., Kodama, T., Smith, S. M. and Yoeman, M. M. (1998) Transformation of chilli pepper (*Capsicum frutescens*) with a phenyl alanine ammonia-lyase gene. *Plant Biotechnol.* 15, 189–193.

Ye, X., Al-Babilli, S., Kloti, A., Zhang, J., Lucca, P., Beyer, P. and Potrykus, I. (2000) Engineering the provitamin A (carotene) biosynthetic pathway to (carotenoid-free) rice endosperm. *Science*, 287, 303–305.

Yeoman, M. M., Meidzybrodzka, M. B., Lindsey, K. and Lauchlal, W. R. (1980) The synthetic potential of cultured plant cells. In: Sala, F., Parisi, B., Cella, R. and Ciffiri, O. (eds) *Plant Cell Cultures, Results and Prospectives*. Elsevier North Hall Publ. Amsterdam, pp. 327–343.

Zenk, M. H. (1965) Biosynthase von vanillin in *Vanilla planifolia. Andr. Z. Pflanzenphysiol.* 55, 404–414.

Zhu, K., Zhang Wen-Jun, Z. L. and Chen Ou-Yang (1996) Transgenic sweet pepper plants from *Agrobacterium* mediated transformation. *Plant Cell Rep.* 16, 71–75.

7 Irrigational aspects of *Capsicum*

L. B. Naik

Capsicum, a shallow rooted crop, is very sensitive to soil moisture variations. Excess or deficit soil moisture must be avoided. More than half of the root system is concentrated in the 5–15 cm layer. For field irrigation, the land should be laid into ridges and furrows to economise water. Under conditions of water scarcity, alternate furrows and widely spaced furrow irrigation are recommended. Generally, eight to nine irrigations may be required to raise the crop depending upon agroclimatic conditions. The irrigation requirement of the crop is found to be 110 cm. Scheduling of irrigation can be made based on cumulative pan evaporation (40 mm), IW/CPI ratio (0.75–0.90), available soil moisture depletion (40–60% ASM) and soil moisture tension (25 kPa). Drip irrigation was found to be superior to sprinkler and furrow irrigation in saving water and reducing the incidence of disease. Spraying 200 ppm alachlor solution as antitranspirant was found to be beneficial.

Introduction

Chillies are shallow rooted crops and are very sensitive to soil moisture variations. Field moisture must be carefully monitored throughout the crop growth. Excess soil moisture particularly following the fruit set must be avoided. When there is a soil moisture deficit, blossom end rot may occur and under greater moisture stress, fruit abortion is possible.

Hot chillies, which are usually grown for red dry chillies, are raised as a rainfed crop in several parts of the world that receive an annual rainfall of around 80–100 cm. Chilli seedlings are usually transplanted with the onset of monsoon in the tropics when the relative humidity is around 90%. This transplantation facilitates quick establishment of the seedlings. The chilli crop is also raised during winter and summer seasons, and also in places where rainfall is not sufficient for the growth of the crop, and so a supplemental irrigation is required. Non-pungent types of chillies are raised under irrigated conditions.

Rooting depth

The depth of rooting of vegetables is influenced by the soil profile. If there is a clay pan/hard pan or other densification, then it is not possible to root the vegetable at its normal depth. Although *Capsicums* are of a shallow to moderately deep rooted vegetable crop (90–120 cm), the root pattern development indicates that more than half of the root system (dry weight) is developed in the 5–15 cm layer (Keng *et al.*, 1979). The root length density of pepper decreases with soil depth, and rapidly so below 20 cm, while horizontal variation in root distribution is relatively small (Morita and Toyota, 1998). *Capsicums* extract 70–80% of water used from a depth of 0–30 cm (Dimitrov and Ovtcharova, 1995).

Methods of irrigation

The land should be laid out either in ridges and furrows or in flat beds for irrigation. Ridges and furrows are better than flat beds or check basins for the economisation of water (Subramanian et al., 1998). Additionally, chillies cannot withstand water stagnation. Seedlings should be transplanted halfway on the ridge for better growth and higher yields. The distance between two ridges depends upon the row spacing of the crop. Hegde (1989a) observed that alternate furrows and a widely spaced furrow irrigation system holds great promise for reducing the irrigation need, as such a set-up has the inbuilt advantage of holding a smaller quantity of irrigation water compared to that of every furrow irrigation.

Matev et al. (1970) compared the different types of irrigation using the cv. Kurtovska Kupija-1619. The lowest yield was obtained with pipe irrigation. Sprinkler irrigation was most suitable for light sandy clay soil, while irrigation by furrows and pipes was most suitable for heavy meadow soils. The efficiencies of rain hose sprinkler and furrow irrigation systems were investigated by Eom and Im (1990) in a sandy loam soil. Cumulative infiltration by furrow irrigation was 25% of the soil infiltrability when the soil water potential was −0.5 bar. The loss of irrigation water by runoff was 27–31% and 58–61% for sprinkler and rain hose, respectively. The application efficiency and storage efficiency of sprinkler irrigation was 61–73% and 52–89%, and that of rain hose irrigation 73–76% and 55%, respectively. The advantage with sprinkler and drip irrigation systems is that pesticides could be mixed with irrigation water to control diseases (Nashev, 1998) and insects (Cabello et al., 1997). In an evaluation of low head drip irrigation, pitcher irrigation and subsurface irrigation using clay pipes, it was observed that in addition to being cheap, simple and easy to use, subsurface irrigation was effective in improving yield, crop quality and water use efficiency (Batchelor et al., 1997).

Adverse effect of water stagnation

Chilli plants cannot withstand water stagnation and excessive moisture at any of the growth stages. If saturated conditions continue for more than 24 hours, the plants may be killed. Saturated conditions inhibit the uptake of nitrogen, and as a result crops receiving heavy irrigation or incessant rains during the rainy season look pale yellow, leading to reduced growth and yield. The application of higher rates of nitrogen was found to overcome the inhibitory effect of high soil moisture on nitrogen uptake (Thomas and Heilman, 1967). Dew and heavy rain at flowering are injurious to the crop, causing flower buds and young fruits to drop off.

Adverse effect of water deficit and critical stages

The principal limiting factor for farmers in arid and semi-arid regions is water. Farmers grow crops that are able to adapt to drought conditions (Muchow, 1989). Mechanisms that permit plants to survive in these unfavourable environments vary between species and include change in stomatal response, osmotic adjustment and a greater movement of photosynthates to the roots in order to increase root length for extracting water at a greater depth. The genus Capsicum exhibits similar physiological responses when confronted with water deficit (Hulugalle and Willat, 1987).

The two most critical stages of moisture stress in chillies are the initial establishment of transplanted plants and the stage prior to blossoming. Chilli plants have two to three peaks of blossoming, depending upon the availability of moisture, nutrient and incidence of pests and diseases. Any moisture stress at blossoming leads to flower and fruit drop. Water stress during the reproductive stage results in a lower yield (17%) compared with stress imposed during the vegetative phase (33%) (Prabhakar and Naik, 1997).

Frequency of irrigation

The frequency of irrigation will depend on the total supply of available moisture reached by the roots and the rate of the water use. The former is affected by soil type, depth of wetted soil and the depth and the dispersion of roots. The latter is influenced by weather conditions like temperature, relative humidity, wind velocity and the age of the crop. In summer, a weekly interval of irrigation is preferred, while in winter 10–12 days intervals may be sufficient in tropics. Generally, eight to nine irrigations may be required to raise the crop depending on rainfall, soil type, humidity and prevailing temperature. Under tropical conditions in chilli planted in monsoon, providing three to five supplementary irrigations whenever a dry spell exceeded seven days resulted in an 80–90% higher yield than in non-irrigated crops (Prabhakar and Hebbar, 1998). In southern peninsular Indian conditions, irrigating once in seven to eight day intervals significantly increased fruit size, seed yield and seedling root length compared with applications at longer intervals of 10–11 or 13–14 days (Vanangamudi *et al.*, 1990).

Depth of irrigation

The depth of irrigation depends upon the soil type and the stage of crop growth. In clay soil the depth of irrigation may be 6–8 cm and for sandy soil the depth may be 4–5 cm for each irrigation. At peak period of growth, namely, 90–120 days after transplanting, the crop's water requirement is more and so also is the irrigation requirement (Hosmani, 1993).

Irrigation requirement

A general rule is that vegetables will need about 2.5 cm of water per week from rain or supplemental irrigation in order to grow vigorously. In arid regions about 5 cm of water is required per week. Irrigation requirement depends on the season of cultivation, weather factors, soil type, types of irrigation, etc. Under the tropical conditions of southern India, the irrigation requirements of chilli have been found to be 110 cm (Sivanappan, 1979). In Cuba, the highest yield of 19.5 t/ha was obtained by irrigating at 85% of field capacity under the second irrigation after flowering, which delivered a total water of 1800 m^3/ha (16 cm) in nine irrigations (Leon *et al.*, 1991).

The maximum evapotranspiration of sweet pepper grown under an untreated green house in Hungary, where the soil water potential was maintained at values higher than −20 kPa, ranged from 0.5 to 4.00 mm per day, with a whole season's irrigation requirement was equivalent to 34.8 cm (Chartzoulakis and Drosos, 1998).

Varietal variations in water use

Hot peppers have more water requirements than non-hot types. In a pot experiment carried out to examine water utilisation between hot and non-hot types in Hungary, it was revealed that hot type SZ-103 produced 1 g of dry matter from 315 ml water, while for non-hot type this value was 298 ml (Somogyi, 1974).

Studies made on three cultivars of *C. chinensis* in Venezuela with the aim of selecting cultivars that were most resistant to water deficit yet had high yields, revealed that fruit production was affected to different degrees depending on the cultivar (24–40% reduction) when irrigation frequency was increased from three to six days.

Irrigation scheduling

The scheduling of irrigation can be made based on weather or soil moisture conditions as follows:

Cumulative pan evaporation

Irrigation scheduling based on evaporation values indicated that irrigation at 40 mm cumulative pan evaporation (CPE) exhibited the highest dry matter production, fruit yield, N uptake and water use efficiency (Prabhakar and Naik, 1997). Pulekar et al. (1990) also reported the highest yield of green chilli when the crop was irrigated at 36 mm CPE.

IW/CPE ratio

The irrigation water (IW) to CPE ratio (IW/CPE) has been found to be a useful criteria for scheduling irrigation. The highest yield of chilli was recorded at IW/CPE ratios ranging from 0.75 to 0.90, depending on varieties and locations (Mary and Balakrishnan, 1990; Censur and Buzescu, 1998; Chartzoulakis and Drosos, 1998; Selvaraj et al., 1998; Subramanian et al., 1998). However, the highest water use efficiency was observed at 0.5 IW/CPE (Palled et al., 1988).

Available soil moisture depletion

While irrigating, enough water is to be applied to bring the soil moisture content of the effective rooting zone up to the field capacity. This is the quantity of water that the soil will hold against the pull of gravity. Add water when the moisture in the root zone has been depleted. In a study on the influence of different irrigation regimes on the off-season Capsicum, Boicet et al. (1989) found that irrigation throughout the vegetative cycle when soil water level had decreased to 85% field capacity resulted in the highest total yield, highest yield of grade one fruits, the largest fruits, and the highest average number of flowers per plant. This treatment also produced the best returns. While for Capsicum cv. California Wonder irrigation at 40% and 60% of available soil moisture gave the highest fruit yield in India (Hegde, 1989b).

Soil moisture tension

Irrigating the field when the soil moisture tension reaches a specified value is a useful method for scheduling irrigation. The soil moisture tension is measured by using tensiometers. Irrigation when soil moisture tension exceeded 1 atm increased the yield of chilli without affecting the number of fruits (Basaccu and Garibaldi, 1971).

Smittle et al. (1994) studied the effect of water regimes on yield and water use of bell pepper. Irrigation regimes consisted of applying water when the soil water tension at 10 cm exceeded 25, 50 or 75 kPa during crop growth. Yield and water use were greatest when irrigation was applied at 25 kPa. Using plastic tunnels with trickle irrigation, the highest yield was produced when irrigation was applied at 15 kPa (Dyko and Kaniszewski, 1989).

Drip irrigation

A large number of experiments have been conducted regarding the beneficial effects of drip irrigation over other methods, like sprinkler or surface irrigation. Advantages of drip irrigation include the saving of water, the efficient utilisation of applied nutrients, water having a higher salt content can be used for irrigation and less weed intensity as a result of less surface area wetted.

Capsicum plants irrigated with water containing 95 mg/L and having electrical conductivity of 0.6 mmhos/cm had a higher content of soluble and diffusible ions in the leaves when irrigated by sprinkler than when irrigated by drip, while there was no significant difference in yield between the two methods (Gornat *et al.*, 1973).

Sprinkler and drip irrigation were compared on green pepper in an arid tract of Jerusalem in Israel. Yield, leaf growth and root development of pepper plants (cv. Califernia Wonder) were all greater with drip than with sprinkler irrigation (Goldberg and Shmueli, 1970).

Pepper plants of cv. California Wonder were drip irrigated at a constant frequency of one to two days with different amounts of water based on evaporation from a class A pan. The amount of water applied was 0.82, 0.95, 1.33 and 1.75 of the pan evaporation. Irrigation at 1.33 of the pan evaporation resulted in the highest yield of pepper (Shmueli and Goldberg, 1972).

An investigation was made in Tamil Nadu, India, to find out the water requirement of chilli crop variety K-1 and its response to drip irrigation. There was a saving of 62% of water by drip irrigation. Yield of the crop increased by 25% and reduced weed infestation by 50% (Sivanappan *et al.*, 1978; Sivanappaa, 1979).

In another study conducted in Israel, yields obtained under a drip system was 74 t/ha, while in a sprinkler system it was 59 t/ha. The yield difference was attributed to lesser nitrate nitrogen in the root zone in drip (60–150 ppm) compared to 250–300 ppm in sprinkler system (Sagiv *et al.*, 1978).

Methods of micro irrigation system

Among the various kinds of micro irrigation systems, namely, rotary, micro sprinkler, stationary micro sprinkler, canewall, drip irrigation, online drippers and drip irrigation microtube tested on a clay soil, the canewall drip tape recorded the highest benefit : cost (B : C) ratio (2.84%) and net extra income. However, the net extra income was the maximum with the stationary microsprinkler with B : C ratio of 2.74% (Shinde and Firake, 1998).

Prabhakar and Hebbar (1998) observed that the micro irrigation system produced 17–29% higher marketable yield and the microtube irrigation system resulted in a significant reduction in investment costs.

In a greenhouse experiment, *Capsicum* plants were grown using either the nutrient film technique (NFT) or rockwool, with drip irrigation. Plants grown using NFT gave a higher yield than those grown in rockwool and took up more water due to a higher leaf area index. However, the higher evaporation from the rockwool meant that total water consumption in the rockwool system was much higher than in the NFT system (Abou Hadid *et al.*, 1993).

Drip irrigation and disease incidence

Drip irrigation increases chilli yield by providing either favourable soil moisture conditions or unfavourable conditions for disease incidence. Jin Haixie *et al.* (1999) observed that drip irrigation created a higher marketable green chilli yield than the alternate row furrow irrigation. *Phytophthora* root rot disease incidence in the infested plots was significantly higher under alternate row furrow irrigation than drip irrigation. There was no disease development in the uninfested plots regardless of the irrigation method. The disease decreased green chilli yield by 55% and the combined yield (green + red chilli) by 36% compared to that in uninfested plots in alternate row furrow irrigation. A similar effect was also observed in the infection of *Phytophthora capsici*, causing mortality in green house grown *Capsicum* (Rista *et al.*, 1995) and in the open production of *Capsicum* (Biles *et al.*, 1992). Cafefilho and Duniway (1996) found that *Phytophthora* root rot of pepper could be reduced in a low rainfall area by positioning the drip emitters away from plant stems with a subsurface location (15 cm below soil). De Qiang and other workers (1996) also observed flood irrigation to be one of the reasons for the incidence of southern *Sclerotium* blight in chilli in China.

Effect of quality of irrigation water

Chilli irrigated with high electrical conductivity (4.2 mmhos/cm) water showed retarded growth, delayed flowering, reduced fruit set and yield, while the plants irrigated with water having electrical conductivity values ranging from 0.58 to 1.59 mmhos/cm exhibited increased flowering, fruit set and yield (Venkatachalam, 1982).

Saline water for irrigation

Salinity and sodicity of the irrigation water affect the growth and yield of crops. Even different varieties of the same crop may give differential responses under such conditions. Hence, the selection of salinity tolerant crops, cultivars and their specific strains are some of the important factors for the proper utilisation of saline irrigation water. Chillies need frequent and light irrigation to meet their water requirements and under such circumstances high amount of salts will accumulate in the root zone. Therefore, it becomes imperative to devise ways and methods for the better utilisation of such waters to obtain optimum yield.

In a pot study to determine the influence of saline water (electrical conductivity ranging from 1 to 6.95 msKm) on the growth of chilli, it was observed that Relative Growth Rate (RGR) decreased with increasing salinity and was related to the retardation of Net Assimilation Rate (NAR). The leaf area ratio did not decrease with increasing salinity. The leaf expansion rate was inhibited by salinity. Salinity did not influence the growth balance between leaves and other organs. The retardation of NAR was related to the obstruction of stomatal conductance and transpiration (Hirota *et al.*, 1999).

Use of different ratios of fresh water, treated urban waste water and saline ground water (mixed to obtain six levels of electrical conductivity (EC) ranging from 0.3 to 15 ds/m) for irrigation of pepper in a mediterranean environment indicated that pepper showed a high growth rate with a lower mortality during establishment (Borin *et al.*, 1997).

In a study on the use of brackish water in raising chillies in red soil, the addition of gypsum (1 t/ha) to the irrigation water showed the best results with the highest pod yield, in comparison to soil applied gypsum and the application of various manures (Sundaravadivel *et al.*, 1996).

At a higher osmotic concentration of saline water (300 mm/20,000 ppm) the germination percentage was lower and the time taken to germinate was longer (Hashem *et al.*, 1991; Palma *et al.*, 1996).

In a field study on the use of non-saline water (0.25 ds/m) and saline water (0.76–1.0 ds/m) for the irrigation of chilli crops grown for dry chillies, it was observed that two applications of non-saline water alternating with one of saline water was the most promising treatment for using saline water to supplement the limited supply of canal water (Srinivas *et al.*, 1991).

Through drip irrigation satisfactory yields of *Capsicum* (up to 2.1 kg/m^2) were achieved with saline (brackish) water, having an EC value upto 6.0 ds/m (Chattopadhyay and Mairi, 1990).

Pitcher irrigation

In pitcher technology the supply of water to the plants was found to be optimum and losses were recorded as negligible because of the controlled and regulated flow of water within the root zone. In a study carried out in arid regions of Rajasthan, India, it was observed that through pitcher irrigation more saline water can be utilised without affecting crop yield since the average yield of dry chillies recorded with normal, saline (EC – 12 ds/m) and high residual sodium concentration (RSC) 13 Me/L water were 581, 561 and 556 kg/ha, respectively. This may be due to the fact that most of the roots are present in the wetting zone (0–60 cm). Vegetable crops such as chillies which require frequent and light irrigation can be grown successfully under arid and semi-arid areas where water is scarce (Vikram *et al.*, 1999).

Magnetised water

The theory claiming that irrigation water passed through a magnetic field (magnetic treated water, MTW) can affect plants indirectly via the soil micro flora/fauna population was investigated using pepper plants. Responses to MTW were observed, but not when the soil was sterilised, thus confirming the theory (Moran *et al.*, 1993). Increased growth of pepper by 8.5% was observed by irrigation with MTW as compared to ordinary irrigation water (Dunand *et al.*, 1989).

Antitranspirants

In arid and semi-arid regions irrigation water is very scarce and the excessive use of water owing to the high rate of transpiration can be reduced by spraying antitranspirants on the foliage. Irrigation at 12 day intervals, along with spraying of 200 ppm alachlor solution as an antitranspirant, was found to be most economical for chilli cv. Pusa Jwala in Andhra Pradesh, India (Suryanarayan *et al.*, 1983).

Temperature of irrigation water

The temperature of irrigation water influences the growth and development of chilli plants. In European countries where paprika is extensively grown (Bulgaria, Hungary) and where stored irrigation water is used, the effect of the temperature of irrigation water is profound. The results of field experiments showed that irrigation with water at 10–12°C interfered with the normal flowering and fruit drop and reduced the yield by 7.6%. Irrigation water at 20–25°C allowed normal growth and development, and increased yield by 10–20% (Ugarcinski, 1964).

References

Abou Hadid, A. F., Elshinawy, M. Z., ElBeltagy, A. S. and Burge, S. W. (1993) Relation between water use efficiency of sweet pepper grown under nutrient film technique and rockwool under protected cultivation. *Acta Horticulturae*, **323**, 89–95.

Basaccu, L. and Garibaldi, G. (1971) The effects of various cultural practices on the yield of capsicum in the presence or absence of attack by *Verticillium dahlia*. *Annali dellafacota di scienze Agrarie della Universita d.gen studi di Torina*, 141–162.

Batchelor, C., Lovell, C. and Murata, M. (1997) Simple micro irrigation techniques for improving irrigation efficiency on vegetable gardens. *Agril. Water Management*, **32**, 37–48.

Biles, C. L., Lindsey, D. L. and Liddell, C. M. (1992) Control of *Phytopathora* root rot of chilli peppers by irrigation practices and fungicides. *Crop Protection*, **11**, 225–228.

Boicet, T., Pujol, P., Duany, J. L. and Verdecia, J. (1989) Behavior of some reproductive indices, yield and quality of capsicum plants grown out of season under different irrigation regimes. *Revista Ciencies Technicas Agropecuarios*, 2, 47–54.

Borin, M., Bonuiti, G., Semerari, A. and Carone, M. (1997) Use of treated waste waters for the irrigation of some vegetables in mediterranean environment. *Irrigation & Drainage*, 44, 20–24.

Cabello, T., Gomez, M., Barranco, P., Lucos, M. and Belda, J. E. C. (1997) Evaluation of Oxamyl against *Homoptera* pests in green house grown pepper applied with drip irrigation. *Tests of Agrochem and Cultivaras*, 18, 2–3.

Cafefilho, A. C. and Duniway, J. M. (1996) Effect of location of drip irrigation emmitters and position of *Phytophthora capsici* infection in roots on *Phytophthora* root rot of pepper. *Phytopathology*, 86, 1364–1369.

Censur, M. and Buzescu, D. (1998) Forecast of watering vegetable crops by using the evaporation conversion coefficient. *Anale Instituttal de cercetn petru Legumicultur floricultur, Vidra*, 15, 301–309.

Chartzoulakis, K. and Drosos, N. (1998) Water requirement of green house grown pepper under drip irrigation. Proc. Int. Symp. on the importance of varieties and clones in the production of quality wine. *Kecskemet*, Hungary, 24–28 Aug, 1997, 3, 175–180.

Chartzoulakis, K. and Drosos, N. (1999) Growth, yield and water use of peppers grown in an untreated plastic greenhouse. *Int. J. Parasitology*, 129, 155–160.

Chattopadhyay, S. B. and Mairi, A. (1990) Application of nonsaline and brackish water for vegetables using drip irrigation. Proc. 11th Int. Cong. on the use of plastics in Agriculture, New Delhi, India. 20-2-90 to 2-3-1990, pp. B-185, B-193.

De Qiang, Huang, Zimon and Chenand LandYu Fang Zhu (1996) Southern *Sclerotium* blight of water caltrap and its control. *Plant protection*, 22, 32–34.

Dimitrov, Z. and Ovtcharova, A. (1995) The productivity of peppers and tomatoes in case of insufficient water supply. Proc. ICID special Technical session on the Role of Advanced Technologies in Irrigation and Drainage system in making effective use of scarce water resources, Rome, Italy, Vol. I. 9.1–9.7.

Dunand, R., Morera, R. and Trujillo (1989) Effect of magnetized water on increased growth in different plant species. *Ciencia Y Technica en La Agriculture, Riegoy Drenoje*, 12, 29–37.

Dyko, J. and Kaniszewski, S. (1989) The effect of soil moisture level on capsicum yield. *Bieuletyn warzywniczy*, Suppl. II, 189–194.

Eom. K. and Im, J. N. (1990) Irrigation efficiency and yield response of irrigation methods in vegetable crop cultivation. *Res. Rep. Rural Dev. Admn. Soils and Fertilizers*, 32, 8–14.

Goldberg, D. and Shmueli, M. (1970) Sprinkle and trickle irrigation of green pepper in an arid zone. *Hort. Sci.*, 6, 559–563.

Gornat, B., Goldberg, D., Rimon, D. and Benasher, J. (1973) The physiological effect of water quality and method of application on tomato, cucumber and pepper. *J. Amer. Soc. Hort. Sci.*, 98, 202–205.

Hashem, M. M., AbouHadid, A. F. and Elbeltagy, A. S. (1991) Studies on the germination ability and seedling growth of pepper (*Capsicum annuum*) growing in Egypt at higher salinity. *Egyptian J. Hort.*, 18, 87–94.

Hegde, D. M. (1989a) Effect of method and volume of irrigation on yield and water use of sweet pepper (*Capsicum annuum* L.). *Ind. J. Hort.*, 46, 225–229.

Hegde, D. M. (1989b) Irrigation and nitrogen requirement of bell pepper (*Capsicum annuum* L.) *Indian J. Agic. Sci.*, 58, 668–672.

Hirota, O., Villavicencio, E., Chikushi, J., Takeuchi, S. and Nakano, Y. (1999) Effect of saline water irrigation at fruit maturity stage on transpiration rate and growth in sweet pepper (*Capsicum annuum*). *J. Fac. Agri. Kyushu Univ.*, 44, 39–47.

Hosmani, M. M. (1993) *Chilli Crop* (Capsicum annuum L.). Publ. by Mrs S. M. Hosmani Near Savanur Nawab Bunglow. Narayanapur, Dharwad 580 008, India. p.112.

Hulugalle, N. R. and Willat, S. T. (1987) Pattern of water uptake and root distribution of chilli peppers grown in soil columns. *Can. J. Plant. Sci.*, 67, 531–532.

Jin Haixie, Cardenal, S. E. S., Sammis, T. W., Wall, M. M., Lindsey, D. L. and Murray, L. W. (1999) Effect of irrigation method on chilli pepper yield and *Phytophthora* root rot incidence. *Agril. water management.*, 42, 127–142.

Keng, J. C. W., Scott, T. W. and Lugo Lopez, M. A. (1979) Fertiliser management with drip irrigation in an oxisol. *Agron. J.*, 71, 971–980.

Leon, M., Denver, R. and Leon, M. (1991) Water requirement of sweet pepper (*Capsicum annuum*) cultivar Medallade Oro grown during a nonoptimal period. *Agrotecnia de Cuba*, 23, 33–41.

Mary, S. S. and Balakrishnan, R. R. (1990) Effect of irrigation, nitrogen and potassium on pod characters and quality in chilli (*Capsicum annuum* L.). *South Indian Hort.*, 38, 86–89.

Matev, T., Vasilev, V. and Dimitrov, Z. (1970) The influence of different irrigation methods on some biological and physiological activities in pepper. *Nauc. Trudov. Viss Sel. – stop. Inst. V. Kolarov. Plovdiv*, 19, 107–114.

Moran, R., Shani, U. and Lin, I. (1993) The effect of magnetic treated water on the development of pepper and melon plants in sterilized soil. *Hassadeh*, 74, 266–271.

Morita, S. and Toyota, M. (1998) Root system morphology of pepper and melon at harvest stage grown with drip irrigation under desert condition in Baja California, Mexico. *Jap. J. Crop Sci.*, 67, 353–367.

Muchow, R. (1989) Comparative productivity of maize, sorghum and pearl millet in a semi arid tropical environment II. Effect of water deficit. *Field Crops Res.*, 20, 207–219.

Nashev, G. (1998) Chemical control through sprinkler irrigation of pepper blight caused by *Phytophthora capsici* Leonina. Proc. Int. Symp. on the importance of varieties and clones in the production of quality wine. Kecskenet, Hungary, 24–28, Aug. 1999, 2, 751–755.

Palma, B., Penaloza, P., Galleguillos, C. and Trujillo, C. (1996) Germination of seeds and development of seedlings of capsicums (*Capsicum annuum* L.) in a constant saline environment. *Phyton.* (Bueno Aires), 59, 177–186.

Palled, Y. B., Chandrasekharaiah, A. M., Kachapcer, M. D. and Khot, A. B. (1988) Yield and water use efficiency of chilli as influenced by irrigation schedules and nitrogen. *Farming System* 4, 25–28.

Prabhakar, M. and Hebbar, S. S. (1998) Water management in chilli under semiarid conditions. Water and nutrient management for sustainable production and quality of spices. *Proc. Nat. Seminar*, Madikeri, Karnataka, India, 5–6 Oct. 1997, 131–135.

Prabhakar, M. and Naik, L. B. (1997) Effect of supplemental irrigation and nitrogen fertilization on growth, yield, nitrogen uptake and water use of green chilli, *Ann. Agril. Res.*, 18, 34–39.

Pulekar, C. S., Patil, B. P. and Rajput, J. C. (1990) Water use, yield and economics of chilli as influenced by irrigation regimes and genotypes. *J. Maharashtra Agril. Univ.*, 15, 247–248.

Rista, L. M., Sillon, M. and Fomascro, L. (1995) Effect of different irrigation strategies on the mortality of pepper by *Phytophthora capsicii* Leonian in green houses. *Hort. Argentina*, 14, 44–51.

Sagiv, B., Bar-Yosef, B., Kafkafi, U. and Mini, A. (1978) Fertilisation and manuring on sprinkler irrigated fields of pepper compared with fertilisation via a trickle irrigation system. Preliminary Report, Aril. Res. Orgn., Inst. of Soil and Water, No. 763, 48.

Selvaraj, K. V., Iqbal, I. M., Dawood, M. S., Krishnasamy, S. M., Muralidharan, V. and Hariharan, M. S. (1998) Irrigation management under differential moisture regime for chilli crop (*Capsicum annuum*) Proc. Nat. Seminar Madikeri water and nutrient management for sustainable production and quality of spices, Karnataka, India, 5–6 Oct. 1997, 161–163.

Shinde, U. R. and Firake, N. N. (1998) Economics of summer chilli production with mulching and micro irrigation. *J. Maharashtra Agril. Univ.*, 23, 14–16.

Shmueli, M. and Goldberg, D. (1972) Response of trickle irrigated pepper in an arid zone to various water regimes, *Hort. Sci.*, 7, 241–243.

Sivanappan, R. K. (1979) Drip irrigation for vegetable crops. *Punjab Hort. J.*, 19, 83–85.

Sivanappan, R. K., Rajagopal, A. and Palaniswamy, D. (1978) Response of chilli to drip Irrigation. *Madras Agrc. J.*, 65, 576–579.

Smittle, D. A., Dickens, W. L. and Stansell, J. R. (1994) Irrigation regimes affect yield and water use by bell pepper. *J. Amer. Soc. Hort. Sci.*, 119, 936–939.

Somogyi, G. Y. (1974) Water demand and irrigation of red pepper. Bull. Veg. Crop. Res. Kecksmet Hungary.

Srinivas, S., Vishwanath, D. P., Hunshal, C. S., Srinivasa, N. and Balikai (1991) Response of chilli to frequency of saline water use in irrigation cycle. *Mysore J. Agric. Sci.*, 25, 199–204.

Subramanian, P., Krishnasamy, S. and Devasagayam, M. M. (1998) Influence of irrigation methods and regimes on the growth and yield of chillies. *South Indian Hort.*, 46, 99–101.

Sundaravadivel, K., Muthuswamy, P., Krishnasamy, R., Ramamurthy, S. and Periaswamy, M. (1996) Use of brackish water in raising chillies in red soil. *Madras Agric. J.*, 83, 37–39.

Suryanarayana, V., Raju, K. T. R. and Rao, D. V. S. (1983) Effect of irrigation frequencies and antitransplants on chilli. *Veg. Sci.*, 77, 83.

Thomas, J. R. and Heilman, M. D. (1967) Influence of moisture and fertilisers on growth by sweet pepper (*C. annuum*) *Agron J.*, 59, 27–30.

Ugarcinski, S. (1964) Effect of irrigation water temperature on the growth and fruiting of capsicums. *Grad. Losar. Nauka*, 1, 85–99.

Vanangamudi, K., Subramanian, K. S. and Baskaran, M. (1990) Influence of irrigation and nitrogen on the yield and quality of chilli fruit and seed. *Seed Research*, 18, 114–116.

Venkatachalam, R. (1982) M. Sc. (Hort.) Thesis Tamil Nadu Agricultural Univ. Coimbatore, India.

Vikram, Chauhan, Singhania, R. A., Singh, A. K. and Ashok Kumar (1999) Impact of saline water by pitcher method on chillies production – a study. *Indian J. Agric. Res.*, 33, 62–66.

8 The cultivation and processing of *Capsicum* in India

K. V. Peter, P. Indira and C. Mini

The cultivation practices of *Capsicum* vary in different countries and the practices followed in India are described here under the subheadings soil and climate, season, nursery preparation and transplanting, fertilizer application, growth regulators, irrigation, intercultivation, weed management, harvesting and yield. Drying the harvested fruits to a safe moisture level, cleaning and grading are the different steps in processing. Different modifications and equipment to speed up the processes have been described.

Introduction

Capsicum native to South and Central America, Mexico and the West Indies continue to be cultivated there and have been introduced, and is now widely cultivated throughout temperate, subtropical Europe, the Southern US, tropical Africa, India, East Africa and China. Indian chillies are medium to highly pungent, while African chillies are the hottest ones. The cultivation practices of chilli in different countries vary greatly, and the typical practices followed in India as reported by Govindarajan (1985) are given below.

Chilli is cultivated from sea level to 1,600 m and grows effectively with an annual rainfall of 600–1,250 mm. The important chilli growing states in India are Andhra Pradesh, Maharashtra, Karnataka, Orissa and Tamil Nadu. Andhra Pradesh leads in the production of this crop.

Soil and climate

The chilli crop is raised extensively in black cotton soils under rainfed conditions. It is also grown in red sandy soils under irrigation and to a limited extent in coastal alluvial soils. The chilli crop raised for ripe dry chilli is mainly concentrated in the black cotton soils of Karnataka, Maharashtra, Andhra Pradesh and Tamil Nadu (Hosmanii, 1993). The less pungent sweet pepper or *Capsicums* are grown in the hilly tracts of Uttar Pradesh, Himachal Pradesh and also in the cooler south Indian plains and hills. A well-drained loamy soil, rich in organic matter, is considered the most suitable for chilli cultivation. Clayey loams that can retain moisture are good for rain-fed crops. Acidic and alkaline soils are not suitable for growing chilli. The germination and early vigour of plants are affected by saline soils. Sandy loam soils with organic matter are ideal for *Capsicums*. Sweet peppers are insensitive to acidic soils and can grow well under a pH of 5.5–7.0 (Thakur *et al.*, 1999).

Chilli is a warm season crop but low humidity and a high temperature result in the shedding of buds, flowers and young fruits. Very low temperature also results in poor growth. Performance of sweet peppers is better under low temperature conditions. Soil temperature below 10°C retards growth and development of chilli, while 17°C is the optimum temperature. Atmospheric temperature ranging from 20°C to 25°C is ideal for chilli. *Capsicums* can be successfully cultivated in a mild temperature of 17–23°C. *Capsicums* cannot tolerate frost.

Season

The main crop for the southern plains of India is sown in May–June, transplanted in five to six weeks and harvested in October. In Punjab, the crop is planted in March–April to avoid damage by frost. In the Gangetic plains, chilli is grown as a cold weather crop, sown in September and harvested in February.

Nursery preparation and sowing

Raised nursery beds are formed in an area of $120\,m^2$ during May–June and December–January. Farm yard manure is incorporated in the nursery beds. Around 1,250 g of seeds would be required to raise seedlings for planting an area of one hectare. The seeds are treated with Bavistin (2.5 g) to prevent incidence of seed-borne diseases. Lines are drawn across the beds at a spacing of 2.5 cm and seeds are sown sparsely along these lines and then covered with top soil. The beds are then mulched with paddy straw and watered with a rose-can. On the 20th day after sowing Furadan, 3 g granules are applied in between the seedling lines across the bed, stirred with the soil and the beds are irrigated (Veeraraghavathatham *et al.*, 1998).

In certain tracts of chilli cultivation, direct sowing is also done. The land is ploughed to a fine tilth. A seed rate of 3–3.5 kg/ha is followed and the seeds are broadcast, either alone or mixed with sand and a shallow harrowing is done to cover the seeds. Seedlings are later thinned to one per hill and gap filling is completed 30 days after sowing.

Transplanting

It is better to harden seedlings before transplanting them by reducing irrigation. Seedlings are transplanted by about sixth week to either ridges or furrows. Farm yard manure is incorporated at 25 tonnes/ha. During summer, seedlings are given temporary shade for two to three days. Transplanting is usually done either in the early morning or late in the evening. At the time of transplanting, seedling treatment with *Azospirillum* culture enhances the establishment and better growth of seedlings. *Azospirillum* culture at 2 kg/10 L of water is used for dipping the root end of seedlings for about 30 minutes.

Plant density depends on many functions, like variety, soil fertility and irrigated or dry land cultivation. In Karnataka and Maharashtra, spacing of 75×75 cm and 90×90 cm are generally followed, whereas in Andhra Pradesh and Tamil Nadu narrow spacing of 45×45 cm or even closer is practised. Wider spacing of 90×90 cm is followed in tall varieties like Bydagi and Sankeshwar in Karnataka and Maharashtra for concentrated dry red chillies; bush varieties are grown in a closer spacing of 60×30 cm. No single spacing is recommended. Under All India Co-ordinated chilli trials, the recommendation is 60×45 cm for varieties and 50×45 cm for hybrids.

Fertilizer application

A wide array of fertilizer recommendations have been reported in chillies. The choice will be based on soil type, irrigated or rainfed cultivation and the variety to be grown. Chilli crops generally respond well to nitrogen and potash application rather than phosphorus. Usually, phosphorus and potash fertilizers are applied in one dose, while nitrogenous fertilizers are given in two or three split doses. The practice in Tamil Nadu is basal dressing with 60 kg of phosphorus (375 kg of superphosphate) and 30 kg of potash (50 kg of Muriate of potash), along with 40 kg nitrogen (87 kg of urea) along one side of the ridge and mixed with the soil. Nitrogen at 40 kg each is used for top dressing on the 30th, 60th and 90th days of transplanting. A fertilizer mixture having 120, 180, 90 kg/ha each of N, P and K, respectively, is recommended for *Capsicums* in

Himachal Pradesh. One-third N along with P and K are given as the basal dose, whereas second and third on top dressing, three to four and six to eight weeks after transplanting (Thakur *et al.*, 1999).

Growth regulators

Growth regulators are used in chilli to induce early flowering, for preventing flower and fruit drop, for uniform ripening of fruits and to increase or decrease seed content in fruits. To improve fruit set, as well as to reduce flower and fruit drop, naphthalene acetic acid (50 ppm) is sprayed before flowering (equal to 1 ml of planofix mixed with 4.5 L of water).

In Tamil Nadu there is a practice of triacontanol application. Triacontanol (2.5 ml of Vipul dissolved in 10 L of water) is sprayed on the foliage of the plants using a hand sprayer. Triacontanol increases the photosynthetic efficiency and thereby the yield.

Irrigation

The field is irrigated at the time of transplanting. On the third day, life irrigation is given and subsequent irrigation is given once a week or ten days depending on the soil moisture conditions. *Capsicum* is grown under irrigated conditions, as the crop is very sensitive to soil moisture. By applying antitranspirants like alachlor (20 ppm) in chilli the irrigation frequency can be reduced.

Inter-cultivation, weed management and inter-cropping

The main objectives of inter-cultivation are (a) control of weeds, (b) soil aeration and (c) soil moisture conservation. It is carried out mainly by hoeing. To prevent cracking and to conserve soil moisture repeated inter-cultivation is performed. By repeated inter-cultivation, the weed intensity in succeeding crops will be of the lower order. Though the cultural methods of weed management are widely used by the farmers, they are becoming laborious, time consuming and expensive due to labour shortage. Under this situation, the option left is the use of herbicides or a combination of both herbicides and cultural methods. Among the herbicides, trifluralin, alachlor, etc. are extensively used for controlling weeds in chilli. As a pre-emergent weedicide, 2 L of Basalin (Fluchloralin 1 L a.i./ha mixed in 500 L of water) is sprayed on the soil surface just before transplanting. This is followed by irrigation. As a phytosanitary measure of disease control, weeding is essential.

Under rainfed cultivation in Tamil Nadu, onion, brinjal and coriander are grown as intercrops. Castor and agathi grown on bunds serve as barrier crops in pest management while the inter-crops give additional income to farmers.

Harvesting and yield

The field-sown crops are grown mainly for ripe fruits, but for the transplanted crop the first harvest is usually picked green to stimulate further flush of flowering and fruitset. The stage of picking depends on market needs also. Flowering begins one to two months after transplanting and it takes another month for the green fruits to emerge. Thereafter, ripe fruits are picked at an interval of one to two weeks and harvesting continues over a period of three months. The number of picking varies from six to ten depending upon the season, cultivar and cultural practices (Muthukrishnan *et al.*, 1986). The ripe and almost red fruits are heaped indoors for two to three days to attain a uniform red colour. They are then dried under the sun by spreading in a thin layer for about 15 days. A chilli crop normally yields 2–2.5 tonnes of dry chilli and 7.5–10 tonnes of green chilli from a hectare.

Fully matured green fruits are harvested in the case of sweet peppers. Usually they attain harvestable maturity in 60–75 days after transplanting. About four to five pickings are possible and the yield is around 15 tonnes/ha.

Primary processing

Primary processing is for readying the harvested fruits for market by drying them to a safe moisture level for transport, storage and further processing.

On harvesting, chillies and *Capsicums* have a moisture content of 65–80%, and this must be reduced to 10% (Pruthi, 1993). Traditionally, this has been achieved by sun-drying fruits without any special treatment. This is the most widely used method throughout Asia, Africa, Central and South America and in the USA.

Sun-drying

In India, ripe red or nearly ripe fruits are harvested, along with their stalks, and heaped indoors for three to four days to develop a uniform colour. This curing is an important treatment, as sun-drying immediately after the harvest leads to bleached and non-uniform colour in the case of fruits, which have not completely turned red. The cured fruits are dried by spreading them on dry ground under the sun for five to fifteen days, depending on day temperature and humidity. Chillies have to be dried in a manner so that they retain their characteristic red colour and lustre. Usually, the chillies are spread in a single layer for drying. After two days of drying in this manner when the fruits are still flaccid, they are trampled upon or are rolled over to enable a greater quantity of the dried product to be packed into gunny bags for storage and transport. The yield of dry chillies is 25–30%, depending on the cultivar (Sasthri, 1959).

The temporary storage of high moisture chillies in bags could start a deteriorative process affecting their colour and microbiological quality. Laul *et al.* (1970) suggested a high loss of quality in areas, where occasional showers create unhygienic conditions. Over-drying and excessive delay in drying results in the growth of microflora and the subsequent loss of quality or total spoilage. Thus, a proportion of chillies produced by sun-drying result in a poor quality product that has a shrivelled appearance and with varying numbers of broken pods.

Improvements

Efforts have been made to develop processes and equipment to speed up the drying process.

Lease and Lease (1956) showed that by delaying the harvest and allowing the ripened pods to wither on the plant and then drying them faster by slicing the pods in forced draft driers gave a superior quality produce. Lease and Lease (1962) suggested 65°C as the optimum drying temperature for good quality produce. Laul *et al.* (1970) suggested that the rate of drying was higher with blanching, reducing the drying time by 40–60%. They found that the drying could be completed in three hours by fluidized bed drying of pricked fruits at 60°C. Luhadiya and Kulkarni (1978) developed a method for producing dehydrated chilli: green chillies, pricked, blanched and dip treated in sodium hydroxide (2%) for 10–20 minutes, washed free of alkali and rapidly dried between 60°C and 70°C. CFTRI, Mysore has developed a chemical emulsion known as Dipsol containing potassium carbonate, refined groundnut oil, gum acacia and butylated hydroxy anisol, which ensures a faster rate of drying, better retention of colour and pungency and a higher yield of superior quality product. Chillies are often smeared with oil of Mahua (*Madhuca longifolia*) to impart glossiness (Kachru and Srivastava, 1990). Solar driers have been designed by Regional Research Laboratory (Jammu) which effects complete drying of the commodity in four to five days with a marked improvement in colour and storage characteristics. In USA the freshly harvested fruits are taken to factories for rapid drying. The harvested fruits are first washed in water, then immersed in dilute hydrochloric acid to remove the pesticide and fungicide residues. Fruits are again washed in water to remove acid, before the fruits are cut mechanically into small strips and loaded on to the trays of the counter-current hot-air driers. With the control of air-flow and humidity, the drying time and colour loss can be reduced (Govindarajan, 1985).

Cleaning and grading

After drying the fruits they are cleaned of extraneous matter, damaged and discoloured pods, etc. before storage or packing. Dried chillies are usually stored in gunny bags. Storage at the farmer's level is essentially a short duration, ranging from one to three months. Storage at trader's level is also done in bags, which are stacked in godowns, where fumigation is done to prevent insect infestation.

For marketing, chillies are graded as first or second sort, mixture, etc. Grades such as special, medium and fair are also adopted.

References

Govindarajan, V. S. (1985) Capsicum – production technology, chemistry and quality. Part 1 – History, botany, cultivation and primary processing, *CRC Crit. Rev. Food Sci. Nutr.* 22(2): 122–129.

Hosmanii, M. M. (1993) *Chilli Crop* (Capsicum annuum L.) 2nd edn, pp. 87–156, Sarasijakshi M. Hosmanii.

Kachru, R. P. and Srivastava, P. K. (1990) Status of chilli processing. *Spice India* 3(1): 13–16.

Laul, M. S., Bhale Rao, S. D., Rane, V. R. and Amla, B. L. (1970) Studies on the sun-drying of chillies. *Indian Food Packer* 24(2): 22–25.

Lease, J. G. and Lease, E. J. (1956) Factors affecting retention of red colour in peppers. *Food Technol.* 10: 368–375.

Lease, J. G. and Lease, E. J. (1962) Effect of drying conditions on initial colour retention and pungency in red peppers. *Food Technol.* 16: 104–106.

Luhadiya, A. P. and Kulkarni, P. R. (1978) Dehydration of green chillies. *J. Food Sci. Technol.* 15(4): 139.

Muthukrishnan, C. R., Thanka Raj, T. and Chatterjee, P. (1986) Chilli and capsicum (In) *Vegetable Crops in India* (ed.) Bose, T. K. and Som, M. G. Naya Prokash, Calcutta, pp. 343–354.

Pruthi, J. S. (1993) Chillies or capsicums (In) *Major Spices of India – Crop Management and Post Harvest Technology*. Publication & Information Div. ICAR, pp. 206–211.

Sasthri, B. N. (1959) Capsicum. *Wealth of India, Raw materials*. CSIR, New Delhi, p. 67.

Thakur, P. C., Joshi, S., Verma, T. S. and Verma, H. C. (1999) Pusa Deepthi – new capsicum hybrid. *Indian Hort.* 43(1): 6.

Veeraraghavathatham, D., Jawaharlal, M. and Seemantthini Ramadas (1998) *A Guide on Vegetable Culture*, 3rd edn, pp. 25–37, Sun Associates, Coimbatore.

9 The preservation and production of *Capsicum* in Hungary

Norbert Somogyi, Moór Andrea and Pék Miklós

Capsicum was introduced to Europe at the beginning of the sixteenth century and became an important crop in several countries, Hungary included. Like other *Capsicum* producing countries, there are two main types grown in Hungary: one for fresh consumption, the sweet *Capsicum*, and the paprika for use as a condiment. The paprika characteristics differ in several ways from the sweet *Capsicum*. Sweet *Capsicum* is mainly grown under controlled environmental conditions and is also field-grown. The paprika is only field-grown. Spice is produced from paprika after drying and milling. The production area of sweet *Capsicum* was about 8,000–10,000 ha during the last three to four decades. This area has decreased dramatically in the last two years; currently, it is about 4,000–4,500 ha. The production area of paprika has remained steady at around 3,000–7,000 ha for a long time. Since sweet *Capsicum* is grown all over the country – except in the cold, rainy areas near the western border – paprika is only grown in the two traditional regions (Szeged and Kalocsa) of the southern counties. Hungary used to be one of the most outstanding exporters of the condiment paprika. Hungary's activity in the world market has been reduced by the emergence of the Southern Hemisphere countries as paprika producers and by the change of consumers' habits (using paprika-based sauces, paprika oleoresins). The Hungarian growers and processors must go through significant changes to meet the requirements of the world market.

The Hungarian sweet *Capsicum* was unique until the last decade. It differed both in colour and shape from the sweet *Capsicum* types grown worldwide. Due to a widened research cooperation and exchanges of genetic material, the typical colour and shape formation produced in the Charpathian Basin is now available from the world's breeding and seed companies. Because of the nature of sweet *Capsicum*, although grown in a large scale (5–6% of the total of the vegetable growing area in Hungary), export possibilities are limited. The only sizeable amount of export is shipped to Germany. Some small quantities are exported to neighbouring countries, and the rest is sold nationally.

Introduction

Sweet Capsicum

The first written records of the appearance of paprika in Hungary are from the middle of the sixteenth century. Paprika was grown as a rarity in the garden of Margit Szechy, the step-mother of a most distinguished General. It can be assumed from the information of the next two or three centuries that *Capsicum* was cultivated only for use as a condiment. The cultivation of the sweet *Capsicum* known today began at the end of the nineteenth century. Bulgarians were the first growers who cultivated *Capsicum* in the southern part of the country (Szentes and surroundings).

At the beginning of the twentieth century production of sweet *Capsicum* spread to other parts of Hungary. It is interesting to note that the sweet *Capsicum* growing regions are not the same as the traditional paprika producing areas. Important sweet *Capsicum* areas were in the middle of the country (Boldog, Nagykőrös, Cegléd) and in the south–southeastern part (Gyula, Baja, Bogyiszlo). But today this typical vegetable crop is cultivated all over the country.

Sweet *Capsicum* growing greatly depends on consumption and market demand. The whole vegetable cropping land of Hungary was 45,000–60,000 ha, and out of this 2–3% was for sweet *Capsicum*. Production significantly increased after the Second World War. It occurred when the processing industry developed and began exporting to the East (mainly to the former USSR). The biggest area sown was towards the end of the seventies when *Capsicum* was grown on 14–15% of the total vegetable cropping land. Due to the changes in technology and improved growing methodology the area sown by *Capsicum* decreased (4,000–5,000 ha) significantly. Nevertheless, the average yield increased and therefore the production increased more than before as a consequence of the improved facilities. Under the improved growing facilities (glass houses, heated and unheated plastic houses), the average yield increased to 80–150 tons/ha. Sweet *Capsicum* grown in the field, yields 25–60 tons/ha, depending on the cultivar and the season.

Paprika

The production of condiment paprika in Hungary is highly labour intensive. It became a significant crop at the end of the nineteenth century, although reports of its cultivation exist from the sixteenth century in Szeged, where the crop was mentioned as one of the crops grown. It was introduced from Turkey by monks who excelled in healing, and was used as an effective medicine against malaria.

Looking at the history of Hungarian paprika production we can distinguish several classical periods:

- Until the middle of the nineteenth century feudalistic family self-sufficiency and the beginning of the production for the market;
- Until the start of the First World War paprika production is market oriented based on free-competition;
- From the end of the First World War until the end of the Second World War a state regulated production order was characteristic;
- Paprika milling was under state monopoly for the period of 1940–90;
- In the last decade of the twentieth century again free market, production and processing are based on competition.

Spice made of paprika is known as "Hungaricum" worldwide, and is an essential element of the Hungarian cuisine. Until the turn of the nineteenth century it was known mainly in public life as medicine. Shepherds of the Great Plain used it only as a spice. From that point, it has become more popular because of its spicy properties, and it became an important export commodity. As a spice, the Hungarian paprika has a purposeful role in the central European "heavy cuisine". First of all because it helps in the digestion and preservation of food (Dunszt, 1939). However, during the last 100 years it has become popular in numerous countries of the world, because of its excellent colouring properties and for its flavour (Figure 9.1).

Paprika is classified according to its pigment contents, fineness of milling and pungency. This classification was used in the past and continues to be used today. When the milled product

Figure 9.1 The fish soup of Szeged that is made with Hungarian style paprika is a typical Hungarian dish.

appeared on the market during the 1850s there was no official regulation regarding the quality. There were only two to three known categories. With the introduction of the steam-mills from the mid-to-late 1800s, four more new quality categories were accepted. For the sake of uniformity it became imminent that quality classification must be regulated by law.

The first act to regulate the grading system of quality was the 1895 XLVI Act forbidding the adulteration of agricultural crops, products and commodities. The Act of 1907, 26.859/VI.3, further refined it. This classified the milled product into four categories: I, II, III and mercantile. It was only allowed to deviate from that during the First World War, because of the economic situation (Obermayer, 1934). According to the present Hungarian standard the applied quality rating system is: special, delicate, sweet-noble, goulash and rose.

During recent years, paprika was produced on 3,000 ha in both the ecological regions of Szeged and Kalocsa. Depending on the season and production technologies the average yield is highly variable. The annual average of 50,000–60,000 tons of raw condiment paprika produces 6,000–10,000 tons of milled product.

Literature review

Paprika (*Capsicum annuum* L.) originated from South America and came to Europe – probably first to Spain – in 1493 after the discovery of the American continent (Pickersgill, 1986, 1989). From there it came to Hungary across the Balkan through Turkish growers. The first paprika plants were planted at the end of the 1500s. At first it was considered as an ornamental plant and was grown for culinary usage at the beginning of the seventeenth century. The plant later enjoyed tremendous popularity around the time of Napoleon (Somos, 1981). The book containing the first detailed description was written by Csapó (1775). According to this book paprika is grown in vegetable gardens and the long red fruit is dried and crushed to a powder. Veszelszki (1798) said around that time the farmers of Fot, Palota and Dunakeszi grew paprika. The first cultivation trials were conducted at the botanical garden of the University of Pest in 1788. Since that time different *Capsicum* varieties were found in the "Index seminum" of the botanical garden (Augustin, 1907). In letters that Count Hoffmansegg sent to his wife about his journey in Hungary, he mentioned: "here I really liked a Hungarian dish, meat with paprika. It must be very healthy ..." (Bálint, 1962). August Elrich, a German traveller did not talk so nicely about the Hungarian paprika in his book of "Die Ungarn wie sie sind" (1831). He called paprika "Diablische Paprika Brühe". He wrote that for people who are not used to it, the effect on the palate is like embers or even worse (Augustin, 1907).

National trade of the milled production of paprika started in the second part of the nineteenth century and the export trade began at the end of the century. Two main growing regions of paprika were established, namely Szeged and Kalocsa. The official quality testing of milled paprika for the protection of commercialised milled products was introduced at the end of the nineteenth century in Szeged, while the breeding work commenced in Kalocsa in 1917 and in Szeged in the 1920s (Szanyi, 1937; Benedek, 1960, 1974). Only Hungarian bred paprika cultivars are grown in Hungary. In contrast, the Hungarian bred sweet paprika occupies only 80–85% of the production, because during the last 15 years foreign breeding and seed companies have gradually produced more cultivars.

Production regions, cultivars, growing and processing technology

Climate of Hungary

Hungary is located on the north latitude 46°–48°. It means Hungary is on the northern border of the paprika growing area. The vegetation period is relatively short. Late spring frost may occur between 15–20 April. In some cases the first autumn frost may come at the end of September, but definitely in the middle of October.

Almost the whole area of the country is suitable for paprika growing given its temperature, precipitation and sunshine-hours. There are no striking differences in climate from region to region, although the sunshine-hours are the highest on the southeast part, and the precipitation is the least (about 2,000 hours and 500 mm per annum, respectively), while the sunshine-hours are less and the precipitation is more (about 1,800 hours and 700–800 mm per annum) on the western part of the country. The biggest growing regions' (South Hungary) meteorological data is shown in Figures 9.2–9.8 (Data provided by National Meteorological Station, Szeged Station).

Production regions

Sweet Capsicum

Sweet *Capsicum* can be grown in any Hungarian region except along the western border of the country where the precipitation is higher and the temperature is lower than the average. Only 8–10% of the country's soil and climate conditions are unsuitable for growing sweet *Capsicum*. Nevertheless, as traditional growing regions evolved, immigrant Bulgarian market gardeners settled at the southern part of the country and started vegetable production. Observing the Bulgarians' success of production and commercialisation of sweet *Capsicum*, the Hungarians

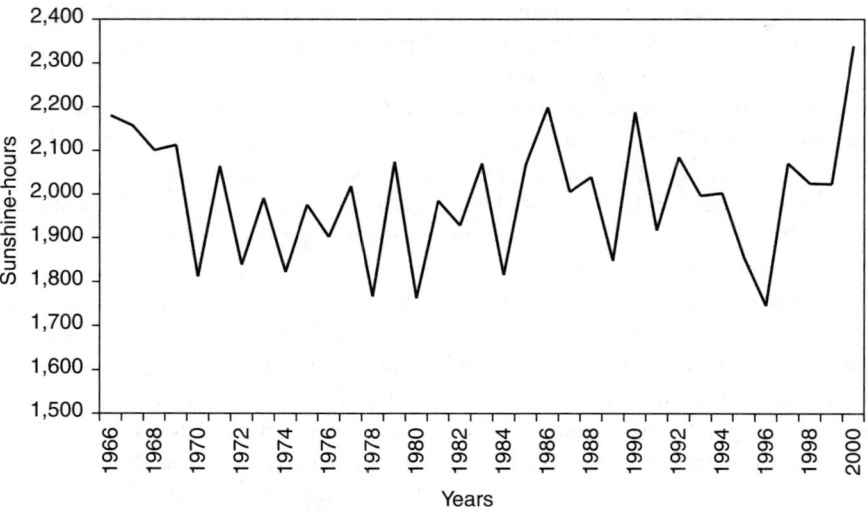

Figure 9.2 Sunshine-hours in Szeged (Hungary) during the period 1966–2000.

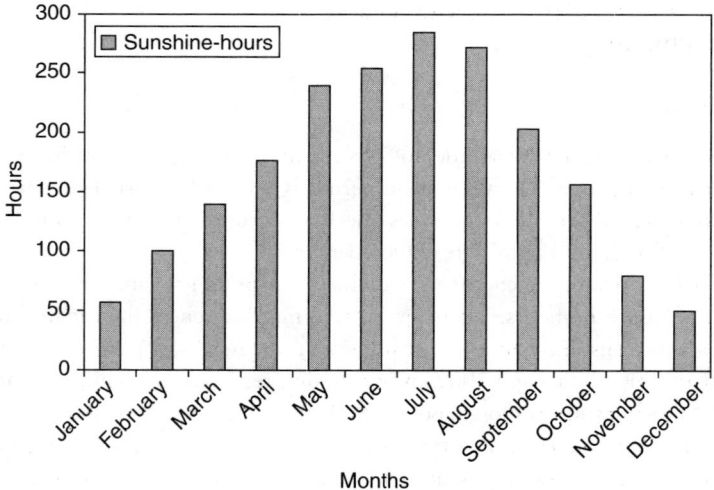

Figure 9.3 Average sunshine-hours in Szeged (Hungary) during the period 1966–2000.

became involved as well. Production regions were established first in the country's southern warmer counties (Bacs-Kiskun, Bekes, Csongrad), which were notable for their easily warming rich sandy humus soil and good water management (county of Jaszsag, Pest). Production was also established on flood plains in the middle of the country, where the soil was of good quality clay. New growing regions developed in the twentieth century (Nyirseg, Hajdusag). Sweet *Capsicum* requires soils with 2–4% humus content with 6.8–8.5 pH. Rainfall should be around 500–600 mm per annum; 50% of that is needed during the vegetation period. This is a

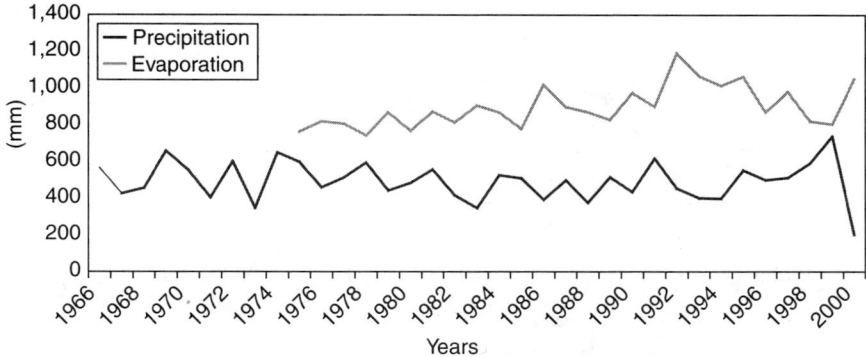

Figure 9.4 Annual precipitation and evaporation in Szeged (Hungary) during the period 1966–2000.

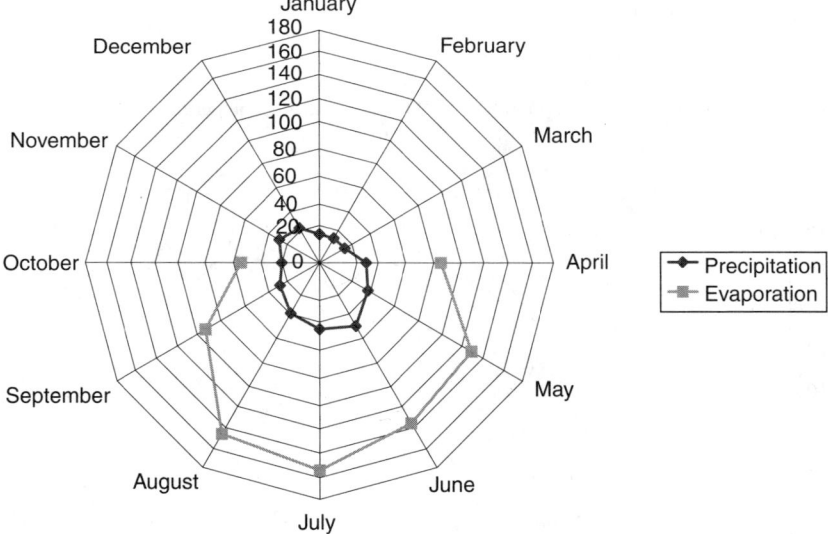

Figure 9.5 The monthly average precipitation and evaporation (mm) in Szeged (Hungary) during the period 1975–2000.

very important aspect of sweet *Capsicum* cultivation, that the water demand of sweet *Capsicum* is higher than that of condiment paprika.

Condiment paprika

Two important growing regions became established in Hungary by the end of the nineteenth century, namely Szeged and Kalocsa. These regions produce the bulk of the Hungarian paprika even today, in spite of the fact that in the last decades people started to grow this crop in the region of Mezohek, and even in the region of Boldog, located in the north of Hungary.

The soil types in the region of Szeged depend on the location: it is mainly medium-heavy, the sub-soil is salty on of the chernozem type. There are also flood-plains near where the rivers Tisza

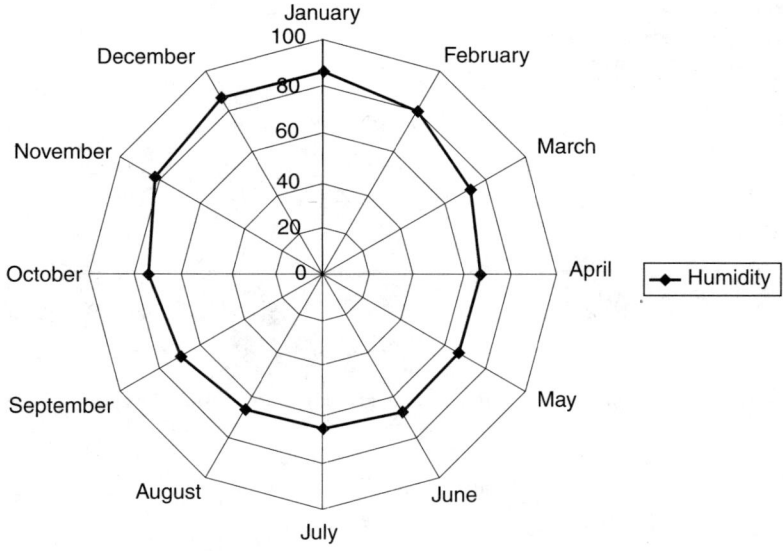

Figure 9.6　The relative air humidity (%) in Szeged (Hungary) during the period 1975–2000.

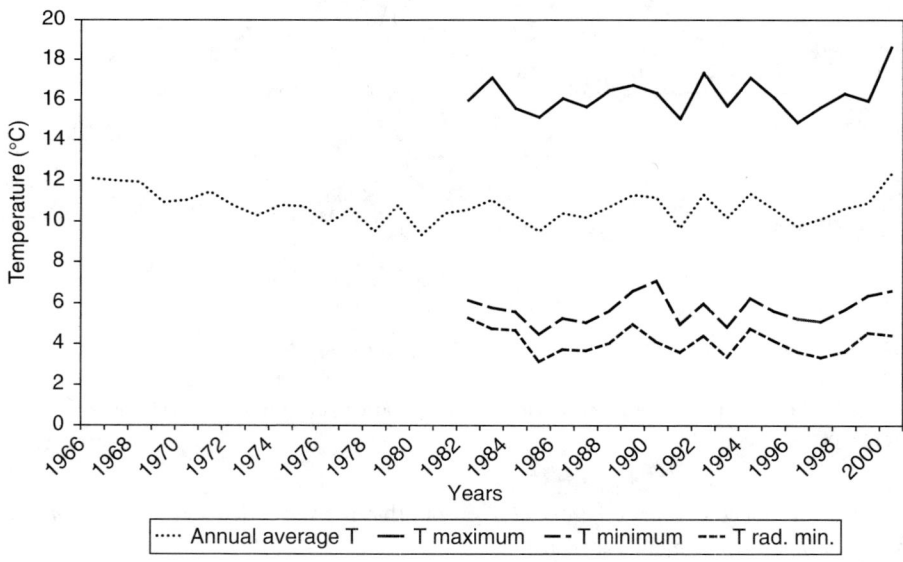

Figure 9.7　Average temperatures in Szeged (Hungary) during the period 1966–2000.

and Maros join. However, in a significant part of the region we can find lighter, sandy loam. These soils have inadequate water management with less humus, however, the cultivation is easier and the soil warms up faster. At the same time, the quality of the condiment paprika is lower than that obtained from the heavier soils. These lighter soils are suitable for intensive sweet *Capsicum* production under controlled conditions. In the region of Kalocsa the soil is mostly heavy, with relatively low humus content on the former flood plains of the Danube river.

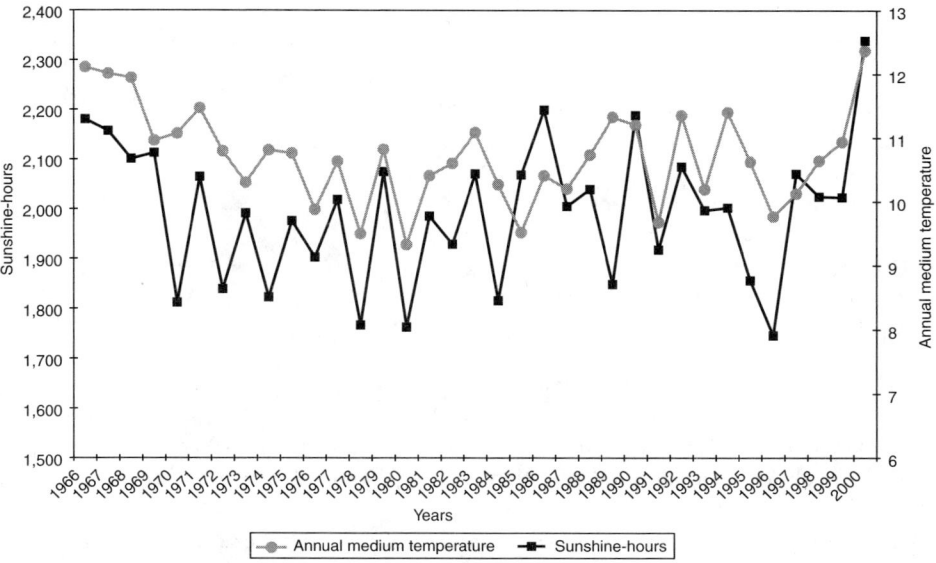

Figure 9.8 Annual sunshine-hours and annual medium temperature in Szeged (Hungary) during the period 1966–2000.

Cultivar types

Sweet Capsicum cultivars

The sweet *Capsicum* cultivars in Hungary changed a lot during the last centuries, especially during the last 50–60 years. In the previous centuries the seed was handed down from father to son and the growers passed the seed on to each other. The seeds of the best and earliest fruits were always kept for further sowing. It can be assumed that on many occasions spontaneous pollination or mutation created new types or varieties. Breeding techniques applying genetics goes back only to the last 50–60 years in Hungary. The first sweet *Capsicum* cultivar which received State registration was "Cecei sweet 3" bred by Angeli (1958). He selected this from a pungent white type and created a non-pungent cultivar. With his work, an active and efficient *Capsicum* improvement started and, consequently, the Hungarian sweet *Capsicum* types and an assortment of cultivars were continuously increased and spread all over the world. The production of sweet *Capsicum* hybrid seed cultivars was initiated in Hungary in the middle of the 1960s (Moór, 1969). Breeding work was conducted by the state-owned Vegetable Crop Research Institute and its predecessors, and by the Horticulture Departments of Universities until the mid-80s. Since then, several private breeders' cultivars received Plant Variety Protection. All the cultivars grown belong to the species of *Capsicum annuum* L.

According to the National Institute for Agricultural Quality Control sweet *Capsicum* cultivars/varieties are categorised by the following groupings:

> White fruit, indeterminate
>> pungent
>> without pungency
> white fruit, determinate
>> without pungency

pale green fruit, indeterminate
  without pungency
hornshaped, indeterminate
  pungent
  without pungency
pointed, hot, indeterminate
pointed, hot, determinate
tomato-shaped, indeterminate
  without pungency
California Wonder type
  without pungency
other
  pungent
  without pungency

Approximately 100 cultivars were registered in the National list of Varieties in 2000. The size of this book does not allow a description of all types, so only the most important ones are mentioned.

One of the most important traits of sweet *Capsicum* cultivation is the sensitivity to lack of light (Zatykó, 1979). This characteristic is important for determining whether the cultivar can be force-grown or not. This means determining whether the cultivar may be grown only in the field or whether its production is economical under green house conditions out-of-season, as well.

1   Ciklon F1 – indeterminate, white fruit ripening to red, sweet, conical, upright fruit, for all production systems. Fruits are 12–15 cm long and 5–6 cm in diameter. Yield is 8–15 kg/m^2 depending on the production technology. It contains Tm2 resistance. Under local light conditions, sowing is done at the end of September, and the first 2 cm long fruits appears after a vegetation period of 125–130 days. This can decrease by 50% under more intense light conditions.

2   Taltos – white fruit turning to red, indeterminate, sweet, conical, blunt, pendulous fruits, grown in the field. Fruits are 10–15 cm long and 5–6 cm in diameter. Potential yield is 30–35 tons/ha.

3   "Pungent apple" – white fruit turning to red, indeterminate, pungent, apple-shaped upright fruit, grown in the field. It is mainly used by the canning industry (to pickle). Fruits are 6–7 cm in diameter and 4–5 cm long. Potential yield is 18–25 tons/ha.

4   Feherozon – white fruit turning red, determinate, without pungency, upright fruits, grown in the field and also under green house conditions. Fruits are 12–15 cm long and 5–6 cm in diameter. Yield in the green house is 6–8 kg/m^2 and in the field is 25–35 tons/ha.

5   Rapires F1 – pale green turning red, indeterminate, pungent, long, conical, pendulous fruits. It can be grown in any type of controlled facilities. Fruits are 15–20 cm long and 3–4 cm in diameter. It contains Tm2 resistance. Yield depends on the technology 7–8 kg/m^2.

6   Tomato shaped green – dark green ripening to red, indeterminate, sweet, flat, round, seamed, pendulous fruits. Fruits are 8–12 cm in diameter and 3–5 cm high. The potential yield of fully ripe fruits is 18–20 tons/ha.

Paprika varieties

The Hungarian condiment paprika's cultivation period is short, there are only $5–5\frac{1}{2}$ months available for the vegetation period. In spite of the short vegetation period, the quality of the harvested crop is excellent in most years. The high pigment content and the high dry matter guarantees a very good base material for milling. The Hungarian varieties' yield can be up to 50% more, with improved attributes, by cultivating them in areas where the vegetation period is longer. This is based on the Hungarian–Spanish (Somogyi *et al.*, 1998), the Hungarian–Portuguese and the Hungarian–Australian (Derera, 2000) cooperative experiments. The full genetic potential of the Hungarian cultivars is limited by the climatic limitations. Between 1993 and 2000 the condiment paprika production area was between 3,000 and 6,500 ha and the raw paprika production was between 26,000 and 65,000 tons. The size of the official growing land was 3,000–6,500 ha in 1993–2000 and the amount of the raw production was between 26,000 and 65,000 tonnes. Consequently, the quantity of the milled product varied between 5,000 and 9,400 tonnes.

All of the condiment paprika cultivars grown in Hungary were bred in Hungary. They belong to *Capsicum annuum* L. covar. *longum* by botanical classification. There are two exceptions (cv. Kalocsai A cherry type, cv. Kalocsai M cherry type). Regarding the growth habit, they are continuous, semi-determinate and determinate. There are two types of orientation of the fruits: erect and pendulous. The categories are indicated below and the pungency is confirmed in brackets:

1 Varieties of continuous growth habit, pendulous fruits: Szegedi 20, Szegedi 80, Szegedi 57–13, Remény, Kármin, Szegedi 178 (pungent), Szegedi 179 (pungent), Szegedi F-03 (pungent), Kalocsai 50, Kalocsai 90, Kalocsai V-2 (pungent), Kalocsai E-15, Csárdás, Folklór;
2 Varieties of continuous growth habit, erect fruits: Kalocsai 57–231;
3 Varieties of semi-determinate growth habit, pendulous fruits: Kalocsai 801, Kalocsai 702, Zuhatag;
4 Varieties of semi-determinate growth habit, erect fruits: Kalocsai M 622, Rubin;
5 Varieties of determinate growth habit, erect fruits: Kalocsai D 601, Kalocsai D 621 (pungent);
6 The cherry type paprika is classified as paprika, but it differs botanically from the rest. They are *Capsicum annuum* covar. *cerasiforme* and not *longum*. They are pungent, their importance is found in gastronomy. If green fruit is harvested it can be pickled or made into salad. When the ripe fruits are harvested they can be used for hot sauce or dried spice (not milled) flakes. The two cherry cultivars that differ in fruit size and growth habit are Kalocsai M and Kalocsai A.

A detailed description of the most typical cultivars of each category is given here.

1 Cultivar of continuous growth habit and, pendulous fruits: Szegedi 80 a sweet cultivar. The fruits are 12–14 cm long, dark red when ripe, pigment content is 8.0–10.0 g/kg after post-ripening treatment. The solids content at picking is 20%. Its yield potential under intensive conditions is 20–25 tons/ha. It has a reasonable tolerance to diseases and can be transplanted or directly seeded. Due to an early ripeness a reasonable yield can be relied upon before the first frosts (Figure 9.9). The Szegedi F-03 is the same type but with pungency (Figure 9.10).
2 Cultivar of continuous growth habit and erect fruits: Kalocsai 57–231, a sweet variety. The bush is 45–55 cm high, its fruits are scattered and 10–14 cm long, slightly bent. They are of fire red colour, or dark red after post-ripening treatment. Its main value lies in its good pigment (8–9 g/kg) and solids content. It is a mid-early maturing variety. Its yield potential is 15–16 tons/ha. It can be both transplanted or directly seeded. It has a good tolerance to diseases.
3 Cultivar of semi-determinate growth habit, with pendulous fruits: Kalocsai 801, a sweet variety. It has a loose spreading foliage. The plant is 40–45 cm high. The fruits are on short peduncles and can be easily picked by hand, they are 10–12 cm long weighing 22–28 g.

Figure 9.9 The "Szegedi 80" non-hot paprika variety.

They are straight, gradually tapering towards a pointed tip, and the colour is dark red when ripe. Their pigment content at picking is 6.0–7.0 g/kg, going up to 8.0–9.0 g/kg after post-harvest ripening. The dry matter content is 18% when ripe. It is an early, intensive variety bringing a high yield as an exchange for watering and good nutriment supply. Its potential harvest is 20–22 tons/ha and has a high field tolerance to viral diseases.

4   Cultivar of semi-determinate growth habit with erect fruits: Kalocsai M 622, a sweet cultivar. The bush is 35–45 cm high with sparse foliage, and has a rigid stem with short intern-odes. Leaves are leathery and thick, so it has a good field resistance to fungal infections. The fruit is 10–15 cm long, gradually tapering towards a pointed tip, and is dark red when ripe. The pigment content at picking is 6.0–8.0 g/kg, increasing to 9.0–12.0 g/kg after post-harvest ripening. If transplanted, the entire crop can be harvested at the same time due

Figure 9.10 The "Szegedi F-03" hot paprika variety.

to its short growing season and early, uniform ripening. It is primarily directly seeded. It is the most widely spread cultivar in the Kalocsa region. It requires intensive agronomic conditions, but the compensation is a high yield. Its yield potential is 20–25 tons/ha. It also has a high tolerance to diseases, which is the basis of secure production (Figure 9.11).

5 Cultivar of determinate growth habit and erect fruits: Kalocsai D 601, a sweet variety with erect fruits appearing in bunches on the stem. The bush is 30–35 cm high and the fruit bunches raise above the foliage and ripen uniformly. Fruits are 10–12 cm long, slightly bent, pointed and ripen to a deep red colour. It contains 6.0–7.0 g/kg pigment at picking and 8.0–9.0 g/kg after ripening. The dry matter content is above average. It has a short growing season and early, uniform ripening. It is primarily recommended for direct seeding. On a large scale it can be harvested by machinery in one operation. The yield potential is 15–16 tons/ha. It can be grown successfully under irrigation on a brown sandy soil with high organic content.

6 Cherry type paprika: Kalocsai M cherry type. It is a cultivar of pendulous fruits and loose foliage. The bush is 40–60 cm high and of continuous growth. The fruit is 3–3.5 cm in diameter, slightly flat and globe shaped with a closed style point. Its surface is smooth and the cross-cut is oval. The colour of the fruits is dark red when ripe. The average weight is 7–9/g. The dry matter content is 20–22% when ripe and the capsaicin content at picking is 120–140 mg/100 g. It has a good tolerance against viral diseases. It has a mid-early, continuous ripening and a good yield. It is primarily recommended for transplanting and requires intensive growing conditions. It needs a soil rich in organic matter that can be easily warmed up (Figure 9.12).

Figure 9.11 The "Kalocsai M 622" non-hot paprika variety.

Technology

Sweet Capsicum

PRODUCTION METHODS

Sweet *Capsicum* is produced under cover in glass or plastic greenhouses (forcing system) and in the field, depending on the environmental conditions (similar to other European countries).

Figure 9.12 The "Kalocsai M cseresznye" cherry type paprika variety.

Forcing system

Growing structures can be: glass covered or plastic covered. According to production periods we recommend the following forcing seasons:

Early forcing: Sowing in September–October, transplanting in November–January to heated glass house.

Mid-early forcing: Sowing in November–December, transplanting in January–March to heated plastic house.

Cold forcing: Sowing at the end of February and transplanting at the end of April to unheated plastic houses.

Autumn forcing: Sowing in July and transplanting in September to heated plastic houses.

Growing in the open field: At first the seedlings are grown in hotbeds or in heated greenhouse usually until the middle of March. Transplanting could commence after the spring frosts approximately at the end of May.

SOWING AND GROWING SEEDLINGS

Using any production technology seedlings are grown under controlled conditions in seedling media (trays, pots, etc.).

TRANSPLANTING

Transplanting varies according to the production technology and cultivars used; 4–5 plant/m^2 are planted for early forcing, or 8–10 plant/m^2 are planted in the field.

PLANT CULTIVATION

The use of natural soil, both in forcing or open field production, is typical of the Hungarian sweet *Capsicum* industry. Growing without soil has only been recently achieved in greenhouses, which represent only 1–2% of the total production area. Because of the above-mentioned tilling, weeding is one of the most important cultivation projects. It is done by small garden machines or by hand in greenhouses. Drip irrigation is used for the irrigation and fertilization in 80% of the growing systems. Micro-dispenser irrigation is used for the rest. Overhead dispenser irrigation system (linear) is typical, but recently several farms switched to drip irrigation to save water.

Sweet *Capsicum* has to contend with two serious diseases when grown in greenhouses; they are the tobacco mosaic virus (TMV) and powdery-mildew (*Leveillula taurica*). Most of the cultivars intended for forced production have TMV resistance. Licensed pesticides are used against powdery-mildew. The cucumber mosaic virus (CMV) and *Xanthomonas* could cause damage in the field. For protection against the last two diseases, pesticides are used as well. Among the numerous insect pests Thrips, Aphids and Whitefly can cause significant damage. Biological and chemical methods are used for protection.

HARVESTING

Harvesting by hand in greenhouses is done on an average every 7–10 days. Fruits are always harvested at market maturity (white, pale green). An exception is the tomato-shaped sweet *Capsicum* that is harvested and sold at biological maturity (red).

Harvesting begins at the end of July or beginning of August in the field. It finishes with the first autumn frost (end of September–middle of October).

Paprika

For a long time paprika was only grown on a small area by a small circle of growers mainly in Szeged–Alsóváros during the 1880s. However, it was a completely new crop in the region of Kalocsa. Originally paprika was directly seeded in the gardens of the villages and towns. Since cultivation technology has improved (transplanting, etc.) it became a field crop. Because of the changes of cultivation technology, direct seeding became an important technology.

Paprika is considered to be an intensive farming system, and a row crop. Paprika is generally known as a transplanted crop today, but it was originally directly seeded. At the early stages the growers had no facilities for growing seedling. At the end of the nineteenth century the technology of growing cold-bed seedlings became popular. At the beginning of the twentieth century a lukewarm-bed or molinos seedling production was introduced, and later the warm-bed (Dutch) seedling production became accepted (Mátray, 1940; Erdei, 1970). As a result of improved technology during the last decades seedling production in plastic houses was widely used, in most cases with complementary heating equipment to protect against sudden cooling. Direct seeding of paprika is practised as a large scale because of logistic and economic aspects of production. Additionally, the increasing cost of labour and energy are a powerful argument for direct seeding. Even though transplanting requires less seeds for sowing, the vegetation period of these plants is a bit shorter, so the production safety and the harvested crop's quality are better. Several authors have studied this issue, namely Szepesy (1974, 1981–82), Kapitány (1978) and Kapitány and Szepesy (1978). They agreed that direct seeding can have the same result as transplantation if early cultivars, perfect agronomy and a "good year" occurs. For these reasons this method should only be used on a portion of the property.

Our observation is that most of the Hungarian paprika cultivars – except the semi-determinate and determinate ones – are able to compensate for the fewer number of plants per hectare by heavier growth and higher fruit production per plant. They could also manage the same production per hectare under suitable climate and production technology conditions. This observation supports the hypothesis that there is justification of growing hybrid paprika if certain conditions are secured in the Hungarian climate.

SOWING AND PLANTING

Considering field production under Hungarian climatic conditions, two kinds of production technology are known and applied: growing seedling and direct seeding. When seedlings are produced the seeds are sown in unheated plastic green houses during the second half of March or in the first few days of April, at the rate of 20–25 g/m². The seeds are then covered by a 1 cm layer of sand. It is advisable to use pre- and post-emergence herbicides before and after sowing. After the seedlings emerge they can be treated, if necessary, with bactericides and fungicides. Weeding has to be done manually. Planting of the hardened seedlings should start in the second half of May, or in the first days of June. Usually, it is carried out by a transplanter ranging from two- to four-row units. The seedlings should be planted in 60–75 cm rows with 18–25 cm between plants by using two to three seedlings in each spot. This process will result in 200,000–250,000 plants per hectare.

To employ the method of direct seeding a suitably prepared land is needed. The sowing can start at the end of April depending on the weather. A pre-irrigation assists in seed emergence. The optimal number of plants is 400,000–500,000 per hectare. After applying pre- and post-emergence herbicides the weeding is later done mechanically or manually.

IRRIGATION AND PLANT PROTECTION

The growers must ensure the availability of an additional water supply because paprika requires more water than is naturally available under the Hungarian climate. This problem is solved with a water-gun. This is not the best for the crop because of the high pressure and the large size of drops. It is better when the irrigation is done by the linear system. There is less mechanical damage and the dispersion is better.

Fumigation of the soil protects against pests in the soil (most of the time nematodes). The damping of seedlings during the early growth needs to be considered. This disease can cause considerable damage and is caused by fungus, namely *Rhizoctonia solani* and occasionally *Phytium debaryanum*. Practical preventive protection includes fumigation or using fungicides for the sterilisation of the soil. After transplanting, and in case of direct sowing, protection is important against *Xanthomonas vesicatoria* pv. *campestris*, especially if it rains a lot around the end of May and in the beginning of June. If cooling follows the precipitation, *Pseudomonas syringae* pv. *syringea* may induce infections. In paprika a symptom called "infectious wilting" is often observed. Most of the professionals agree that this symptom is caused by several pathogens (e.g. *Verticilium* sp., *Fusarium* sp., etc.). A significant loss in yield can also be caused by Stolbur disease. Various cicada species spread a mycoplasma called *Stolbur mycoplasma* that causes this disease. During the second half of the vegetation period different sucker pests (aphids, cicada) may induce heavy damage on crops. They are vectors for CMV and Alfalfa Mosaic Virus. CMV's typical symptom is the so called "new-faith", with a large number but very small fruits. "Vetesi bagolylepke" may cause notable damage.

Yeilds vary widely in Hungary, but are generally between 8 and 20 tonnes. The harvested fruit's dry matter content is between 15–18%. The large farms use a modified bean-harvester in addition to manual labour. The disadvantage of using mechanical harvesting is that the percentage of damaged fruits is higher, the possibility of second infection and loss during storage could be higher and consequently, the quality of the milled product will deteriorate. Hand harvesting is done only on small parcels of land. Usually it is done in several steps. The advantage is that pickers harvest only the healthy, red fruits. The quality of fruits picked by hand is better when compared to the mechanical harvesting, and the mechanical damage is less. Traditionally harvest started in the last days of August, but it is preferable to begin after September 8. Red and semi-ripe fruits were only harvested in three to four steps before the first autumn frost (Bálint, 1962). After two to three days of wilting, the harvested crop is stringed and exposed to four to six weeks of after-ripening. During this time the fruits undergo a useful biochemical change. The total red pigment content is increased, the sugar content is lowered and significant moisture loss occurs.

At the first picking, 60–65% of the crop could be harvested, depending on the cultivar. Ideally, 75–80% of the crop should be harvested at the first picking. The general belief is that after the first picking the remaining fruits will ripen faster. Due to this belief some growers start the first picking earlier than necessary. This way the second harvest can be done earlier and the total yield will be higher, contrary to a delayed first picking and a milder, risky second picking. To have a good quality product the aim is to harvest as much fruit as possible at the first picking.

Processing technology

Selected fruits go through a four to six week after-ripening period. Ripping, the removal of the calyx, drying and milling is to follow. This is the base of the traditional processing technology. The processing methods as known today developed in parallel with industrial development. In the eighteenth century, the general procedure was that harvested crops were stringed (Figure 9.13). Following the after-ripening period they were dried, bagged and crushed in bags and sifted. A foot driven mortar and pestle later became the usual practice employed for the crushing stage. In the second half of the nineteenth century it became common practice to remove the calyx before drying. The paprika was then crushed, milled by using water-, wind- or steam-mills before sifting.

By the end of the nineteenth century ripping became an accepted step because it satisfied all the requirements of preserving the important biological traits of the fruit, and a really good milled product could be made without any external material (and of course without calyx, which has no spicy taste at all). Ripping ensured the possibility of producing the completely sweet powder before the appearance of sweet varieties. The job of processing developed in parallel with ripping in Szeged. Ripping became a trade and provided a livelihood separate from growing.

After removing the calyx, they ripped the shorter side of the fruits, taking out the placenta (containing septum and seed). They stringed the pericarpium and kept them in a heated place until they became bone dry. The seeds were then separated from the placenta and the septum and washed until they were free of capsaicin. The dried seeds were added to the dried, crushed pericarpium in the required amount and then milled.

Figure 9.13 The traditional pre-drying and post-maturation method of paprika.

Traditionally, milling – by mortar grinding – took place in a couple of two stone water-mills, and later in wind-mills. These kind of mills were forced out rapidly by the use of steam-mills. Due to their specialisation of paprika milling they were used right up until the end of the nineteenth century. By that time newly built steam-mills working with 10–14 couple of stones won the battle over them (Bálint, 1976, 1977).

Unfortunately, the large processors cannot take into consideration the biochemical processes which affect the quality of the milled product.

In Hungary, one can only find a good quality milled product using the traditional processing technology, such as home milling (Figure 9.13), and not with the large scale operators. The large processors buy raw fruits and after a temporary storage, without the necessary after-ripening and hand selection, dry it immediately. This practice can cause problems such as the half-product may not be suited to make a good quality milled product after six to eight months of storage.

One of the deficiencies of modern harvesting and processing methods is that the problem of after-ripening is not solved which is the key to good quality.

References

Augustin, B. (1907) *Historisch-kritische und anatomisch-entwicklungsgeschichtliche Untersuchungen über den Paprika*. Rosner, Németbogsán.

Bálint, S. (1962) *A szegedi paprika* (The paprika of Szeged). pp. 1–131. Budapest: Akadémiai Kiadó.

Bálint, S. (1976) *A szögedi nemzet* (The nation of szoged), *A szegedi nagytáj népélete* (The public life of the Szeged region), A Móra Ferenc Múzeum Évkönyve 1974/75–2. 1. rész. pp. 608–611. Szeged: Móra Ferenc Múzeum.

Bálint, S. (1977) *A szögedi nemzet (The nation of szoged), A szegedi nagytáj népélete (The public life of the Szeged region)*, A Móra Ferenc Múzeum Évkönyve 1976/77–2. 2. rész. pp. 174–175, 211–212. Szeged: Móra Ferenc Múzeum.

Benedek, L. (1960) A szegedi fűszerpaprika-kutatás története (The history of paprika research in Hungary). In: *Délalföldi Mezőgazdasági Kísérleti Intézet Közleményei*. Mezőgazdasági Kiadó, Budapest, 1962, pp. 167–178.

Benedek, L. (1974) A fűszerpaprika (The paprika). In: *A Gabonatermesztési Kutató Intézet Jubileumi Évkönyve*. Szeged, pp. 73–84.

Csapó, J. (1775) *Új füves és virágos magyar kert*. Posony: Landerer.

Derera, N. F. (2000) Recent developments in condiment paprika recearch. *Australian Agri-Food 2000 Research Forum*. Melbourne, 17th August 2001.

Dunszt, K. (1939) A fűszerpaprika használatának módjai a konyhatechnikában és alkalmazása az élelmiszeriparban (Methods of using paprika in gastronomy and applying in food industry). *Kísérletügyi Közlemények*, **XLII**(1–3), 1–8.

Erdei, I. (1970) *A fűszerpaprika palántaelőnevelése (Seedlings growth of paprika)*, 1–32.

Kapitány, J. (1978) Ültetett és helyrevetett fűszerpaprika-növényállomány össze-hasonlító értékelése (Correlation of transplanted and direct-sowing paprika vegetation). *Zöldségtermesztési Kutató Intézet Bulletinje*, 13, 25–35.

Kapitány, J. and Szepesy, K. (1978) Állománysűrűség és termőképesség összefüggései hely-revetett fűszerpaprikánál (Correlation between paprika plant number and yield). *Zöldségtermesztési Kutató Intézet Bulletinje*, 17, 33–42.

Mátray, S. (1940) A szántóföldi fűszerpaprika langyosági (molinós) palántanevelése. [Seedling growth in lukewarm (molinos) bed]. *A szegedi M. Kir. Növénytermesztési Kísérleti Állomás közleménye*, 1–7.

Moór, A. (1969) A paprika hibridmag termesztése (Hybrid seed production of paprika) Thesis. Hort. Univ. of Budapest.

Obermayer, E. (1934) A magyar paprika áruismerete (Knowing the Hungarian paprika as a product). In: *Időszerű kérdések a gazdasági földrajz, vegytan, áruismeret köréből*. Budapest, Pátria, 1935.

Obermayer, E. (1938) A magyar fűszerpaprika termesztése (Production of Hungarian paprika), Pallas, Budapest.

Pickersgill, B. (1986) Evolution of hierarchical variation patterns under domestication and their taxonomic treatment. In: B.T. Styles (ed.) *Intraspecific Classification of Wild and Cultivated Plants*. Systematics Association, Special Volume no. 29. Clarender Press, Oxford, pp. 191–209.

Pickersgill, B. (1989) *Genetic Resources of Capsicum for Tropical Regions*, AVRD Publication No. 89–317, pp. 1–9.

Somogyi, N., Garcia Pomar, M.I. and Pék, M. (1998) Applied spice pepper processing in Spain and Hungary. *Bulletin of the Vegetable Crops Research Institute*, 28, 78–96.

Somos, A. (1981) *A paprika (The paprika)*, pp. 1–396. Akadémiai Kiadó, Budapest.

Szanyi, I. (1940) Fontos feladatok a magyar fűszerpaprika termesztése terén (Important subjects of the production of Hungarian paprika). *Mezőgazdasági Közlöny*, 1940 **március**, 1–12.

Szepesy, K. (1974) A szegedi fűszerpaprika fajták optimális ültetési sűrűsége (The optimal plant density of Szeged varieties). *Zöldségtermesztési Kutató Intézet Bulletinje*, 9, 151–154.

Szepesy, K. (1981–82) Az állománysűrűség hatása a fűszerpaprika termésmennyiségére és minőségére helyrevetéses termesztés esetén (The effect of the plant number on paprika yield and quality in direct-sowing). *Zöldségtermesztési Kutató Intézet Bulletinje*, 15, 53–55.

Veszelszky, A. (1798) *A növevény-plánták országából való erdei és mezei gyűjtemény, vagy-is fa és fűszeres könyv*. Pesth: Trattner.

Zatykó, L. (1979) *Paprikatermesztés*. Budapest: Mezőgazdasági Kiadó 1896. évi 38.286 FM rendelet.

10 Post-harvest handling and processing of *Capsicums*

V. Prakash and W. E. Eipeson

The genus *Capsicum* encompasses a number of species differing in size, shape, colour and pungency. Due to these differences some of them are used as vegetables, while most others are valued as condiments and culinary supplements. The post-harvest handling and processing technologies for *Capsicum* have developed considerably as a consequence of the increased production and newer applications of this crop. Bell peppers and a sizable quantity of chillies are consumed fresh and their harvesting indices and scientific handling protocols have been standardized. Chillies and paprika are initially dried and stored in preparation for processing. The accumulated scientific evidence shows the role of various processing factors on the quality of the processed products like chilli powder, oleoresin and colour extract. In keeping with this knowledge, process parameters have been modified to develop new technologies for obtaining superior products. The emergence of the industrial food processing sector along with newer food applications requiring tailor-made ingredients have also introduced more stringent demands for *Capsicum* products. R&D and the industry are poised to face the challenges.

Introduction

Tropical South America is believed to be the original home of *Capsicum*, where the different varieties were known by different names in regional dialects, one of them being "chilli". They were introduced into Spain in the fifteenth century by Columbus who named it red pepper. *Capsicum* is believed to have been introduced into India by the Portugese. The different species of *Capsicum* which were established in India, China and South Asia are the long and thin varieties of moderate pungency which are locally called "chillies". The highly pungent varieties having small fruits also called chillies (bird chillies) are more common in Africa. The highly coloured, mildly pungent to sweet varieties known as "paprika" were developed in Europe.

Chillies, red peppers, sweet peppers, cayenne, paprika and most other cultivated varieties of varying pungency belong to the species *Capsicum annuum*. A few chilli varieties, such as bird chillies (African chillies) and tabasco chillies, are classified under *C. frutescens*. However, five major cultivated species of the genus *Capsicum* namely, *C. annuum*, *C. baccatum*, *C. frutescens*, *C. chinense* and *C. pubescens*, have been recognized the world over. Regardless of the term *Capsicum* used by taxonomists, the scientific literature in different countries have used different terms to describe the type of chillies/*Capsicums*. This practice has lead to considerable ambiguity. The term "chilli" is used in India, the UK, Africa and the countries in the East, but not generally in the US. The British Standard Specifications differentiate between chilli and *Capsicum* based on the degree of pungency. The International Standard Organization, however, recognizes only two types namely, "chillies" and "paprika". These types cover the easily perceptible to strongly pungent types and the big fleshy vegetable, and the sweet or just recognizable pungency type of *Capsicums*, respectively. Commercially, chillies with moderate pungency are grown in tropical countries like India, China, Pakistan, Nigeria, Malaysia, Turkey and Japan; high pungency varieties are grown in Africa and paprika with milder pungency is grown in European countries like Hungary, Spain, Bulgaria, Romania and Poland.

Capsicum annuum is the most important among the species grown all over the world. India is the largest producer of chillies (dry) in the world with an estimated annual production of 1.066 million tonnes during 1996–97 (Annual report, Spices Board 1998–99). Chillies, both fresh and dried, are used as condiments and culinary supplements for their pungency and characteristic pleasant flavour. Bell pepper (*C. annuum* var. *grossum*) is used in stews, salads, pizza, meat loaf and also as a vegetable and culinary supplement. Bell pepper is also known by different names, such as green pepper, sweet pepper and in India as Simla mirch.

Over the years, ripe chillies and paprika have become very important raw materials for processing into a variety of products like chilli powder, curry powder, chilli oleoresin and chilli colour. They are also used to a limited extent for canned, frozen, pickled and fermented (e.g. Kochujang) products. Due to their importance in world trade, various developments have taken place in crop improvement, storage conditions, processing technologies as well as the packaging and storage of *Capsicums*. The large volumes of scientific information generated in the recent years on the beneficial pharmacological properties of capsaicinoids (Govindarajan and Sathyanarayana, 1991) have further boosted the commercial importance of *Capsicums*.

This chapter looks into some of the post-harvest technological aspects of this important spice and attempts to shed light onto the future prospects of processing them for value addition.

Handling and storage of fresh produce

Fresh chillies

Chilli fruits are harvested for use as green vegetables when they are fully mature but before they change colour from green to red. The chilli crop takes about 40 days to flower after transplanting and a further 30 days to develop fruit suitable for green harvest (Rajput and Parulekar, 1998). Fresh green chillies are usually marketed as such and are only occasionally cold stored. Therefore, very little published information is available on this aspect. The available information as well as the studies carried out at the Central Food Technological Research Institute (CFTRI), Mysore, India, show that the optimum cold storage temperature for chillies is $10 \pm 2°C$. Under the storage conditions, the storage life of chillies is extended up to 10–12 days. Prepackaging in polymeric film or paper bags improves the storage life and retains the freshness of chillies. It has been reported that the decay in stored fresh chillies caused by *Erwinia carotovora* is hastened by high relative humidity (Rajput and Parulekar, 1998). Chillies meant for making dried red chillies are ready for harvest after about 45–50 days of flowering when they are fully mature and red or nearly ripe red.

Bell pepper

Some of the important varieties of bell pepper produced in India are: Bullnose, California Wonder, Chinese giant, HC-201, HC-202 and HC-213. They differ in size, shape, flesh thickness and other quality factors. Bell peppers used as a vegetable are usually harvested when they are of market size and still green in colour. Those used for dehydration and sauce are allowed to ripen on the plant before being harvested. Due to their brittle structure, bell peppers should be handled with care. Transpirational losses in bell peppers are very high, limiting their storage life. New Mexican type peppers become flaccid in three to five days (Lownds and Bosland, 1988; Miller *et al.*, 1986) at 20°C (7–10% weight loss). The post-harvest bio-chemical changes associated with the softening phenomenon in *C. annuum* fruits have been studied (Priyasethu *et al.*, 1996). They showed that the fruit texture decreased with an increase in polygalocturonase activity. Prepackaging of the fruits in perforated polyethylene packages has been shown to reduce water loss rates by 20 times or more (Lownds *et al.*, 1994). Prepackaging also reduces colour development (red colour) across nine cultivars on storage (Boswell, 1964). Based on a Instrumental Sphere data logging tool, potential damage causing operations on the bell pepper packaging line have been identified and remedial measures suggested (Marshall and Brook, 1999).

Bell peppers are susceptible to chilling injury at temperatures below 7°C. Chilling injury also predisposes them to *Alternaria* rot (McColloch, 1962) and *Botrytis* rot (McColloch and Wright, 1966). However, pre-harvest spray with growth regulators like paclobutrazol, uniconazole and mefluidide has been shown to alleviate chilling injury development in bell pepper on storage for 28 days at 2°C (Laurie *et al.*, 1995). Waxing of the fruits has also been found to be beneficial in extending their storage life (Hartman and Isenburg, 1956; Hardenburg *et al.*, 1986). Other edible coatings (protein, cellulose and oil based) also have been investigated to increase the storage life of green bell peppers. Among them, oil based coating gave the best results in reduction of respiration rate and the preservation of good colour (Lerdhangangkul and Krochta, 1996). As ethylene accelerates senescence in bell pepper, it is not advisable to store them along with other ethylene-producing fruits. Evaporative Cool Storage (ECS) of bell peppers has been shown to extend their storage life to 12 days, compared to two days under ambient conditions. The ECS was maintained at 20–24°C with 95% relative humidity (RH) (CFTRI, 2003).

Pre-cooling of bell peppers before cold storage is essential to obtain good storage life. This is best achieved by forced air cooling. Dipping bell peppers in hot water for four minutes at 53°C and packaging in polyethylene bags was found to be most effective for maintaining bell pepper quality. They remained hydrated and green and had a lower rate of respiration and less chilling injury (Gonzalez *et al.*, 1999).

Low oxygen atmosphere retarded ripening and respiration during transit and storage (Pantastico *et al.*, 1975). The storage life of bell peppers has been found to be prolonged under controlled atmosphere storage conditions of low oxygen and high carbon dioxide (Hughes *et al.*, 1981; Luo and Mikitzel, 1996).

Processing of *Capsicums*

Standardization, grading and storage of dry chillies

The Bureau of Indian Standards has outlined specified standards for dried chillies based on physical characteristics, as well as on other factors such as total ash, acid insoluble ash, non-volatile ether extract and fibre content. Under the Prevention of Food Adulteration Act (1954), minimum purity standards are laid down for chillies. Agmark specifications for the grading of dry chillies for export takes into account various physical, chemical, sensory and microbiological parameters (Agmark, 1962).

Microbial and insect infestations are serious problems during the storage of dried *Capsicum*. Ethylene oxide fumigation in bulk is recognized as the best treatment to achieve practical commercial sterility (Govindarajan, 1985). Methyl bromide and phosphene are used as

Table 10.1 Breeding objectives for major fruit quality traits in various market types of pepper (*Capsicum* spp.)

Market type	Important fruit quality traits
Fresh market (green, red, multi-colour) whole fruits	Colour, pungency[a], shape, size, lob number, flavour, exocarp thickness, endocarp seed ratio, vitamin A and C
Fresh processing (sauce, paste, canning, freezing)	Colour, pungency, flavour, pericarp thickness, endocarp seed ratio
Dried spices (whole fruits, powder)	Colour, pungency, flavour, dry weight, low crude fibre, endocarp seed ratio
Oleoresin extraction	Essential oils (colour, pungency)
Ornamental (plants or fruits)	Colour, pungency, shape, dry weight

Note
a Qualified for non-pungency or quantified for degree of pungency.

fumigants for insect control. Ionizing radiation with a dosage of 10 kGy has been shown to destroy both microorganisms and insects. A dosage of 7.5 kGy has been shown to be sufficient for eliminating fungal populations, and oleoresin yields have increased from 24.45% to 31.61% by irradiation due to the enhanced extractibility (Onyenkwe and Ogbade, 1995).

As mentioned earlier, a large number of products are produced from chilli and paprika. The raw material quality requirements for different end uses vary. In order to satisfy the precise requirements, extensive work has been carried out on chilli breeding. Some of the breeding objectives are given in Table 10.1 (Peter, 1998). For the genetic upgrading of oleoresin quality and deeper red colour in Indian chillies, Pusa Jwala was crossed with IC 31339. The new strains attained superiority over parental varieties by possessing 20% and 27.5% more capsaicin and oleoresin, respectively, and a higher colour value (CV) of the order of 22,000 and 27,200 CV, respectively (Wealth of India, 2000).

Drying and dehydration

Dried chillies and paprika are the raw materials for the commercially important products of chilli powder, oleoresin and colour. Therefore, the most important primary processing operation for chillies and paprika is drying.

Traditionally, chillies are sun dried to a safe moisture level (about 10%) mainly for trans-portation, storage, distribution and further processing. Over the years, with the growing demand for speciality products with the required pungency, colour and particle size, along with the information generated on the significance of curing and drying conditions on the functional properties of *Capsicum*, have lead to the development of improved methods for their dehydration. The major factor contributing to the product quality is the cultivar. The chemical quality char-acteristics (oleoresin, colour and ascorbic acid content) of some important varieties of chillies grown in different regions of India are given in Table 10.2 (Papalkar *et al.*, 1992; Pruthi, 1998).

In addition to the intrinsic factors, varietal characteristics have also been found to affect the stability of colour in chillies. Lease and Lease (1956) showed that under similar conditions of processing and storage of red peppers as powder for six months, one variety retained 78% of the initial colour while in another, there was almost complete loss. The role of harvesting maturity to colour retention in red chilli powder has been reported (Isidoro *et al.*, 1995). Subsequent

Table 10.2 Chemical quality of some varieties of chillies

Variety	Oleoresin (%)	Total extractable colour (Asm[a]/ ASTA[b])	Ascorbic acid (mg/100 g)
X235[a]	8.19	8,550	158.46
Musulwadi[a] selection	7.46	7,235	64.50
NGP 12[a]	8.50	7,260	168.90
NGP 22[a]	8.30	8,417	168.75
Loni Budruk[a]	8.67	7,119	163.07
Parbani tall[a]	6.27	7,425	145.46
BR Red[b]	8.60	1,248	58.20
Pant C1[b]	9.80	1,144	62.80
Jwala[b]	10.60	2,542	86.40
CO 2[b]	11.60	256	60.80
SP 47[b]	8.40	846	28.90

Notes
a Papalkar *et al.* (1992).
b Pruthi (1998).

curing of chillies has also been shown to ensure maximum colour formation in paprika (Vinkler, 1973). An important finding on paprika was that there was a more rapid colour deterioration of the low moisture powder stored at high temperature when the harvested fruits were allowed to wither on the plant than when harvested fully mature and ripe and cured outside (Kanner *et al.*, 1977). This observation is in accord with the practice of harvest and curing practiced in Hungary and Spain and the continued increase of red pigment concentration during curing outside recorded by many investigators there (Govindarajan, 1985). Storing bell pepper after pre-packaging in perforated (0.064–0.42%) polyethylene bags at 3°C gave good colour development which retained the firmness. Under these conditions, the fruits did not develop chilling injury, even at the lower temperature (Meir *et al.*, 1995).

The other important factors are the drying conditions. Sun drying normally takes about 15 to 20 days to reduce the moisture content to 10–15% and under uncontrolled conditions can lead to bleaching and dull colour formation in chillies. Improvements in the sun drying of chillies through pre-treatments, such as pricking, blanching, treating in alkaline solution (dipsol) and drying in preforated trays both in shade and directly in the sun, have been reported by Laul *et al.* (1970). By this method, the sun drying time is reduced from 15–21 days to almost a week. Colour and pungency are better retained and a more hygienic quality product is produced. A further improvement on sun drying is the development of the solar drier which effects the complete drying of the commodity in four to five days with much better colour and storage characteristics (Pruthi, 1998).

Mechanical dehydration under controlled conditions of temperature, air velocity and humidity has further improved the quality of dried chillies. Different types of driers, such as the cabinet drier, tunnel drier, multistage belt drier and fluidized bed drier, can be used for the purpose. Controlled rapid drying of plant-withered and sliced fruits at temperatures below 80°C, preferably 60–70°C, has been shown to give the highest colour and colour retention (Lease and Lease, 1962). In the US, popular chilli varieties are dried in belt driers by a forced draft of air at a temperature of 80°C to a moisture content of 7–8% (Feinberg, 1973). A two stage dehydration, involving initial drying to 12–15% moisture and storage at 0°C and redrying when required for grinding to 7–8% moisture, is also practised. This two stage process has the advantage of better colour and pungency retention. It has been observed that there is a continuation of light induced carotenogenesis in *Capsicum* after harvest, followed by a light induced degradation of the pigments. By combining a first step of illumination and a second step of darkness during dehydration, it was possible to obtain dry peppers having 20–40% more pigment concentration (Mosquera Minguez *et al.*, 1994). Even under controlled conditions of dehydration certain quality changes take place. Luning *et al.* (1995) have reported, after hot air drying of *Capsicum* a decreased levels of odour compounds (Z)-3-hexenal, 2-heptanone (Z)-2-hexenal, (E)2-hexenal and linalool which have green vegetable like, fruity and floral notes. Therefore, it is necessary to optimise the dehydration conditions to minimize the loses.

Dry chillies are usually packed in jute bags of up to 100 kg. In order to prevent breakage of dry chillies while compressing the packs, the moisture content is kept around 10%. Among the various factors affecting dry chillies during storage, darkening due to temperature effect is the most predominant. The present commercial practice is to store dry chillies in cold storages.

Ground chillies

In recent years, the global demand for ground chillies has steeply increased, mainly due to their convenience in use. Ground chillies offer an additional advantage in that it is also possible to get the required functional properties for specific end uses by blending different varieties of chillies. Even though spice grinding is essentially a size reduction unit operation, the requirement of high quality products has prompted considerable engineering improvements. Different types of machines are used for spice grinding. They include attrition mills, impact mills, roller mills, vibro energy mills, fluid energy mills and bowl mills (Ramesh, 1989). Modern spice mills, which handle large volumes, have optimized material flow, closed circuit pneumatic conveying, dust collection aids and noise reduction fixtures (Russo, 1976). In order to reduce volatile loss and other quality deterioration due to the heat generated during the grinding operation, cryogenic grinding has also been attempted (Wistreich and Schaffer, 1962; Besek *et al.*, 1985).

Hungarian paprika is a specialty product valued for its colour. Fresh paprika is also a rich source of ascorbic acid (vitamin C) which can be as high as 300–400 mg/100 g. Szent Gyorgi, the Hungarian scientist, was awarded the Nobel Prize in 1937 for isolating ascorbic acid from paprika. Different grades of paprika powder are produced from carefully graded paprika which are not stored for more than one year. The fruits are freed from the calyx and peduncle which dilutes the colour and adds fibre. The placenta, which carries pungency stimulants and seeds which dilute colour and also contribute to the potentially oxidizable fat leading to the risk of colour deteriorations are also removed. The seeds whose addition is required to facilitate grinding are washed in cloth bags to free them from adhering tissues, which may be rich in capsaicinoids. For the pungency grade, the fruits and other parts removed or rejected from other grades are used (Govindarajan, 1986a). The CFTRI, Mysore, has developed a new technique to fractionate chillies into three grades: (i) Capsaicin rich powder; (ii) Skin; and (iii) Seeds. In the process, the chillies are dried and ground in a suitable mill wherein all three grades are separated. Data from a typical batch show *Capsicum* rich powder, skin and seeds as 5–6%, 35–40% and 40–44%, respectively (CFTRI Process No. CPS-3560). These fractions can be separately extracted to obtain capsaicin rich oleoresin, colour concentrate and fixed chilli seed oil.

Chilli powder requires suitable packaging to retain its functional properties during storage. These properties are known to be adversely affected by the absorption of moisture, effect of light, oxidation by air and storage temperature.

Capsicum oleoresins

The term "Oleoresin" is used to describe desolventized total extracts by a specified solvent. It sometimes refers to a product obtained by benefication of one or more functional components in the total extracts by some amount of fractionation. In its use as a food additive, the best oleoresin of *Capsicum* is that which contains the colour and flavour components (pungency, aroma and related sensory factors) and that which truly recreates, when appropriately diluted in food formulations, the sensory qualities of fresh materials (Govindarajan, 1986a). Being a solvent extracted product, the purity and residual amount of solvent are to conform to the specifications of international and national food laws. Different solvents have been used for extracting oleoresins. The early systematic studies reported by Tandon *et al*. (1964) have generated the laboratory data for producing chilli oleoresin. They reported the distribution of the functional components like colour and capsaicinoids in chillies. Among the solvents used for oleoresin extraction, they found that alcohol gave the highest yield of oleoresin, but the lowest extraction of colour, while ether gave good yields of both. Mathew *et al*. (1971) undertook studies to develop a process for producing chilli oleoresins. They found ethylene dichloride to be a better solvent compared to alcohol or hexane. Their studies also showed that 94% of the colour and 90% of the capsaicinoids were obtained by extraction of the chilli pericarp that was freed from seeds and stalks.

The CFTRI has pioneered the developmental work for the establishment of a spice oleoresin industry in India. The process for production of chilli oleoresin developed at CFTRI has been extensively used by the spice processing industry in India. In the process chilli is ground to the required particle size and extracted with a suitable solvent system. The Regional Research Laboratory, Thiruvananthapuram, India, has also developed a process for the production of chilli oleoresin, including separating fractions of highly pungent oleoresin, colour and seed oil. Counter-current extraction reduces the solvent requirement to 2.5 volumes from 4 volumes for straight runs. During the removal of the solvent from the miscella foaming can be controlled by using anti-foaming agents or air jet control which reduces material loss by entrainment. The last traces of the solvent are removed by steam injection. A vacuum is used in the final stages, as there are very little volatiles in chillies. The spent solvent-free and dried solids contain about 28% protein, 36% carbohydrate and 29% fibre, which could be used in animal feed composition. In the earlier years, ethylene dichloride was a preferred solvent for oleoresin extraction. However, due to their suspected toxicity (Blum and Ames, 1977) poly-halogenated hydrocarbons are discouraged for use in food processing. Presently hexane, acetone and ethyl acetate are the preferred solvents.

There are three types of commercial *Capsicum* oleoresins (Govindarajan, 1986a). Oleoresin of *Capsicum* is mainly used as a food colouring in meats, dairy products, soups, sauces and snacks. Red pepper oleoresin is used for both colour and pungency, mainly in canned meats, sausages, smoked pork, spreads and soups and in a dispersed form in some drinks such as gingerale and in some snacks. Oleoresin from African *Capsicum* is more pungent and is used for its counter-irritant properties in plasters and some pharmaceutical preparations.

Capsicum colour

In view of the increasing evidence on the carcinogenic potential of synthetic colours, there has been extensive research into the isolation, characterization and use of natural colourants. *Capsicum* colour (red) is one which has been investigated extensively and produced commercially. Govindarajan (1986b) has exhaustively reviewed the chemistry of *Capsicum* colours, and Francis (1995) has described the use of carotenoids as colourants. More recently, Deli *et al*. (1996) have made a detailed HPLC analysis of the carotenoids from *Capsicum*; out of the 56 peaks, 34 have been identified. Capsanthin and capsorubin are the most important colouring pigments in *Capsicum. Capsicum* colour has assumed great commercial importance as a food colourant. In the past, the application of *Capsicum* extract was restricted to savoury products due to its spicy flavour. Purified paprika extracts in which the flavour compounds have been significantly reduced, are now available as colours used in sweet preparations such as sugar confectionery. The extract is available in both oil soluble and water miscible forms (Henry, 1998).

Due to the commercial importance of natural colours, most of the process information on *Capsicum* colour is covered under patents. However, some published literature gives useful information. The early report of Todd (1957) gives details of production of *Capsicum* colour concentrate from fresh red fruits. In the process, red *Capsicums* are broken mechanically, the peri-carp and seed portion are separated and the pericarp is converted into a puree. The puree is passed through a fine sieve to separate juice and pulp. The pulp is reslurried and passed through the sieve to recover maximum colour. The colour components are recovered as a precipitate by the denaturation of the proteins having the sugars, gums and other undesirable components in the solution. The precipitated colour is centrifuged to reduce the water content, drum dried and extracted into acetone. The acetone extract is vacuum distilled to recover the solvent. The resul-tant ruby-red oil is mixed with permitted antioxidant to prevent colour loss during storage. Preparation of water dispersible *Capsicum* colour has been described by Elshikiny and Abd-El-Salam (1970).

Specifications for Capsicum *oleoresins*

The Essential Oils Association (EOA, 1975) of USA has detailed specifications for three types of *Capsicum* oleoresins (Table 10.3). The three types are oleoresin *Capsicum*, oleoresin red pepper and oleoresin paprika. The pungency standards are determined by diluting an alcoholic solution of the oleoresins in 3–5% sugar solution. This solution is then tasted by panel members and the first perceptible stinging sensation constitutes standard protocols. The strength of the dilution agreed by three of the five panelists gives the Scoville Heat Units (Govindarajan *et al.*, 1977). The Scoville test run shows a very high correlation to total capsaicinoids content (1,50,000 Scoville units = 1% capsaicin). The colour value in the EOA specifications is determined by measuring the absorption of a 0.01% solution of the oleoresin in acetone at 458 nm. The absorp-tion is then multiplied by a factor 61,000.

Geographical appellation and trademarks

Capsicum belongs to that class where certain varieties whose colour, flavour and properties are unique owing to their respective geographic origin. These unique traits may in future become trade mark issues. We need to address this issue very carefully and ensure that proper documen-tation, both at field and laboratory level, ensures that intellectual crop properties are assigned to the country in question. This issue needs the attention of policy makers, scientists, technologists and others alike.

Future prospects for *Capsicums*

Being the largest producer of *Capsicums* in the world, India deserves to have a dominant position in its value addition and export. At present, the country's export of dry chillies and other

Table 10.3 EOA specifications for *Capsicum* oleoresins

Title	Oleoresin Capsicum (African chillies)	Oleoresin red pepper	Oleoresin paprika
Number	EOA No. 244	EOA No. 245	EOA No. 239
Botanical source	*C. frutescens* L. or *C. annuum* L.	*C. annum* L. var. *longum* Sendt.	*C. annuum* L.
Preparation	Solvent extraction of dried ripe fruit, with subsequent removal of solvent	Solvent extraction of dried ripe fruit, with subsequent removal of solvent	Solvent extraction of dried ripe pods and subsequent removal of solvent
Appearance and colour	A clear red, light amber or dark red, somewhat viscid liquid with characterisitc odour and very high bite	A deep red liquid with characteristic odour and high bite	A deep red somewhat viscid liquid with characteristic odour
Scoville Heat Units (for pungency by described method)	480,000 min.	240,000 min.	—
Colour value (by described method)	4,000 max.	20,000 max.	As stated on label (generally 40,000 to 100,000)
Solubility			
Alcohol	Partly soluble with oily separation	Partly soluble with oily separation	Partly soluble with oily separation
Benzyl benzoate	Soluble	Soluble	Soluble
Fixed oils	Soluble	Soluble	Soluble
Glycerine	Insoluble	Insoluble	Insoluble
Mineral oils	Insoluble	Insoluble	Very slightly insoluble
Propylene glycol	Insoluble	Insoluble	Insoluble
Residual solvent (EOA No. 1-1D-3-1 gas chromatographic method)	Meets with Federal, Food, Drug and Cosmetics Act Regulations; methelene chloride, trichloroethylene, individually or collectively not more than 30 ppm; isopropyl or methyl alcohol not more than 50 ppm; hexane not more than 25 ppm		

products has not reached the desired levels. Therefore, various improvements in the whole chain of crop production, post-harvest technologies, processing and marketing are required.

Even though under the Coordinated Vegetable Development Programme of the Indian Council of Agricultural Research (ICAR), New Delhi, India, considerable progress has been made in increasing productivity and resistance to diseases, there is scope for further improvements. Today, aflatoxin in dry chillies is a major export issue, which calls for the development of aflatoxin free lines.

Due to various reasons, natural plant colourants are acquainting great commercial importance. Among them *Capsicum* red colour is the most promising, considering its wide application. In India, the promising *Capsicum* varieties having good colour value and low pungency are Bydagi, Arka Kabir, the Warrangal chilli and KT-P1–19. They are, however, not comparable to Hungarian paprikas of high colour values and negligible pungency. Therefore, it will be very useful to develop such varieties so that the growing demand for chilli colour can be met better.

The demand for tailor-made chilli products to suit diverse product applications and large scale food processing is increasing. These applications call for the provision of products with the required levels of aroma, pungency and colour, in addition to good storage stability. The accumulated scientific information indicates the importance of the harvesting stage and the primary processing conditions on product quality, especially the colour. This knowledge could be properly utilised to develop new technologies.

It is rather difficult to develop chilli varieties suitable for different product applications. The blending of chilli powders and oleoresins with the help of modern sensory techniques could be standardized to obtain various formulations.

Chilli extracts are presently produced mostly by solvent extraction. Even though the industry is now able to maintain the specified limits of solvent residue levels, there is scope for developing cost effective alternative technologies, such as supercritical gas extraction, which have the additional advantage of selective extraction of components.

Microbial and insect infestations are serious problems in chillies. In addition to the development of pre- and post-harvest protocols, controls in various critical processing steps have to be introduced. In addition to the fumigation and chemical sterilisation practised, irradiation could be further investigated and popularized.

It is needless to state that packaging of the products is critically important for maintaining their storage quality, as well as to aid in marketing. Continuous improvements are possible in this area, along with the development of newer packaging materials having functional properties.

Conclusion

Capsicums, which were used as culinary supplements essentially for their colour, flavour and pungency, have acquired much more importance as a source of natural food colour and valuable pharmacological compounds. In order to maximize the beneficial properties of *Capsicum*, concerted R&D work is required to develop newer varieties that have higher levels of these compounds. Additionally, technologies have to be continuously upgraded to produce end-product specific *Capsicum* formulations.

Along with the ever growing nutrition and health conscious consumer, the demand for *Capsicums* and their products are expected to grow steadily.

References

Agmark (1962) Chilli Grading and Marketing Rules Agricultural Marketing Directorate, Govt. of India, Nagpur.

Annual report (1998–99), Spices Board, Cochin, India.

Besek, C. A., Wilson, L. A. and Hammond, E. G. (1985) Spices quality: effect of cryogenic and ambient grinding on volatiles, *J. Food Sci.*, 50, 599.

Blum, A. and Ames, B. N. (1977) Possible hazards in the use of poly halogenated compounds in general, *Science*, 195, 17.

Boswell, V. R. (1964) Pepper production, *USDA Inf. Bull.* No. 276.

CFTRI (2003) Annual report, under preparation.

Deli, J., Malus, Z. and Tolh, G. (1996) Carotenoid composition of fruits of *Capsicum annuum* cv Szentesi Kozzarvu during ripening, *J. Agric. Food Chem.*, 44(3), 711.

Elshikiny, S. and Abd-El-Salam, M. H. (1970) Preparation and properties of colour from *C. frutescens* var. California wonder and safflower., *J. Food Sci.*, 35, 875.

Essential Oils Association (1975) Specification for oleoresin paprika – EOA No. 239, Oleoresin red pepper, No. 244, Essential Oil Association, New York.

Feinberg, B. (1973) Vegetables. In: WB Van Arsdel, M. J. Copley and A. I. Morgan (ed.), *Food Dehydration*, Vol. 2, 2nd edn, AVI, West Port, Conn.

Francis, F. J. (1995) Carotenoids as colorants, *World of Ingredients*, September/October, 34, 37.

Gonzalez Aguilar, G. A., Cruz, R., Buez, R. and Wang, C. V. (1999) Storage quality of bell peppers pretreated with hot water and polyethylene packaging, *J. Food Quality*, **22**(3), 287.

Govindarajan, V. S. (1985) Capsicum – production, technology, chemistry and quality. Part I. History, botany, cultivation and primary processing, *CRC Crit. Rev. Food Sci. Nutr.*, **22**(2), 109.

Govindarajan, V. S. (1986a) Capsicum – production, technology, chemistry and quality. Part II. Processed products, standards, world production and trade, *CRC Crit. Rev. Food Sci. Nutr.*, **23**(3), 207.

Govindarajan, V. S. (1986b) Capsicum – production, technology, chemistry and quality. Part III. Chemistry of the colour, aroma and pungency stimuli, *CRC Crit. Rev. Food Sci. Nutr.*, **25**, 185.

Govindarajan, V. S., Shanti Narasimhan and Dhanaraj, S. (1977) Evaluation of spices and oleoresins. II Pungency of capsicum by Scoville heat units, standardized procedure, *J. Food Sci. Technol.* (India), 14, 28.

Govindarajan, V. S. and Satyanarayana, M. N. (1991) Capsicum – production, technology, chemistry and quality. Part V. Impact on physiology, pharmacology, nutrition, structure, pungency, pain and desensitisation sequences, *CRC Crit. Rev. Food Sci. Nutr.*, **29**(6), 435.

Hardenburg, R. E., Watada A. E. and Wang C. Y. (1986) The commercial storage of fruits, vegetables and florist and nursery stocks, US Department of Agriculture, Hand Book No. 68, 65.

Hartman, J. D. and Isenburg, F. M. (1956) Waxing vegetables, *NY, State Col. Agric. Cornell Ext. Bull.*, 1.

Henry, B. (1998) Use of capsicum and turmeric as natural colours in food industry, *Indian spices*, **35**(3), 7.

Hughes, P. A., Thompson, A. K., Pumbley, R. A. and Seymour G. B. (1981) Storage of capsicum (Capsicum *annuum* L.) under controlled atmosphere, modified atmosphere and hypobaric conditions, *J. Hort. Sci.*, 261.

Isidoro, E., Colter D. J., Fernandez, G. C. J. and Southward G. M. (1995) Colour retention in red chilli powder as related to delayed harvest, *J. Food Sci.*, **60**(5), 1075.

Kanner, J., Harel, S., Palevitch, D. and Ben Gera, I. (1977) Colour retention in sweet red paprika (*Capsicum annuum* L.) powder as affected by moisture content and ripening stage, *J. Food Technol.*, **12**, 59.

Laul, M. S., Bhale Rao, S. D., Rane, V. R. and Amla, B. L. (1970) Studies on the sun drying of chillies, *Indian Fd. Packer*, **24**(2), 22.

Laurie, S., Ronen, R. and Aloni, B. (1995) Growth regulator induced alleviation of chilling injury in green and red bell pepper fruit during storage, *Hort. Sci.*, **30**(3), 558.

Lease, J. G. and Lease, E. J. (1956) Factors affecting the retention of red colour in peppers, *Food Technol.*, **10**, 368.

Lease, J. G. and Lease, E. J. (1962) Effect of drying conditions on initial colour, colour retention and pungency in red peppers, *Food Technol.*, **16**, 104.

Lerdhangangkul, S. and Krochta, J. M. (1996) Edible coatings effects on post harvest quality of green bell pepper, *J. Food Sci.*, **61**, 176.

Lownds, N. K. and Bosland, P. W. (1988) Studies on post harvest storage of pepper fruits, *Hort. Science*, **23**, 71.

Lownds, N. K., Banaras, M. and Bosland, P. W. (1994) Post harvest water loss and storage quality of nine pepper (capsicum) cultivars, *Hort. Science*, **29**(3), 191.

Luning, P. A., Yuksel, D., Vuurst-de-vries-R, Van der, and Roozen, J. P. (1995) Aroma changes in fresh peppers (*C. annuum*) after hot air drying, *J.Food Sci.*, **60**(6), 1269.

Luo, Y. and Mikitzel, L. J. (1996) Extension of post harvest life of bell peppers with low oxygen, *J.Sci. Food Agric.*, **70**(1), 115.

Marshall, D. E. and Brook R. C. (1999) Reducing Bell pepper bruicing during post harvest handling, *Hort. Technology*, **9**(2), 254.

Mathew, A. G., Lewis, Y. S., Jagadishan, R., Namboodri, E. S. and Krishnamurthy, N. (1971) Oleoresin from capsicum, *Flavour Industry*, **2**(1), 23.

McColloch, L. P. (1962) Chilling injury and alternaria rot of bell peppers, *US Dept., Agric. Market Res. Rept.*, 536, 16.

McColloch, L. P. and Wright, W. R. (1966) Botrytis rot of bell peppers. *US Dept. Agric. Market Res. Rept.*, 754, 9.

Meir, S., Rosenberger, I., Aharon, Z., Grinberg, S. and Fallik, E. (1995) Improvement of post harvest keeping quality and colour development of bell pepper (cv. Maor) by packaging with polyethylene bags at a reduced temperature, *Post Harvest Biology and Technology*, **5**(4), 203.

Miller, W. R., Risse L. A. and McDonald R. E. (1986) Deterioration of individually wrapped and non-wrapped bell peppers during long term-storage, *Trop. Sci.*, 63, 1.

Mosquera Minguez, M. I., Jaren Galan, M. and Garsido Fernandez, J. (1994) Competition between the process of biosynthesis and degradation of carotenoids during the drying of peppers, *J. Agric. Food Chem.* 42(3), 645.

Onyenkwe, P. C. and Ogbade, G. H. (1995) Radiation sterilization of red chilli pepper (C. frutescens), *J. Food Biochem.*, 19(2), 121.

Pantastico, E. B., Chattopadhyay, T. K. and Subramanyam, H. (1975) Storage and commercial storage operations. In: Pantastico (ed.) *Post Harvest Physiology, Handling and Utilization of Tropical and Subtropical Fruits and Vegetables*, The AVI Publishing Co. Inc., Westport, CT. p. 314.

Papalkar, J. S. P., Sayed, H. M. and Kulkarni, D. N. (1992) Physico chemical qualities of some varieties of chilli, *Indian Cocoa, Arecanut and Spices Journal*, 15(3), 76.

Peter, K. V. (1998) Recent advances in chilli breeding, *Indian Spices*, 35(3), 3.

Prevention of Food Adulteration Act, 1954 and Rules 1955, Ministry of Health and Family Welfare, Government of India, New Delhi.

Pruthi, J. S. (1998) *Chillies or Capsicums* In: *Major Species of India – Crop, Management and Post Harvest*, ICAR, New Delhi, p. 221

Priyasethu, K. M., Prabha T. N. and Tharananthan R. N. (1996) Post harvest biochemical changes associated with softening phenomenon in *Capsicum annuum* fruits, *Phytochemistry*, 42(4), 961.

Rajput, J. C. and Parulekar, Y. R. (1998) "Capsicum". In: D. K. Salunkhe and S. S. Kadam (ed.), *Hand Book of Vegetable Science & Technology – Production, Composition, Storage and Processing*, Marcel Dekker, New York, p. 203.

Ramesh, A. (1989) Equipments for processing of spices, In: *Proceedings of seminar on "Recent trends and developments in post harvest technologies for spices"*, 24th August 1989, CFTRI, Mysore, India, p. 39.

Russo, T. R. (1976) Cryogenic grinding "Carousal" materials handling, *Food Eng. Int.*, 1(8), 33.

Tandon, G. L., Dravid, S. V. and Siddappa, G. S. (1964) Oleoresin capsicum (red chillies) – some technological and chemical aspects, *J. Food Sci.*, 29, 1.

Todd, E. C. (1957) New extraction process improves paprika colour, *Food Eng.*, 29, 81–82.

Vinkler, M. (1973) Changes in pigment content and pigment composition during the storage of intact peppers, konzerv-es, Paprikaipar, 3, 99, *Food Sci. Technol. Abstr.*, 5, 3J425.

Wealth of India (2000) *A Dictionary of Indian Raw Materials and Industrial Products, First Supplement Series (Raw Materials)*, Vol. 1, A-Ci; National Institute of Science Communication, Council of Scientific and Industrial Research, New Delhi, p. 212.

Wistreich, H. E. and Schaffer, W. F. (1962) Freeze grinding versus product quality, *Food Engineering*, May, p. 62.

11 Advances in post-harvest processing technologies of *Capsicum*

J. S. Pruthi

Capsicums, like all other agricultural commodities, contain a high moisture content and vary considerably in shape, texture, size, colour and pungency. Hence, their pre-treatments – curing, cleaning and methods of processing – also vary to some extent. During their post-harvest processing, they are subjected to different types of unit operations such as washing, curing, drying, clearing, grading and packaging, until they are ready for the consumer or for the terminal market. Such post-harvest processing technology should ensure the proper conservation of the basic qualities of the chillies for which they are valued, namely aroma, flavour, pungency or bite factor and colour. All these aspects are briefly covered in this chapter.

Introduction

Capsicum or chilli belongs to the genus *Capsicum* and the family Solanaceae. There is some confusion in the nomenclature of *Capsicum* species. Although the following five species have been globally recognized, *Capsicum annuum* continues to be the most popular and commercially most important species of the genus *Capsicum*, the next being *C. frutescens* (Anu and Peter, 2000).

Species of Genus Capsicum	*Common English names*
Capsicum annuum	Chillies, *Capsicum*, Red pepper, Paprika, Cayenne
Capsicum frutescens	pepper, Bell pepper, Bird chillies or Tabasco pepper
Capsicum baccatum var *Pendulum*	
Capsicum chinense	
Capsicum pubescens	

This classification is exclusively based on chilli fruit characteristics like colour, shape, size, pungency and end uses. According to this classification, there are five major groups under *C. annuum*, as tabulated above, and only one group in *C. frutescens* (Purseglove *et al.*, 1981). The paprika of commerce comes from the following important producing, processing and exporting countries: (1) Spain; (2) Hungary; (3) Morocco; (4) Bulgaria; (5) US; (6) Yugoslavia; (7) The Czech Republic; (8) Romania and (9) Portugal.

Economic importance

India is one of the major chilli producing countries of the world, producing about 1,00,000 tonnes annually. Chilli is also the second largest export earner for India, being next to black pepper. Thus, during the year 1999–2000, India exported about 64,776 tonnes of chillies (Anon., 2000, Spices Board, Govt. of India). However bird chilli is not grown commercially at present in India but it is in great demand in several other countries because of its high pungency (1.0% to 2.0%). Bird chilli is very small in size and has bright red colour. Looking to its great potential in world trade, it is now being cultivated in some rates in India (Anu and Peter, 2000).

Post-harvest technology

Introduction

It is the prime requisite that all *Capsicum* species are harvested at the correct stage of maturity without much physical damage, after which they are processed properly for the market. In most cases, they are sun dried at the farm and transported to an appropriate centre for further processing. *Capsicums* or chillies, like all other agricultural commodities, invariably contain a high moisture content (60–85%) at the time of harvest, which must be brought down to 8–12% moisture. Furthermore, chillies of different varieties vary considerably in shape, texture, size, colour and pungency. Hence, the pre-treatments, curing, cleaning and methods of processing also vary to some extent. During their post-harvest processing, they are subjected to different types of unit operations such as washing, curing, drying, clearing, grading and packaging, until they are ready for the consumer or for the terminal market. Such post-harvest processing technology should ensure proper conservation of the basic qualities of the chillies namely, aroma, flavour, pungency or bite factor and colour, for which they are valued. These post-harvest unit operations are collectively termed as post-harvest technology. These aspects are briefly covered: by Pruthi (1980, 1991, 1992).

Pre-treatments

Correct harvesting and curing

Ripe fruits are plucked together with their stalks and cured at the farmers/growers level itself. The harvested produce is heaped indoors for three to four days so that partially ripe fruits, if any, ripen fully and the whole produce develops uniform red colour. Partially ripe fruits, if dried without this curing treatment, develop white patches and such fruits have less market value. However, in some parts of the country, namely the Tarai region of Uttar Pradesh, the drying of green chillies is started on the same day when the crop is harvested. In case the drying operation does not start on the same day, the green chillies have to be spread under the shade (not heaped) overnight, as keeping chillies even for 24 hours causes considerable damage, leading to total spoilage within four days if left in the heap form. In such parts of the country, it is observed that the maximum time lag between harvesting and the storage of chillies should be 24 hours, even when the green chillies are spread. This places a restriction on the transport of green chillies to distant places for marketing or drying. Because of the poor storability of green chillies, farmers themselves dry the chillies. A small fraction of the total produce of chillies in India is dried by the middlemen who purchase the green chillies from the local markets, brought by the farmers on the very day of their harvesting. Baskets and bags are used for the material handling of the chillies (Pruthi, 1993a,b).

The pungency, initial colour and colour-retention properties of chilli fruits are closely related to maturity. Pods left to ripen and to partially wither on plant, are superior in these three qualities to those picked when fully coloured but which are succulent. The use of partially withered ripe pods is the single condition required for high quality, secondary only to the intrinsic properties of the cultivar grown, and has a greater influence on the final product than drying or storage methods. However, it should be noted that care must be taken over the extent of withering permitted prior to harvesting, since, if prolonged, it can sometimes result in a product with a grey colour (Pruthi, 1992).

Pricking

Pricking the skin of chillies longitudinally helps to reduce the total time of drying. It also helps to retain colour and overall quality. The same applies to the *Capsicums*, red pepper or paprika (Laul *et al.*, 1970). ARI has developed a portable pricking machine which has been described in fair detail on p. 186 (Ilyas *et al.*, 1987a,b).

Alkali treatment

The attractive red colour of chillies/*Capsicums*/paprika can be stabilized to a great extent by steeping them in 2% sodium carbonate solution for 10–12 minutes. Such alkali treatment has been found to be useful in the drying of chillies, particularly in conjunction with olive oil (Laul *et al.*, 1970).

Antioxidant treatment

The attractive red colour of *Capsicums* or chillies, which is mostly due to carotenoids (notably capsanthin, capsorubin, etc.), is stabilized to a great extent by treatment with a suitable antioxidant (Van Blaricum and Martin, 1945; Lease and Lease, 1956b). Van Blaricum and Martin (1945) and Lease and Lease (1956a) have also studied the effects of other factors, such as initial composition, light, air, temperature, condition (whole or ground), harvesting practices and pre-drying treatments on the retention of colour in cayenne pepper during drying and storage.

Grading

(a) Importance/Advantages: Standardization and grading is a prerequisite for development of the modern marketing and trade of any commodity in the country and abroad. Grading serves to classify goods according to the end use, and eliminates waste that would otherwise occur. In general, it simplifies the marketing system by making it possible for buyers to procure goods easily that meet their particular requirements. It facilitates easy price comparison among the different markets, thereby giving buyers and sellers better information on which they can base their decisions.

(b) Organizations for standardization and grading: The Indian Standards Institution has considered almost all spices, including chillies, for specifying standards based not only on their physical characteristics but also on other additional chemical factors, such as total ash, acid-insoluble ash, non-volatile ether extract and fibre content. Under the Food Adulteration Act, minimum purity standards are laid down for chillies from the point of health. In addition, there are standards prescribed by the Directorate of Marketing and Inspection (DMI) under the Grading Act 1937, for chillies as briefly discussed next. DMI is the most pioneering organization in grading.

(c) Grading activities of DMI: Grading activities of DMI can be classified into three broad categories, namely (i) compulsory grading under Agmark for export; (ii) grading under Agmark for internal trade or voluntary grading; and (iii) grading at producer's level or commercial grading. These aspects have been dealt with in fair detail by Pruthi (1993a,b) and hence shall not be covered here again.

Sun-drying technique – traditional or conventional

Drying of chillies is mostly done by spreading the fruits on clean, hard dry ground or on a concrete floor under the sun. Mud floor, roof top or wooden cots are also used for this purpose. As mentioned earlier, chilli is a highly perishable material at the time of harvest due to its high moisture content, which is usually 70–80% (wb), whereas the limit of moisture content suitable for the safe storage of chillies is about 10% (wb). Therefore, chillies must be dried in a manner that preserves their characteristic red colour and lusture. Excessive delay in drying results in the growth of micro-flora and subsequent loss of quality or total spoilage. Also, dirt and dust may deposit on the chillies during open-yard sun drying. Moreover, it involves excessive handling, irrecoverable shatter and drying mass, rendering the chillies vulnerable to weather hazards. Usually, the chillies are spread in a single-chilli-thick layer for drying. After two days of drying in this manner, when the fruits are still flaccid, they are trampled upon or are rolled over for flattering. This treatment enables a greater quantity of the dried product to be packed into gunny bags for storage and transport. In many parts of the country, the drying operation takes 3–15 days for the reduction of the moisture content from 70–80% to 10%, depending upon climatic conditions. If there is cloudy weather and there are intermittent rains during the drying period, damages as high as 50% are reported. Such unfavourable conditions also lead to the discolouration of the dried chillies. On an average, about 25–30% of the fresh weight of chillies are produced after successful drying by conventional methods. The manpower requirement for such drying techniques are two persons for every 10 bags. Except for a seed separation of about 10 kg/tonne of drying chillies, no visible damage, colour loss or glossiness is generally reported during drying in good, sunny weather. On the other hand, drying under unfavourable conditions results in loss of colour with white spots over the chilli surface, and glossiness and pungency is also influenced. Unripe chillies are, however, sometimes boiled or blanched and dried for domestic consumption. Chillies are often smeared with oil of mahuwa (*Madhuca longifolia*) to impart glossiness. However, coconut oil and generally sesame seed oil are not used for this purposes, as they lead to mould growth and discolouration (Shivhare *et al.*, 1987).

Chillies or *Capsicums*, on harvesting, have a moisture content of 65–80% depending on whether they were partially dried on the plant or harvested while still succulent; this moisture content must be reduced to 10% in order to prepare the dried spice. Traditionally, this has been achieved by sun drying which remains the most widely used method throughout Asia, Africa, Central and South America. Even in the United States, where artificial drying is practised by virtually all commercial processors, sun drying is still used by many growers of small areas.

In Nagpur (Maharashtra), chillies or *Capsicums* are dried on open spaces, mostly roadsides, without the benefit of mattings or concrete platforms, and remain exposed to weather for the whole of the drying period (14–21 days). Consequently, fruits become contaminated with dust and dirt, damaged by rainfall, animals, birds and insects. Laul *et al.* (1970) have reported that under these conditions, losses can be as high as 70–80% of the total quantity of spice taken for drying.

In the southern United States, small growers sun-dry chillies or *Capsicums* by. spreading the fruits on drying racks, on roofs and even on the ground sometimes they may be tied together into bundler weighing 5–12 kg and hung on the walls of houses, fences and even on clothes lines. This method of drying may take several weeks (Pruthi, 1993a,b, 1998).

In Japan, the sun drying of chillies is followed by mechanical drying, dressing and then more sun drying before packing. The entire stem or plant with the fruits is cut and hung from bars and then exposed to the sun for partial sun drying. After the removal of 80% moisture, the parts are further dried in a dryer to the critical moisture level. The dried parts are removed with the help of a dressing machine and finally dried in the sun again before packing. This efficient combination of sun drying and dehydration may work well in such cases. The main problem is the heterogeneity in the product with reference to size, colour and quality. Grading may help in getting better results. Improved agronomic practices are also required to get a more uniform product at picking time (Pruthi, 1980).

Sun drying of chillies has also been studied in some detail in India. Whole chillies spread on perforated rectangular aluminium trays (5 kg/m) took about 15 days to dry at a room temperature of 20–25°C and a relative humidity of 34–50%. Pricking the chillies longitudinally reduced the drying time to 12 days, and blanching reduced it to 7 days. In the case of 'checking', the drying time was seven days after dipping in a 2.5% potassium carbonate solution, and it was further reduced to six days in the presence of deodorized olive oil (Laul *et al.*, 1970). Shivhare *et al.* (1987) have described the conventional processing methods of chillies and have also reviewed the research and developmental studies carried out in India and abroad to develop suitable post-harvest technologies for this important cash crop.

Disadvantages of traditional sun drying

The traditional methods of harvesting and sun drying chillies or *Capsicums*, outlined above, have many disadvantages, not least in regard to product quality. Poor handling of fruits prior to drying can result in bruising and splitting. Bruising shows up as discoloured spots on the pods and splitting leads to an excessive amount of loose seeds in a consignment; there is a considerable loss in weight if dried fruits are sold without seeds. Neglect in provision of adequate drying platforms and protection from rain and pests can result in high losses and contamination. If the fruits are not dried properly, they lose their colour, glossiness and pungency (Pruthi, 1992, 1998).

Improved CFTRI method of sun drying

Laul *et al.* (1970) have investigated ways of improving sun drying methods for Indian chillies or *Capsicums*. Drying in a single- and multi-tier tray system, in sun and in shade, and also blanching and pricking pre-treatments were studied. A four-tier system of wire-mesh trays and a single tray of perforated aluminium both took 14 days in the sun to dry fruits having a moisture content of 72–74%, reducing it to about 6%, the normal Indian commercial traditional sun drying methods takes about three weeks to achieve a moisture level of 15–20%. Primary processing of chillies essentially consists of drying and despiking. Better retention of colour and a higher yield of finished product, avoiding breakage of pods and loss of seeds are achieved by adopting CFTRI improved technology for sun drying of chillies over traditional methods.

The improved method essentially consists of dipping fresh chillies in 'Dipsol' for five minutes and then drying on racks that have multi-tier wire net trays (Central Food Technological Research Institute, CFTRI, 1979).

The advantages of the improved CFTRI technology

1   The rate of drying is fast; the time needed for drying is only a week as compared to 15–21 days in the traditional method of sun drying.
2   Requires less space.
3   Helps in better retention of colour and pungency.
4   Gives a more hygienic and superior quality product.
5   Gives higher yield of finished product (2% more) due to minimum breakage and less loss of seed.
6   The improved method adds extra material cost to finished product i.e., less costly.

The field demonstrations, using the 'Bhivapur' variety of chillies, have shown that red chillies sun dried by the improved method brings an additional economic benefit.

The improved technology know-how (CFTRI)

1   Dip fresh chillies in an emulsion (Dipsol) for a short time (approximately five minutes).
2   Drain the excess emulsion.
3   Spread the chillies on to racks having multi-tier wire net trays at the rate of 5–10 kg/m^2 of tray area depending on ambient temperature.

The treated materials dry to the commercial level moisture content in about a week's time.

Preparation of the emulsion 'Dipsol'

'Dipsol' is a water-based emulsion containing potassium carbonate (2.5%), refined groundnut oil (1%), gum acacia (0.1%) and butylated hydroxy anisole (BHA) (0.001). Thus, 100 kg of 'Dipsol' contains the follows ingredients:

Potassium carbonate:	2.5 kg
Refined groundnut oil:	1.0 kg
Gum acacia:	0.1 kg
BHA:	0.001 kg

Dissolve the potassium carbonate and gum acacia separately in water. Dissolve the BHA in refined groundnut oil. Mix water-phase solutions and BHA dissolved in groundnut oil slowly while stirring. The mixture is passed through a homogenizer twice at 200 kg/cm^2 (Laul *et al.*, 1970). Fifteen litres of emulsion are required to treat 100 kg of fresh chillies. A scheme for sun drying chillies by this improved technique is available from CFTRI, Mysore (CFTRI, 1979).

The Pantnagar drying technique

The Pantnagar Centre of All-India Co-ordinated Post-Harvest Technology Scheme of Indian Council for Agricultural Research (ICAR) has investigated drying characteristics of chilli (variety 'Pant C-1'), both for sun drying and hot-air drying. Studies have been conducted for evaluating the performance of cement concrete floor, with black tar coating, a wire-mesh elevated to 0.5–1.0 m high stand, with the thickness of the chilli bed varying between 1.11 and 1.19 cm. Heated air drying studies were conducted with and without stalks at two hot air temperatures (323°K and 328°K) and at three different air velocities (1.00, 1.25 and 1.50 m/s) at 10 cm bed thickness. Drying rates were computed using the numerical differentiation technique. Concrete floor with tar coating was reported to be the best surface for sun drying as it gave higher drying rates. The total average time to reduce the moisture of chillies to 5% (db) was about 20 h of sun exposure for the concrete floor with tar coating and 37 h of sun exposure for elevated wire-mesh platforms. It was observed that about 1% of seeds got separated during the sun drying process. In heated air drying, with the air at 328°K and 1.5 m/s velocity, chillies without stalks achieved maximum drying rates with 12 h required to reduce the moisture from 55.4–16.5% (bd). Chillies with stalks dried at a slower rate. In another similar study at two air temperatures of 50°C and 52.5°C and at air velocity of 1.5 m/s, it was observed that at 50°C air temperature, the moisture content of chillies with stalks was reduced from 76.4% to 50.36% in 11 h. It was concluded that a drying temperature of 50°C and air velocity of 1.5 m/s could be satisfactorily used for the mechanical drying of chillies without influencing the pungency or quality of chillies.

Bhopal (M.P.) sun-drying trials: The Bhopal Centre of All-India Co-ordinated PHT Scheme has also conducted drying studies for chillies on seven different surfaces, namely: mud floor, concrete floor, jute cloth, white canvas, tarpaulin, black-and-white polyethylene sheets and black poly-ethylene.; A saving of 21% time was achieved with tarpaulin compared to the mud floor (Singh and Alam, 1982).

Solar-drying

Recently, attempts have been made to develop solar equipment to improve upon sun-drying techniques, which lead to the following advantages:

(a) better use of available solar radiation;
(b) reduction in drying time; and
(c) cleaner and better quality product, free from dust, dirt and insect infestation.

This equipment is called 'Solar Drier'. The Regional Research Laboratory (RRL), Jammu, India has devised a solar drier for drying chillies in Jammu and Kashmir state (RRL, 1978). Red chillies of Kashmir are very popular throughout the country, as they impart an attractive bright red colour to dishes. Chillies are produced in substantial quantity in Kashmir valley and it is a common scene to find chillies strung together in thread and hung on walls and doors or spread on roof tops. Commercially, plants with fruits still unplucked are harvested and spread out on the ground for about a week for partial drying. Thereafter, the fruits are plucked by hand and spread in the field for final drying. The entire operation takes about 15 days, during which, chillies are exposed to dirt, dust, fungus attack, all of which are in addition to the uneven drying of chillies.

A solar drier has been installed near Pampore in Jammu and Kashmir, which effects complete drying of the commodity in four to five days, with a marked improvement in colour and storage characteristics. The gadget is very simple and is made of mud, stone, pebbles and glass panes and is specially suited for rural areas. It can be conveniently constructed by village artisans. A unit of 2.5 × 2.5 m size can dry 80 kg chillies per batch.

Types of solar driers

Encouraged with the results of sun drying studies, the following solar driers (Figures 11.1 and 11.2) have been designed and developed in India for the sun drying of chilli (Kachru and Srivastava, 1990a,b).

1   *CAZRI solar drier:* The Central Arid Zone Research Institute, Jodhpur Centre of the PHT Section Scheme in India is reported to have developed a solar drier using low-cost materials for construction. A PVC sheet is used as a transparent cover for the drier, providing a green-house effect. Bajra-husk is used as an insulator while other materials include bamboo and MS sheet. The drier is furnished with an aluminium chimney which is painted black to serve as an exit for hot air. The performance evaluation of this drier during the winter sea-son (at Jodhpur) showed that the drier took nine days to reduce the moisture content of chillies from 83.6% to 3.5%, in comparison to the 21 days required to remove moisture from the same quantity of chillies using the open drying method.

2   *Bhopal solar-cabinet drier:* The Central Institute of Agricultural Engineering, Bhopal, India has developed a solar-cabinet drier (Figure 11.1) for the drying of perishable, semi-perishable and wet-processed food materials.

The drier is simple in design and does not require any mechanical prime mover or elec-tricity. It is free from fire hazard and can be fabricated from materials such as wood, glass,

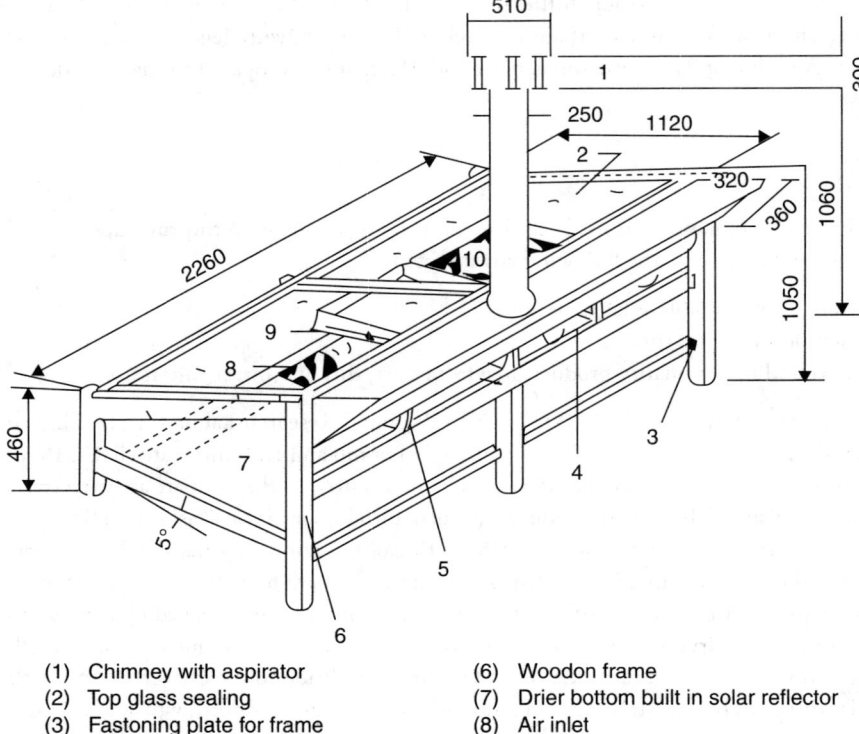

(1) Chimney with aspirator
(2) Top glass sealing
(3) Fastoning plate for frame
(4) Sliding rack
(5) Vertical support for glass sealing
(6) Woodon frame
(7) Drier bottom built in solar reflector
(8) Air inlet
(9) T-type guide for rack
(10) Rack bottom built in wire mesh

Figure 11.1 Solar-cabinet drier (Central Int. Agri. Engng, Bhopal, M.P. (India)).

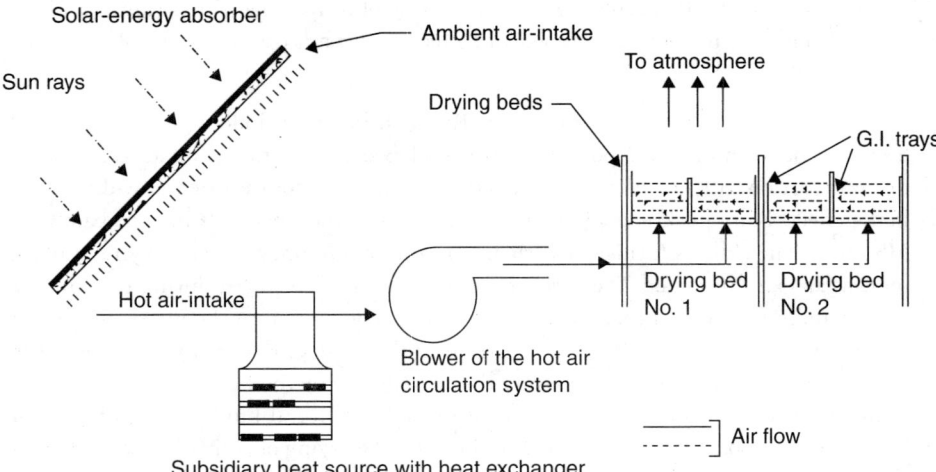

Figure 11.2 A schematic drawing of the improved solar drier for spices (Kachru and Srivastava, 1990a,b).

plywood, wire-mesh and a MS sheet. Specifications, test results of the drier and economics for chilli drying by this equipment are as follows:

(a) Specification for the Bhopal solar drier
   Type: Tray type allowing natural convection
   Overall dimensions (size):   2,260 × 1,440 × 2,410 mm
(b) Test results
   Suitability for crops: Chillies, potato chips/cubes, cauliflower and
                                  leafy vegetables
   Capacity for-solar drying: 30–50 kg per batch
   Labour requirements: 2–3 man-hours/day
   Total drying time: 4–7 days/batch

With the extensive use of such solar driers, sizeable quantities of red chillies and other dried vegetables of improved quality can be produced in rural areas.

Another improved partially mechanized solar drier is illustrated in Figure 11.2.

3   *PKV Akola waste-fired solar drier*: The Akola Centre Scheme has modified a waste-fired drier, developed by the same centre for drying different agricultural commodities (e.g. sorghum cobs, grains and pods) for the artificial drying of chillies.

The modified drier is reported to be suitable for drying 200 kg of fresh red chillies per batch within a total period of 16 h by reducing the moisture content of chillies to 16.5% (wb) from an initial moisture content of 73.40% (wb). The drier is connected to a blower operated by 1 hp electric motor.

Artificial drying or mechanical drying or dehydration of *Capsicums*

In order to avoid dependence on the vagaries of weather and also to reduce microbial contamination, natural convection driers or forced-draft driers can advantageously be used to get a better quality product. An air temperature of 60°C gives a drying time of about 5–9 h, depending on the commodity, particle size, thickness of the layer and method of drying (Pruthi, 1980).

Artificial drying has the advantage over traditional sun drying in producing a better and more consistent quality product, taking less time and minimizing crop losses. Artificial drying has been used for many years in the United States for the pungent forms of *C. annuum* grown in North and South Carolina and Louisiana (finger peppers, which may be 10–30 cm in length), and for the very pungent tabasco chillies (from *C. frutescens*), which are also grown in Louisiana. In South Carolina, tobacco-barns have often been used for this purpose.

Commercial processors in the United States bring harvested fruits to drying centres in bulk, after which they are washed, inspected and spread out on trays, either as whole pods or sliced into 2.5 cm lengths. The fruits are dried in heated buildings or, more usually, in tunnel driers or stainless steel continuous-belt or belt-trough driers, exposing fruits to a forced current of air at temperatures of 50–60°C, thereby reducing their moisture content to 7–8%. Some processors use a two-stage method, first drying fruits to a moisture content of 12–20% and then storing at 0°C; when required for grinding the drying is continued until pods contain 7–8% moisture or less. A flow diagram for optimum dehydration of chillies is presented in Figure 11.3.

Factors affecting the quality of the dehydrated product

There are several factors affecting the quality of the dehydrated product, such as (i) quality and nature of the raw material (Lewis and Natarajan, 1980); (ii) method of preparation (whole, sliced or dried); (iii) pre-treatments like pricking, alkali treatment or antioxidant and treatment with other

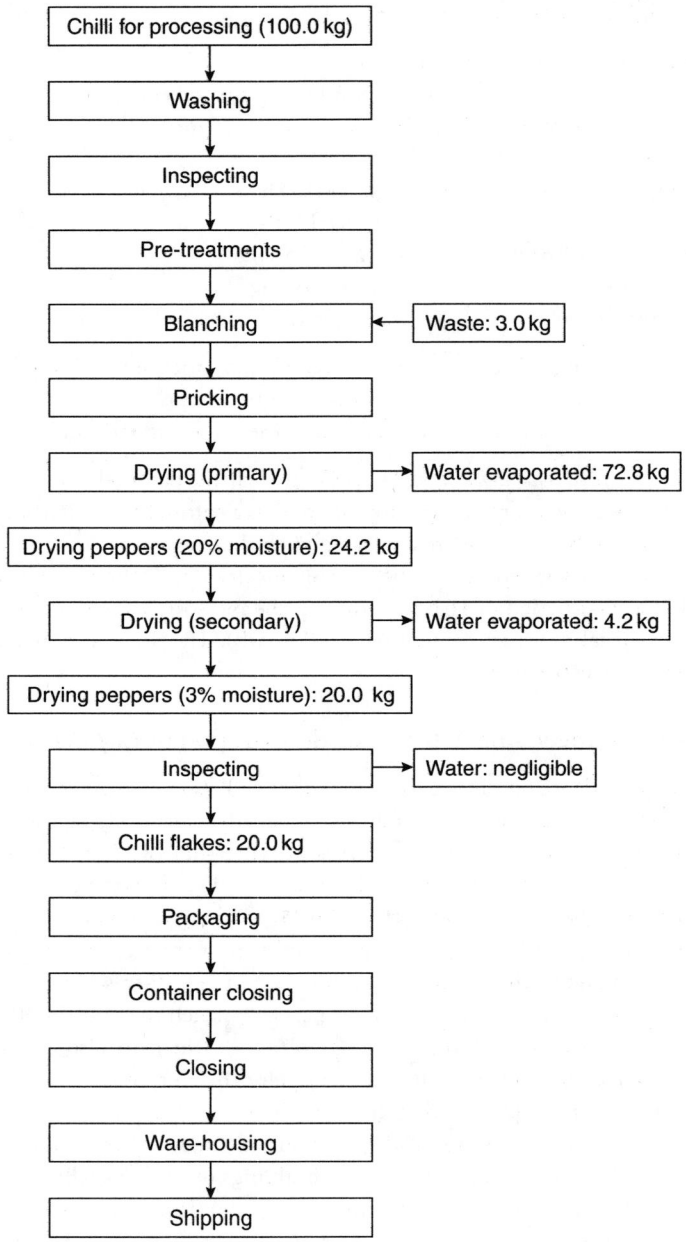

Figure 11.3 Flow-sheet on optimum artificial drying/dehydration of chillies (USA – Commercial Process).

chemicals; (iv) density of lading; (v) the time, temperature and method of dehydration; and (vi) critical temperature of dehydration. Of these, the temperature of the air used during dehydration processes greatly affects both the total time taken for dehydration and the quality of the finished or dehydrated product (Pruthi and Lakshmi Shankar, 1968; Pruthi *et al.*, 1967).

Optimum conditions for artificial drying/dehydration

Artificial drying has advantages over traditional sun drying in that the product is more consistent and of a better quality, the time taken is less and crop losses are also minimized (on the basis of 100 kg fresh red pepper/chillies, the shrinkage ratio is 5 to 1). Lantz (1946) reported that slicing pods reduced the drying time by half and that there was no loss of initial colour on drying for 72 hours at 60–75°C. Lease and Lease (1962) carried out a more comprehensive study of the effects of forced air-drying on Carolina hot pepper (from *C. annuum*). The drying of sliced pods was found to have considerable advantages over whole pods; the drying time was much reduced (by 50%) and a superior initial colour was obtained. The optimum drying temperature for a good quality product was found to be 65°C; single layers of succulent whole fruits dried to 8% moisture content in 12 hours, and in 6 hours when in sliced form. Pods harvested when partially dried (withered) on the plant took even shorter time for curing and exhibited even better quality characteristics. Extended drying was considered permissible at 50°C and at 65°C, but at 80 °C, a lowering of pungency, of initial colour and colour-retention properties was encountered (Figure 11.3).

In 1961 the CFTRI suggested that chillies could be dried in a forced draught of 2,500 cubic feet (71 m³) per minute at 60–70 °C; tea-driers with appropriate modifications are suitable. However, freshly harvested chillies could be dried on the same day in thin layers at lower temperatures. Misra (1972) reported that chillies dried at 60°C turned black and lost part of their pungency and glossiness; he recommended a range of temperature 45–50°C, as suitable.

Dehydrated paprika, Capsicums/*bell peppers*/*chillies* – *a global overview*

While studying the effect of canning and drying on the carotene and ascorbic acid content of chilli, Lantz (1946) found that slicing or slitting the pods of chilli reduced drying time by half and that there was no loss of initial colour on drying for 72 hours at 140°F to 168°F. Lease and Lease (1956a) reported that the colour retention in peppers is affected by the stage of ripeness at harvest. By adding the antioxidant BHA to the ground pepper, the retention of colour is improved (Lease and Lease, 1956b). While studying the effect of drying conditions on initial colour, colour retention, and pungency of red peppers, Lease and Lease (1962) concluded the following:

(1) Drying or curing the sliced pods of Carolina hot pepper at 150°F was optimum for quality.
(2) The BHA antioxidant, when added, markedly improved the colour retention of cured pods and also of freshly harvested pods.
(3) Drying at 70°C led to a lower initial colour, colour retention and pungency.
(4) No correlation was found between the moisture content and the above quality factors.
(5) Aired peppers retained more colour when stored whole than when ground following curing. However, the colour of whole, cured pods was lost on grinding.
(6) Another variety, Louisiana Sports, lost colour faster than did the Carolina variety.
(7) Autoclaving pepper before drying increased the rate of colour loss and suggested that colour breakdown could not be attributed directly to enzyme activities.

Daoud and Luh (1967) reported that the colour of red bell peppers is preserved by freeze drying without blanching. Temperatures of less than 20°C are best for quality retention. The loss in aroma and flavour at 30°C is due to deteriorative chemical changes and enzymatic reactions for example, the formation of brown water-soluble pigments, the formation of cysteic acid and taurine from cysteine and the loss of amino acids and carotenoids.

Chen and Gujmania (1968) showed that the deterioration of extractable colour pigments of dehydrated ground chilli peppers during storage was due to an autoxidative process having the kinetics of a second-order reaction. Consequently, the reaction rate constant, K_2, was used to evaluate the effect of a number of variables, such as moisture content, storage atmosphere and ethoxyquin treatment. It also provided a means for comparing the relative colour stability of different pepper varieties.

De la Mar and Francis (1969) studied carotenoid degradation in bleached paprika. Nearly 96% of the total extractable pigments expressed as β-carotene were lost during sunlight bleaching of paprika, of which 33 and 21, respectively, were definitely or tentatively identified. Sixteen known pigments were found in both samples. Pruthi (1969) also reported the bleaching or degradation of colour (capsanthin and capsorubin) in Hungarian paprika powder during storage.

Phillip and Francis (1971a,b) reported on the isolation and chemical properties of capsanthin and derivatives, as well as on the nature of fatty acids and capsanthin esters in paprika. Rosebrook (1971) reported a method for determining the extractable colour in paprika and paprika oleoresin.

Chilli pricking or punching machine to hasten drying

Drying operations of chillies can be made more economical by reducing the drying time through punching the chillies. It is reported that in the manual method of punching with pins, the operators are unable to do the punching job continuously as they experience fatigue and pain in the hands and eyes. Besides, the output is also very low. Realizing the importance of this technique, a chilli-punching machine has been developed at the Indian Agricultural Research Institute, IARI, New Delhi, India (Figure 11.4), which is capable of punching

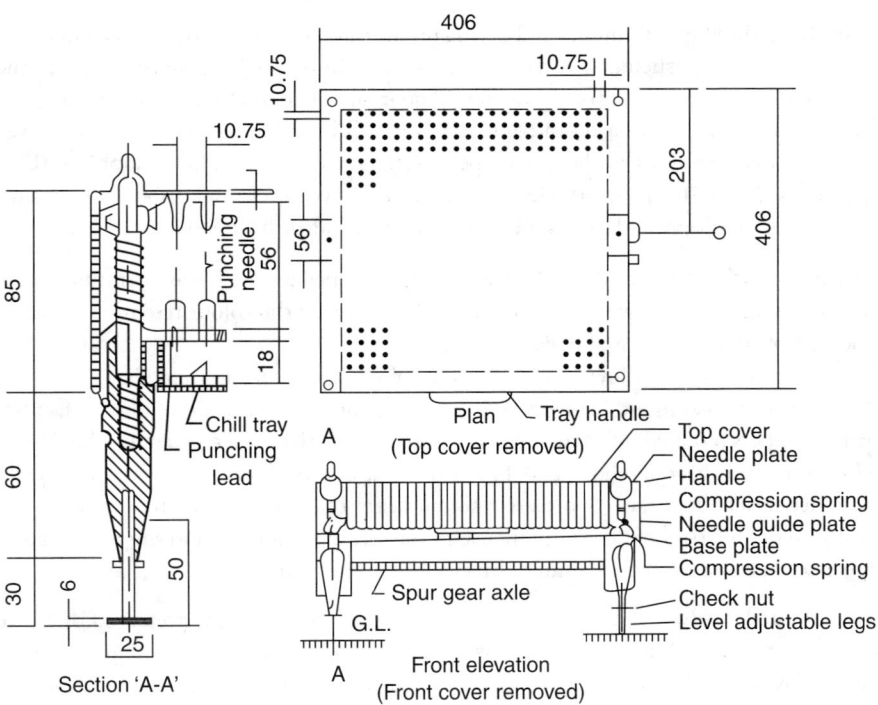

Figure 11.4 IARI chilli pricking or punching machine (courtesy: Ilyas *et al.*, 1987a,b).

Table 11.1 Specifications of the punching machine

Number of needles:	1,024
Number of rows:	32
Machine size:	0.406 × 0.406 × 0.20 m
Tray size:	0.40 × 0.40 × 0.01 m
Batch capacity:	0.25 kg
Diameter of needle:	1 mm
Length of needle:	56 mm

Source: Ilyas *et al.* (1987a,b).

chillies at the rate of 10 kg/h as compared to 1 kg/h by manual method (Ilyas *et al.*, 1987a,b). The IARI machine is portable and easy to handle with a provision for adjusting the gap between needles and tray.

The specifications of the punching machine are given in Table 11.1. Four punching operations made, on an average, 20 holes per chilli and the holes punched by the machine are reported to be uniformly spread over the entire surface area of the chilli.

Chillies seed extractor

The Tamil Nadu Agricultural University, Coimbatore, India, has developed an equipment for the extraction of seeds from dried chillies (Figure 11.5).

The dried chilli fruits are cut into small pieces and the seeds are separated from the fruit. This seed extractor is a continuous type which can be easily operated by a farmer. Scorching and pungent smells endured by labourers in a conventional practice could be eliminated by using this machine. The specifications, test results and economics of this seed extractor are given in Table 11.2.

Storage of chilli seeds

The storability of chilli seed has been studied in relation to packing material and storage period in three types of packages, namely cloth bag, polyethylene bag (300 gauge) and moulded plastic jar at room temperature. The poly bags were heat-sealed and plastic jars were closed with tight fitting lids. The storability of seeds in various packages was studied by evaluating the moisture content on a wet basis and the percentage of germination after one month, three months, six months, nine months and one year of storage. The changes in the moisture content of seeds packed in cloth bags were reported to be more abrupt after a storage period of three months. However, the deviation in germination values, because of packages, was not found to be statistically significant. Seeds stored in polyethylene bags and moulded jars maintained a better quality than those stored in ordinary cloth bags. The safe moisture content for chilli seeds for yielding a minimum 60% germination has been reported to be 7.5%.

The GBPUAT (G. B. Pant Univ. Agri. and Technology), Pantnagar, India has however, reported cases of severe infestation of chilli seeds kept in polyethylene bags with *Lasioderma* spp., where 60% of the seeds were damaged, turning some of them into floury dust.

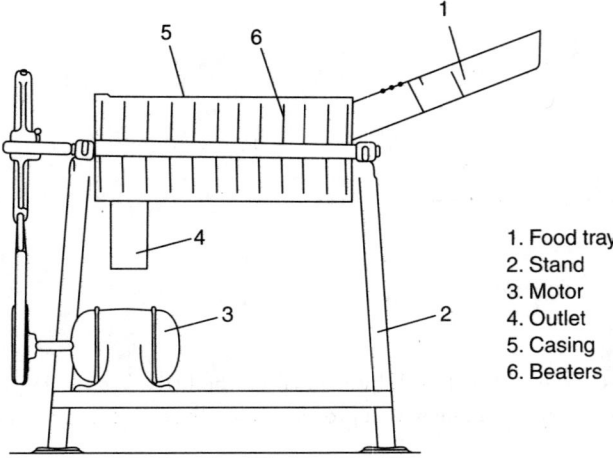

Figure 11.5 Chilli seed extractor.

Source: Tamil Nadu Agri. Univ., Coimbatore, India.

Table 11.2 Specifications of chilli seed extractor

(a) Specifications	
Type:	Power operated, continuous
Overall dimensions:	1,090 × 580 × 910 mm
(b) Test Results	
Suitability for crops:	Extraction of seed from dried chillies
Capacity:	4 quintal dried fruits/day of eight hours
Power requirement:	0.5 hp electric motor
Labour requirement:	One

Source: Tamil Nadu Agri. Univ., Coimbatore, India.

Cleaning, sorting, packaging and storage

After drying or curing of the fruits, they are cleaned of extraneous matter, and of damaged and discoloured pods, before their storage or packaging. Dried chillies are usually stored in gunny bags. Each bag contains 25 kg of dried chillies. The storage at the farmer's level is essentially a short-duration storage ranging from one to three months. Storage at the trader's level is also in bags, which are stacked in godowns where they are fumigated to prevent insect infestation.

Preparation of the dried chillies for the market

Preparation of the produce for sale in the market is an important operation. The condition and the quality of the produce in which it reaches the market greatly influences its price. It is, therefore, of utmost importance that the different operations such as harvesting and drying are performed in a manner which will reduce losses to the minimum and ensure the arrival of the produce to the market in good condition.

The trade usually classifies chillies into three or four grades depending upon the purity, colour and quality of the produce. In some markets, thickness of the skin, pungency and seed content are taken into consideration when classifying the produce. The main quality factors in chillies are (i) colour, (ii) size, (iii) shape, (iv) seed content, (v) pungency, (vi) presence of dirt and other foreign matter, (vii) extent of damage and (viii) moisture content.

No grading is practised by the producers. They do, however, remove discoloured, damaged and rotten chillies during the drying process. Village merchants do not grade either. On the other hand, they adulterate the produce. The wholesale merchants generally discard damaged and discoloured chillies before selling the produce or sending it out to the markets.

Chillies are usually packed in gunny bags. 'Toddy mats' are used in Andhra Pradesh and 'Andkas' in Maharashtra, India. Green chillies are sometimes brought to market in baskets made of split-bamboos and wicker.

Storage studies on dried chillies

The GBPUAT, Pantnagar, conducted some entomological studies on chillies stored for household consumption and also as seed for the next crop, with a view to investigate the possibilities of the development of *Sitophilus oryzae* (L.) and *Rhizopertha dominica* (F.) on dried chillies. It was concluded that both of the pests are unable to develop on dried chillies due to their pungent odour and low moisture level. Studies conducted by this university on the biochemical properties of chilli affected by storage have revealed that the capsaicin content decreases by 60–70% in the case of chillies dried at different conditions and temperature and stored in non-airtight packages for one year in comparison to sealed samples which maintained 0.4% capsaicin content. Samples procured from the local market had a capsaicin content which was 50% less than those in the sealed samples stored for one year, indicating that the effect of storage conditions and the packaging is more pronounced than the temperature at which the chillies are dried. This implies that chillies should be stored in airtight packages.

Disinfestation of whole dried chillies with gamma irradiation

Whole and ground chillies infected with microbes, as well as four species of well-known storage insects, were exposed to cobalt-60 gamma irradiation at 10 kGy. Storage for three months at an ambient temperature (28–30°C) indicated that this radiation dose of 10 kGy is effective for microbial decontamination of chillies as well as to destroy insect pests (Desai *et al.*, 1987). However, irradiation of food products is a highly skilled operation only to be performed by licensed organizations. It is necessary to check whether food products have been irradiated or not. Heide and Bogal (1987) have measured the thermoluminescence and chemiluminescence intensities of chilli samples to determine whether their techniques could be used to detect if the food product has been irradiated (10 kGy) and stored after extraction or not.

Packaging and storage of green chillies

The shelf-life of fresh green chillies is reported to be only three days at room conditions and 9–10 days in cold storage conditions. It can be increased to five to six days and 14 to 15 days, respectively, when the green chillies are packed in 200-gauge polyethylene bags.

Composition and quality of dried chillies

The comparative composition of Indian green and red chillies, paparika and American peppers are given in Table 11.3 (Pruthi, 2001).

Quality attributes of paprika

The varieties of *C. annuum* which are used to produce paprika pods in one growing area may differ in shape and appearance to those in other areas. Some have a fairly round shape with a pointed end; others are elongated. In general, they are medium to small, and quite fleshy. They grow on small and bushy plants. When ripe, they are picked and spread out to dry naturally, or dehydrated in specially constructed tunnels, depending on the area, where they are produced. It is a unique fact in the spice industry that paprika is always developed into a ground product. It is processed into a powder wherever it is grown, whether overseas or domestically. Conversely, most other spices are shipped from the source in their whole form.

Paprika peppers are selectively bred for their colour and flavour. Their breeding can be further controlled to a certain extent by the way in which harvested pods are processed. The seeds and veins have a negligible red colour, therefore, the more of these materials that the processor removes, the more intense the red colour of the ground product. Removal of vein material may also affect pungency, since the pungency present in paprika pods originates in its veins or placenta.

Table 11.3 Composition of chillies, paprika, red pepper and Indian green pepper

Physico-chemical characteristics	Indian green chilli (fresh)	Indian red chilli (dry)	Indian paprika (dried)	American pepper (chilli)	American red pepper
Moisture (%)	82.6	10.0	7.9	6.5	6.2
Protein (%)	2.9	15.9	13.8	14.0	16.0
Fat (%)	0.6	6.2	10.4	14.1	15.5
Fibre (%)	6.8	30.2	19.2	15.6	26.0
Carbohydrates (%)	6.1	31.6	41.1	42.6	28.0
Total ash (%)	1.0	6.1	7.6	7.2	8.0
Calcium (%)	0.03	0.16	0.2	0.1	0.1
Phosphorus (%)	0.08	0.37	0.30	0.32	0.32
Iron (%)	1.20	2.30	0.23	0.01	0.01
Sodium (%)	—	—	0.02	0.01	0.01
Potassium (%)	—	—	2.40	2.10	2.10
Thiamine (mg/100 g)	—	—	0.60	0.59	0.52
Riboflavin (mg/100 g)	1.18	—	1.36	1.66	0.93
Niacin (mg/100 g)	0.5	—	15.3	14.20	13.60
Ascorbic acid (mg/100 g)	111	50	58.8	63.70	29.41
Vitamin A (IU/100 g)	54	576	4,915	6,165	3,530
Caloric value (cal/100 g)	41	246	390	415	420

Source: Pruthi (2001).

One of the paprika's most interesting quality attributes is its content of vitamin C (ascorbic acid). The Hungarian scientist, Dr Szent-Gyorgyi, who won a Nobel Prize in 1973 for his work on vitamin C, found paprika pods to be one of the richest sources of ascorbic acid.

Paprika is used for its colouring and flavouring properties. The end-use determines which of these factors is the most important and, therefore, which paprika is best for a particular buyer. In general, a high 'colour extraction rating' enhances the value of paprika, but in many cases, this can also be the most economic product to use, since only small quantities may be needed. The term colour extraction rating (the amount of colour extracted by an appropriate solvent) is used because surface colour is not always a reliable indication of paprika's colouring properties. Occasionally, a paprika, which looks richly red to the naked eye, will deliver less colour than expected in a finished food product. This is because the surface colour can vary with the fineness of grind, amount of heat temperature developed during processing and moisture content. Freeze grinding helps in the retention of flavour and red colour. Storage temperature and humidity of the raw material at the time of grinding may also affect this outward appearance. This is not to say, however, that the surface colour of paprika be disregarded; it should be considered, along with the fineness of grind and the colour extraction value of paprika.

Quality attributes of dried chillies

The important quality factors of dried chillies are variety, colour, size, shape, seed content, pungency, flavour, freedom from dirt, dust, mould, insects, foreign matter (both organic and inorganic), damage and moisture content. The physical parameters and chemical quality attributes of 16 varieties of chillies grown in different regions of India are presented in Tables 11.4 and 11.5. The physico-chemical parameters of chillies (whole and ground) have been specified under the National Standards (Agmark/ISI/PFA). The quality requirements depend upon their

Table 11.4 Physical parameters of chilli cultivars (dry red pods)

Cultivar	Size	Visual colour	Filter no. and name	Length (mm)	Breadth (mm)	Bulk density (g/ml)	100-pod weight (g)	Seed weight (g)
'BR Red'	Large	Dark red	14 ruby	97	35	0.206	75.3	34.50
'G 4'	Large	Bright red	6 primary red	83	38	0.232	61.7	35.18
'Pant Cl'	Small	Light red	34 golden amber	55	30	0.520	41.3	17.99
'SP 14/5'	Large	Light red	33 deep amber	86	43	0.262	97.7	34.97
'Jwala'	Large	Bright red	6 primary red	100	35	0.202	64.0	23.79
'Musalwadi'	Medium	Bright red	6 primary red	68	30	0.420	52.0	20.05
'CO 2'	Small	Bright red	6 primary red	42	61	0.498	97.3	55.85
'CA960/1-3'	Medium	Light red	33 deep amber	73	34	0.408	56.0	24.40
'S/118-2'	Small	Dark red	18 magenta	42	40	0.560	56.7	21.58
'SP 47'	Large	Light red	34 golden amber	99	30	0.236	71.0	37.92
'K1'	Large	Bright red	6 primary red	98	29	0.208	61.0	26.67
'K2'	Large	Dark red	13 magenta	80	40	0.212	80.7	32.73
'618-126'	Large	Dark red	14 ruby	86	38	0.342	82.4	29.98
'CO 1'	Medium	Bright red	6 primary red	79	34	0.352	76.4	34.74
'CO 3'	Small	Bright red	6 primary red	58	36	0.476	54.2	23.61

Source: Pruthi (1993a,b).

Table 11.5 Chemical quality parameters of chilli cultivars

Cultivar	Moisture (%)	Capsaicin (%) (MFB*)	Total extractable colour (ASTA units) (MFB)	Ascorbic acid (mg/100 g) (MFB)	Oleoresin (%)
'BR Red'	8.8	0.62	1,248	58.2	8.6
'G 4'	9.0	0.44	896	54.6	8.4
'Pant Cl'	10.2	0.42	1,144	62.8	9.8
'SP 14/5'	8.5	0.36	1,052	32.4	12.2
'Jwala'	8.5	0.68	2,542	86.4	10.6
'Musalwadi'	9.2	0.28	1,500	32.4	12.4
'CO 2'	10.0	0.28	256	60.8	11.6
'CA 960/1-3'	8.8	0.42	158	42.4	6.2
'S/118-2'	9.2	0.40	394	52.6	6.2
'SP 47'	10.2	0.50	846	28.9	8.4
'K 1'	8.6	0.27	278	28.1	8.6
'K 2'	10.4	0.32	258	32.7	9.2
'618-126'	10.0	0.42	245	61.6	9.8
'CO1'	9.6	0.42	262	58.5	11.3
'CO3'	9.8	0.30	158	51.7	10.5
Mean	9.39	0.409	749	49.6	9.7
SED	0.04	0.062	0.188	0.096	0.045
CD at 5%	0.09	0.135	0.380	0.201	0.052

Source: Pruthi (2001).

Note
*MFB = Moisture-free basis.

end-use. For flavouring and colouring of foods, chillies should be rich in colour (both surface and extractable) and flavour but mild in pungency. However, for oleoresin extraction, they must be highly pungent and be less in red colour.

Overall, good quality is based on (i) a good pungency level, (ii) a bright red colour, (iii) a good flavour, (iv) medium-sized fruits with a moderately thin pericarp, (v) a smooth glossy surface, (vi) a few seeds in the fruit, (vii) a firm stalk and (viii) freedom from damage, dirt, foreign matter, mould and insects.

The total colouring matter content of commercially important varieties and types of Indian chillies, as determined by the author and expressed as capsanthin content, follows:

Variety	Average colour capsanthin content (g/kg) (moisture-free basis)
'Madras Sannam'	1.46
'Madras Sannam B'	1.40
'Madras Mundu'	1.08
'Guntur Sample'	1.88
'Warangal'	1.92
'Dondaicha'	1.71

Source: Pruthi (1969, 1970).

Ground chillies or chilli powder

Spice milling is an ancient industry, akin to the cereal milling industry, with the difference that in spice grinding there are additional problems of the volatility of essential oils naturally present therein (Pruthi, 1980).

There is a considerable amount of international trade in ground spices, in which black and white pepper powder are by far the most important. They are available in a variety of different grinds. The most common grinds are cracked, coarse ground, table grind and fine grind or pulverized. Generally speaking, the finer the grind, the more immediately available the flavour, but the shorter its shelf-life (American Spice Trade Association, 1964). Other popular ground spices include chilli or *Capsicum*, turmeric, onion, garlic, cinnamon and coriander.

The optimum size of the grind for each spice depends on its ultimate end-use. Standard ground spices are usually ground to allow them to pass through US standard sieves ranging from No. 20 to No. 60 mesh. Spices are also ground to microscopic fineness with a particle diameter of 50 microns, which is one-twentieth of the size of 60-mesh particles.

Proponents of this fine grind argue that extreme fineness contributes to the unlocking of natural flavours, aiding quick and thorough dispersion, and permitting constant control of flavour intensity. When they are used in food products, the uniform distribution of microground spices avoids hot spots. Further, because of the minute particle size, no dark specks can be noticed in the finished product (Neal and Klis, 1964). Also, the colouring components are extracted into the carrier medium or the food product within a short period of cooking. Grinding improves the aesthetic appearance of the product and also enables users to effect greater economy in use, time and labour.

The other advantages of ground spices, according to Heath (1972) are as follows: (i) slow flavour release in high-temperature processing; (ii) ease of handling and accurate weaning; and (iii) no problems in labelling. However, ground spices are known to lose a measurable fraction of their volatile oil or flavouring components owing to the heat generated during grinding (Miller, 1951; Pruthi and Misra, 1963). Pruthi and Misra (1963) reported total losses of 0–1.15% of volatile oil and 0.5–3.0% of moisture from different spices during grinding. They also reported that the product temperature during milling ranged between 42°C and 95°C in different spices, when ground individually. Water cooling of grinding machines (Parry, 1945), the use of liquid nitrogen (Miller, 1951), and the use of 'amulin' during grinding (Grimme, 1954a,b) are some of the means practised in some countries to reduce such losses of volatile oil. According to the Griffith Lab, Inc., conditioning or chilling of spices prior to their grinding curbs the loss of flavour volatiles during both milling and storage. In spice milling, controlled storage temperature and humidity is a must. This system saves labour, time and space. Conditioning the spices for milling also increases throughput up to 20% and minimizes shrinkage during milling. The spices pass through the mill easily, without smearing, blinding or coating the mill. Refrigerated storage of spices also reduces loss in weight. Black pepper when stored in a cooler for three months lost only 0.3% weight, whereas its counterpart stored in a conventional warehouse lost 3.6% weight.

Chilli milling or grinding equipment

In order to reduce flavour loss in ground spices due to the excessive heat produced during grinding, cooling arrangements by means of air blown through the grinders or jacketed water-cooled units are necessary. Closed-circuit grinding with vibratory screens allows a variety of particle sizes.

For size reduction or fine grinding, air-swept beaters, double rollers, cages and hammer mills are used. Hammer mills are not very satisfactory for fine grinding. Plate mills are used for small-scale and domestic purposes, and pin mills are employed for very fine grinding and higher capacities (Pruthi, 1980).

The fat content of spices is also a problem in milling. Particle size, product yield, product uniformity, freedom from contamination, economy and dust from the operation are other factors to be considered in the selection and operation of the grinders (Pruthi, 1980).

Cryomill process of freeze-grinding of chillies/spices

The working principle and description

Wistreich and Schafer (1962) have described a novel 'Cryomill process' of freeze-grinding spices, which is accomplished by the controlled injection of liquid nitrogen (which acts as a direct-contact refrigerant) directly into the mill's grinding zone. A temperature controller, which monitors the flow of liquid through a solenoid valve, maintains the desired product temperature. Part of the exhausted stream of cold nitrogen gas is recirculated to the spice hopper for the pre-cooking of spices (Figure 11.6).

The instantaneous evaporation of the liquid refrigerant quickly chills both the spice and the mill. It also absorbs the frictional heat of grinding. Thus, the temperatures in the grinding zone generally are well below $-100°F$ (Pruthi, 1980).

Advantages of the Cryomill process

This novel process has several advantages over conventional grinding processes: (i) it cuts down the oxidation of spice oils because, as the liquid nitrogen evaporates in the grinding zone, it tends to expel any air in the mill. The flavour is thus much better retained; (ii) it also permits extremely fine grinding because the spice oils solidify at low temperature, thereby making the spices very brittle; (iii) such finely ground spices disperse flavour uniformly throughout the final product; (iv) they virtually eliminate specking problems, as in sausages; (v) in liquid preparations, the settling rate of freeze-ground spices is noticeably reduced; (vi) it curbs the usual loss of spice aromatics and moisture. The ground products retain their original flavour strength and weight. On an average, seven ounces of freeze-ground spices have the same flavouring power as ten ounces of their conventionally milled counterparts; (vii) freeze-ground spices have proved to

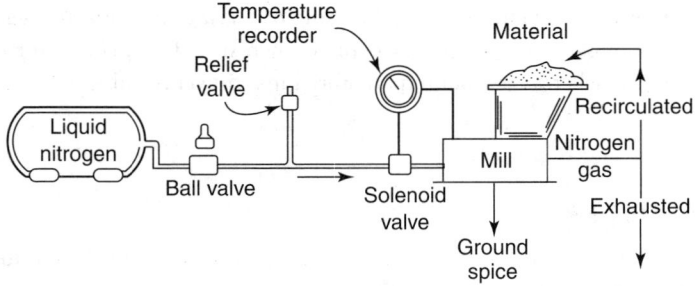

Figure 11.6 Schematic drawing of a freeze-grinding (Cryomill process) installation (Wistreich and Schafer, 1962).

be considerably more stable than conventionally ground products. Possibly, the spices absorb or in some way retain some of the nitrogen. (viii) low-temperature treatment lowers the microbial load on the spices; (ix) grinding rates are increased because the mill's low-temperature operation minimizes the 'gumming up' of grinding surfaces and screws. In a 25-horsepower mill refrigerated with liquid nitrogen, 1000 pounds of spices can be ground per hour; (x) the actual costs of Cryomill spices are lower than those for conventional products when the increased flavour strength is taken into account. In addition, fine grind and greater stability are bonuses to food processors; and (xi) the process is applicable to a variety of other foods, such as cocoa, coffee, tea, coconut, vanilla and dehydrated meats (Pruthi, 1980).

The chilli powder

The most important ground spice item exported from India is chilli powder increasing from 7,100 to 15,900 t during the last five years (1994–95 to 1998–99), registering an average growth of 31% per annum. The value of exports ranged between Rs. 21 crores and Rs. 66 crores. The main markets for Indian chilli powder (based on 1998–99 data) are given below. Countries which buy more than 1,000 t in a year are UAE: 3,070 t, USA: 2,825 t and UK: 1.109 t. The other important markets which buy 200–1,000 t, are Canada: 496 t, Saudi Arabia: 476 t, South Africa: 310 t Pakistan: 386 t, Singapore: 302 t and Bangladesh: 232 t.

Disadvantages of ground spices

Compared with spice oleoresins, ground spices, according to Heath (1972), have the following disadvantages:

1   Variable flavour quality;
2   Variable flavour strength;
3   Microbial contamination;
4   Possible contamination by filth;
5   Easy adulteration by less valuable materials;
6   Presence of lipase enzymes in some spices;
7   Flavour loss and degradation on storage;
8   Undesirable appearance characteristics in end products;
9   Poor distribution of flavour, particularly in thin liquid products (sauces);
10   Discolouration due to tannins;
11   Usually a hay-like aroma in herbs;
12   Dusty and unpleasant bulk handling;

Packaging and storage of dried whole and ground chillies

Dried chillies (whole)

According to the CFTRI process (1961), chillies should be conditioned to 10% moisture and compressed at 2.5 kg/cm^2 by using the Baling process. Wooden crate dunnage with a layer of matting is suggested for storage in suitable godowns or warehouses.

For retail or consumer packing of chilli powder, packing in Mylar-Saran-PF plastic laminate and aluminium combination (AFC) pouches under nitrogen gas is suggested. Such packages can be stored in a cool, dark and dry place for about a year.

Particular care is required to ensure that stocks of chillies received for storage in warehouses or godowns are adequately dry. Detachment of stalks from pods results in the bleeding of seeds from within the pods, leading to a loss in pungency. It is difficult to ensure the stability of stocks due to the low density of the commodity. To obviate this, iron frames have been introduced in some warehouses to facilitate the building of stable stocks. As this commodity emits a strong odour, it should be stored in a separate compartment, as far as possible. Chillies can be attacked by spice beetles and cigarette beetles during storage (Harels *et al.*, 1977).

Ground chillies or chilli powder – consumer package designing/development

Studies carried out at the CFTRI, Mysore, for designing a suitable consumer package for ground and whole chillies, have revealed that (i) a moisture content higher than 15% is critical with respect to mould growth; (ii) the discolouration of the red pigment of chilli during storage is greatly influenced by moisture and temperature; (iii) 300-gauge low density polyethylene film pouches are suitable for 100 g consumer unit packs for ground chilli powder to give a shelf-life of three and six months under accelerated and normal conditions of storage respectively; and (iv) under tropical conditions, 200-gauge low and high density polyethylene films are suitable for packaging of whole chillies in units of 250 g each.

Factors affecting colour retention during storage of dried whole and ground chillies

Storage has a marked influence on the colour of dried chillies although it has little effect on their pungency. Since colour is one of the main determinants of the price which a producer receives, and it may be months before the dried ground product reaches the consumer, the problem of loss of red colour during the necessary storage period is one that needs to be considered.

The greatest influence on colour retention is not their storage conditions but rather the variety of *Capsicum* or chilli grown. Over a six-month storage period for ground material, one variety can retain 78% colour, whereas another can suffer complete colour loss.

The controllable factors from harvesting onwards that are significant with regard to colour retention of the dried spice are as follows:

1   Delaying harvesting until the pods are partially withered on the plant and then curing the sliced pods is considered to provide a product with superior colour-retention properties. The initial colour of the dried product, however, does not appear to have a marked influence on the subsequent rate of colour loss.

2   The colour of ground chillies deteriorates faster than whole or sliced pods on storage but the latter quickly lose colour on eventual grinding.

3   Exposure of dried chillies to air and light accelerates the rate of bleaching, and so storage in airtight containers away from sunlight is desirable. Studies on storage in glass jars and tin containers, however, showed no significant difference in colour retention. Samples in glass bottles showed surface bleaching and those in tins faded more uniformly.

4   When ground chillies were stored with 9–10% moisture, the colour retention was better than in samples with moisture below 7%. Krishnamurthy and Natarajan (1973) studied the storage of whole chillies in cans and found that samples with a moisture of 11.0–12.9% retained a higher colour, expressed as β-carotene content, than in samples with moisture below 9%. However, samples stored at 11–12% moisture turned black and those stored below 7% turned pale.

5   The storage temperature has a greater influence on colour retention than does light, air, kind of container or whether the spice is stored in the whole or ground form. Storage at 37°C was found to accelerate colour loss but it paralleled that occurring at 25°C. Refrigerated storage at 5°C shows a rate of discolouration but it does not affect the inherent colour stability of a particular *Capsicum* cultivar. On regaining an ambient temperature, the refrigerated material suffers a quicker colour loss than corresponding unchilled material. This rapid colour loss possibly corresponds with the shelf-life time of retailer.

6   Rancidity of samples is believed to be correlated with colour loss, and the application of fat-soluble antioxidants has been found to improve colour retention. The addition of anti-oxidants is more effective after curing than before and in the ground spice rather than in whole pods.

Finally, it should be mentioned that rats have a great liking for chillies in spite of their pungency, and therefore care should be taken in storage to protect chillies from the depredations of this noxious animal.

Packaging and storage of fresh bell peppers

Using perforated polyethylene bags of 150–200 gauges can prolong the shelf-life of green peppers. Ventilation of packages should be adequate to avoid off-flavour development and moisture condensation in packages.

The lowest temperature range recommended for storing green bell peppers is 7–10°C for up to two to three weeks. At temperatures below 7°C, bell peppers are subjected to chilling injury. Peppers having a large surface to volume ratio are particularly susceptible to water loss. They must be held in high relative humidity of 90–95% or else they will rapidly become wilted. They are well adapted, therefore, to selective film packaging (Pruthi, 1985, 1987a,b).

For controlled atmospheric storage of bell peppers, the recommendations are 4–8% oxygen and 2–4% CO_2 at 13°C. Oxygen concentrations below 2% combined with 10% CO_2 may cause injury.

Chilli processing technology

Chillies are processed into a number of commercially viable products like chilli paste, chilli puree, chilli powder, dried/dehydrated chilli, chilli oleoresin and chilli seed oil (fixed) which are briefly discussed in this chapter. Additionally, green chillies are processed into pickles, paste or even dehydrated as explained later (Pruthi, 1991).

Dehydrated green chillies

There are reports on the standard conditions and pre-treatments required to produce the best quality dehydrated green chillies (Luhadia and Kulkarni, 1978).

Dried chillies generally contain about 6% stalks, 40% pericarp and 54% seeds. The important constituents of colour and capsaicin are concentrated in the pericarp. About 90% of the capsaicin in chillies has been noticed in the placenta, connecting seeds with pericarp. Placenta which represent only less than 4% of total weight of pod has a capsaicin content of about 7% (Table 11.6).

Drying yield of chillies

Aiyer (1944) reported an average yield of dried chillies for a rainfed crop in India as 280 kg/ha although a good crop may amount to three times this quantity or more. The average yield from an irrigated crop is about 1,650 kg/ha, increasing in good crops to 2,800 kg/ha.

Table 11.6 Analysis of different parts of chilli

Variety	Stalks (%)	Seed (%)	Pericarp (%)	Dissepiment (%)	Pericarp		Dissepiment	
					Capsaicin (%)	Colour value	Capsaicin (%)	Colour value
'Mundu'	6.2	55.6	36.7	1.5	0.17	39,650	6.6	5,978
'Jwala'	5.1	49.7	42.1	3.1	0.58	41,480	7.7	6,100

Source: Pruthi (1993).

Table 11.7 Comparative data on the drying yield of different types of chillies

Items	Indian chilli	'Tabasco'	Bird chilli
Yield of dry fruits (kg/ha)	4,486	5,663	5,549
Percentage conversion wet-to-dry	25	32	37
Number of dry chillies per kg	1,617	6,718	21,385
Picking time for fresh fruits (kg/h)	3.6	0.6	0.15
Mean length of fruit (cm)	7.0	3.0	1.3

Source: Solomon Island Report on Drying of Chillies (1973).

In the Solomon Islands, yields and other useful data have been reported which are presented in Table 11.7.

The dried forms of the fruits of *Capsicum* species, which are for trade, fall into three groups: (i) highly pungent 'chillies'; (ii) moderate to mildly pungent '*Capsicums*'; and (iii) 'paprika' which may be sweet or mildly pungent. Paprika is always traded as a ground product, and chillies and *Capsicums* in whole or ground form. All of the three types are also extracted with solvents to prepare their oleoresins. Blends of ground chillies and *Capsicums* are marketed as cayenne and red pepper or mixed with other spices for the preparation of 'chilli powder'. Related products include larger-fruited, sweet or mildly pungent varieties of *C. annuum*, used in their fresh state as vegetables or in preserves.

Quality requirements for processing

In *Capsicums* and chillies which are used in food preparations, quality is of much importance, and is based on a good pungency level, bright-red colour, good flavour, medium-sized fruits with moderately thin pericarp, smooth glossy surface, few seeds in fruit and a firm stalk.

Medium-sized fruits are preferred, because in storage they remain intact better than longer pods, which tend to break at distal ends. A fairly thin pericarp is necessary as drying is more easily accomplished. On drying, fruits with thick pericarps show a wrinkled surface and dull appearance. Chillies and *Capsicums* with a bright-red colour command higher prices than those which are fully red or even orange or yellow in colour, or deep-red fruits tend to retain their colour in storage longer than those which are of a lighter shade (David, 1982).

Fractionation of red chillies (CFTRI)

The CFTRI, Mysore, has developed a new technique to fractionate red chillies into three fractions or compounds: (i) capsaicin-rich powder, (ii) skin and (iii) seed. The capsaicin-rich fraction

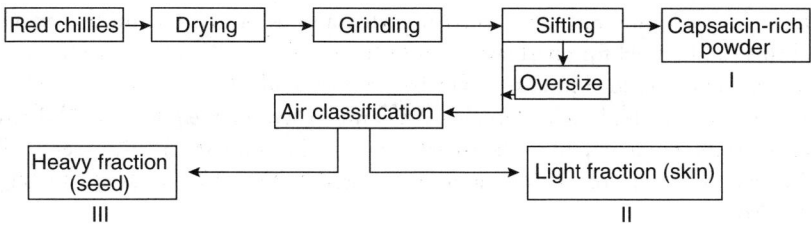

Figure 11.7 The flow diagram for the process of fractionation of chillies developed at the CFTRI, Mysore.

can be used as a substitute for whole chilli powder in oleoresin extraction the skin can be used for colour and low pungent oleoresin, and the seed will be raw material for fixed oil extraction.

(a) *The CFTRI Process*: Chillies are dried to make them crisp and then ground in a suitable mill wherein all three fractions are separated. The material is first sifted to separate the inner sheath or pungent part and then air-classified to separate the seed and skin (Figure 11.7).

(b) *Product yield*:

Products yield per day (basis 300 kg red chillies)	Expected yield (kg)	Range (%)
Capsaicin-rich powder	18	5–6
Skin	120	35–40
Seed	132	40–44
Losses as moisture, etc.	30	8–10
Total	300	

(c) *Raw material*: The main raw material is red chillies with a good colour and pungency, which is available indigenously.

(d) *Equipment*: The main equipment required is a drier, a grinder, a sifter and an aspirator. All are available indigenously.

(e) *Capital outlays*: The CFTRI has suggested that the capacity of the minimum economic unit should be 300 kg of red chillies per day (one 8-hour shift).

(f) *Land required*: The land required is 500 m^2; the area of the building is 180 m^2.

(g) *Safety considerations*: Fine chilli dust may be problematic for workers. Proper exhaust fans and closed processing can avoid this problem.

(h) *Material and energy required*: The following data are for per day of production (one shift):

Red chillies	300 kg
Electric power	250 kW
Raw water	500 gallons

Fractions from Indian chilli oleoresin

The oleoresins produced from Indian chillies are not suitable for pharmaceutical purposes and export since their pungency is very low. In view of this, the Regional Research Laboratory, Trivandrum, developed a process which aims to separate the oleoresin of chillies into two fractions, a highly pungent fraction and a natural chilli colour fraction, free from pungency. The main product is highly pungent oleoresin having a value of over one million Scoville Heat Units and this can be used for pharmaceutical purposes. The chilli colour-free from pungency can be used for colouring food products. The by-product, chilli seed oil (fixed), can be used for edible purposes and can also be added in pickle-making, where a little pungency is also required, as seed oil has some pungency too. The solvent extracted residue from chilli seed (chilli seed-cake) has a protein content of 27–29% and can be used as a fertilizer or as an animal feed too (Pruthi, 1989a).

World demand

Many countries (including India) are applying more and more restrictions on the use of artificial colours and recommend the use of natural colours. The red colour fraction obtained in the process can be very well used as paprika oleoresin. The present world demand for highly pungent chilli oleoresin is not known, but it is assumed to be 100–200 tonnes per annum. As chilli colour is a new product, its future demand is not yet known. But it is expected that as coal-tar based artificial colours are banned and being replaced, the chilli colour may have a good demand as a substitute for synthetic colouring in the food industry. For example, it could be used in tomato sauce/ketchup, paste or pureed soup.

The NRDC Process

Dry chillies with 8–10% moisture are fed into a pin-mill, crushed and then fed into a sieve separator where the chilli pericarp along with its stalks are separated from the seeds. The chilli seed which contains 3.5% of pericarp is separated by air classification (Figure 11.8).

The chilli pericarp is dried at a controlled temperature, ground to the required mesh in a disintegrator and then packed in a battery of percolators where oleoresin is extracted by a counter-current extraction method with solvent continuously fed from the top (see Figure 11.9). A part of the extract coming down is pooled up for desolventization. The remaining extract is fed, crushed and then processed for solvent extraction. The oil obtained is free of all solvent by distillation. The recovery of the solvent from the extractor is accomplished by steam distillation (Figure 11.10).

The product obtained after being subjected to control examination can be either packed in high-density polyethylene carboys or in canned tins.

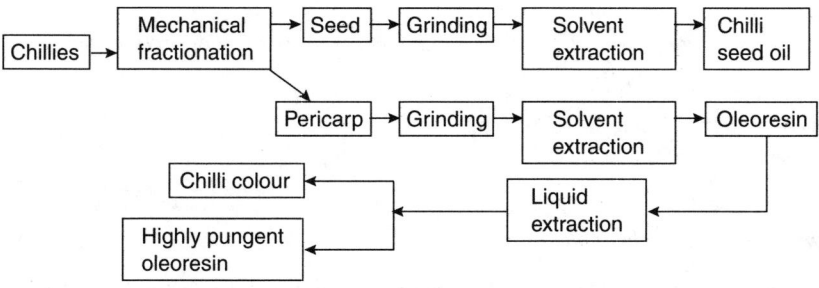

Figure 11.8 Flow-sheet for separating a highly pungent oleoresin fraction and a natural red colour fraction free from pungency (NRDC, India).

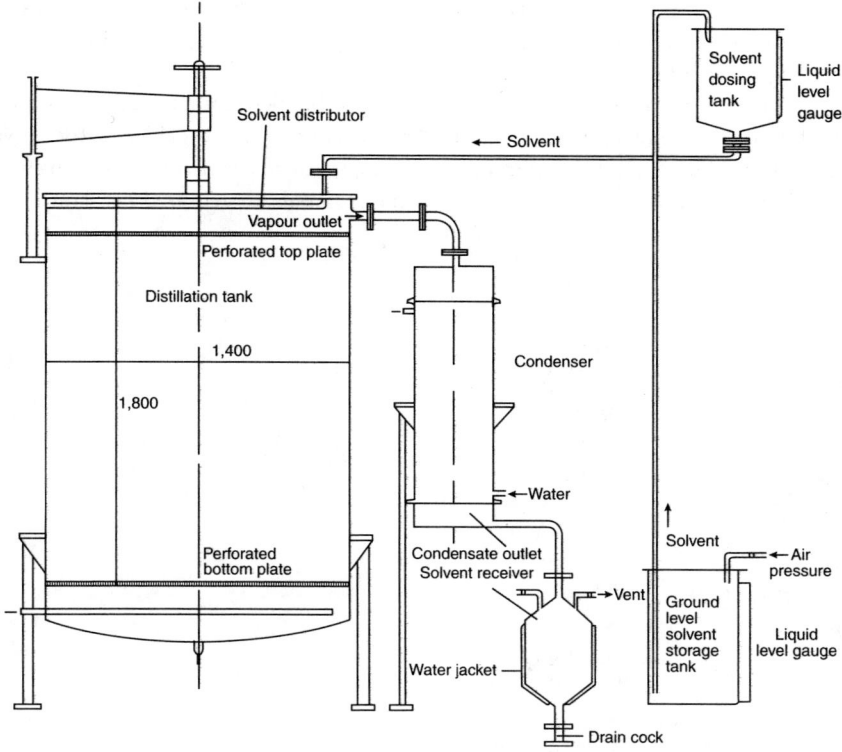

Figure 11.9 Preparation of oleoresin in a solvent extraction plant (NRDC Process).

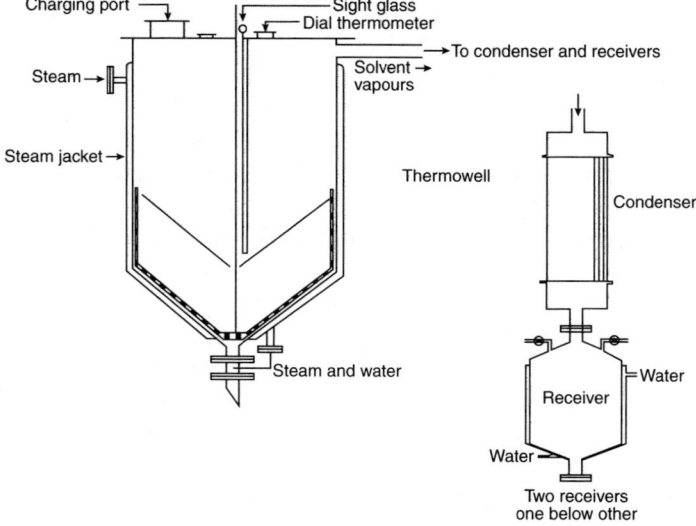

Figure 11.10 Solvent recovery system of oleoresins (NRDC Process).

(a) *Scale of investigation*: The laboratory has carried out work on 20 kg of whole chillies.
(b) *Input*: The main raw materials required are good quality dry red chillies, ethylene dichloride and hexane alcohol which are available indigenously.
(c) *Equipment*: The main equipment required is pin-mills, percolators, a disintegrator, an air classifier, driers, condensers, sieve separators, expellers, centrifugal extractors, a vacuum pump, stirrer, boiler, chilled-water unit and pumps. All are available indigenously.
(d) *Raw material and utilities required*: The following materials are required per day:

Whole dry red chilli:	1
EDC:	63 kg
Hexane:	71 kg
Alcohol:	6 kg
Electric power:	800 kW
Steam:	1.2 t
Fuel oil:	70 L
Process water:	90 gals

(e) *Yield per day*:

Colour 70%:	25.9 kg
Pungent fraction:	11 kg
Oil:	86 kg
Oil-cake:	420 kg

(f) *Plant parameters*:

Production capacity:	One tonne of whole chilli;
Number of shifts:	Two per day of eight hours each.

Manufacture of oleoresin Capsicum *or paprika – a global overview*

As early as in 1949, Goldman described methods for the preparation of spice oleoresins, extracts of red pepper and paprika, by extraction with Me_2CO or EtOH, filtration and removal of the solvent by distillation. Soaking the cake before extraction improved the yield. A diagram of a commercial-size extraction apparatus based on a Soxhlet method is given in his report. Berry (1935) reported the examination of the extractives of *Capsicums*. The oleoresins of *Capsicum* vary in appearance, solubility and degree of pungency according to the solvent used for extraction. Et_2O and alcohol extract much of the non-pungent matter from chillies.

Ferns (1961a,b) compared various methods of extraction of oleoresin from *Capsicum*. A laboratory-scale counter-current apparatus was found to be the best method. Acetone was selected as the best solvent for the extraction of oleoresins from *Capsicum*.

Tandon *et al*. (1964) from CFTRI discussed the preparation of oleoresin of *Capsicum* (red chillies) as well as its technological and chemical properties.

Szabo (1969, 1970) discussed the manufacture of paprika oleoresin and Mathew *et al*. (1971a) covered in detail the preparation and quality control of oleoresin *Capsicum*. There are other reports on oleoresin from Hungarian paprika and there is a Japanese patent on cayenne pepper extract.

Blazovich and Spanyar (1969) described a method for the determination of capsaicin in oleoresin and some other preparations with a high capsaicin content, while the gas chromatographic detection of the acetone content of paprika oleoresin has also been reported. Govindarajan and Ananthakrishna (1970) described the separation of capsaicin from *Capsicum* and its oleoresin, and Mathew *et al.* (1971b) developed an improved TLC method for the estimation of capsaicin in *Capsicum* oleoresin.

Advantages of oleoresins (over whole or ground spices)

1   They are more hygienic than whole or ground spices; they are free from bacteria.
2   They can be easily standardized for strength of flavour.
3   They contain natural antioxidants.
4   They are free from any enzymes.
5   They have a long shelf-life under ideal conditions of storage.
6   They have less bulk in storage.
7   They have less weight in shipping.
8   There is no colour disturbance or speck formation.
9   There is no danger of any mould attack as there are in spices (whole or ground).

Factors affecting the quality and yield of oleoresin

The following factors significantly affect both the quality and recovery or yield of oleoresins from pulverized spices.

1   Selection of proper variety of *Capsicum*.
2   Conditioning or drying of the spices properly before use.
3   Preparation of material for pulverizing.
4   Choice of a proper solvent.
5   Conditions of solvent extraction – time and temperature of extraction.
6   Proper solvent stripping from the miscella.

Factors affecting the proper location of oleoresin plant

The location of a solvent extraction plant and the solvent recovery plant, whether at the source or near the market, affects the marketing and economic aspects of the oleoresins. The following are the major considerations to be taken into account at the time of locating a new plant:

1   The availability of the contemplated raw material.
2   Reasonable cost and good quality of raw material.
3   Scope for the marketing of the oleoresin.
4   Efficient plant design with minimum losses in recovery.
5   Availability of suitable technical expertise and management.
6   Reasonable facilities for labour transport utilities.
7   Assurance of reasonable annual profits.
8   Suitable government regulations regarding labour and transport.
9   Availability of sufficient financing.
10   Facility for economic utilization/disposal of the spice-waste.

The ultimate success of the plant lies in the entrepreneur's ability: (i) to interpret the facts correctly; (ii) to translate ideas into action; and (iii) to achieve the goals set for the enterprise. Systems and checklists should be used in organizing routines, but in the final analysis, the key to success lies in the people behind the undertaking. This is true for every human endeavour and so also for spice extraction, whether at the source or at the market (Pruthi, 1980) careful, regular and correct evaluation of each batch of product is absolutely necessary to achieve success.

Forms of oleoresins (tailor-made)

In some food applications, essential oils (tailor-made) and/or oleoresins may be used 'as is'. However, because of their limited solubility and also the minute quantities required to produce a strength flavour equivalent to that of the parent spice, the manufacturers further process the extractives into forms that are more convenient to use from the standpoint of both limited solubility and flavour strength:

1   Liquid oleoresin: (a) oil-soluble type, (b) water-soluble type;
2   Dry oleoresin: (a) dry, soluble, (b) spray-dried.

Both these forms are soluble or dispersible in the media in which they are intended to be used (Pruthi, 1980).

Types of Capsicum oleoresins

There are three types of oleoresins prepared from the dried fruits of *Capsicum* species.

(i)   Oleoresin *Capsicum* (Bird chillies – *C. frutescens*): The bird chilli is also known as 'African *Capsicum*'. Oleoresin is prepared from the most pungent, small-fruited bird chillies. This oleoresin has a very high pungency and is used exclusively for official pharmaceutical work; it is also employed to impart pungency to some foods and beverages. It is evaluated solely on its content of capsaicin as its colour value is low. Commercial *Capsicum* oleoresins are usually supplied in pungency ratings between 500,000 and 1,800,000 Scoville units (approximately 3.9–14% capsaicin w/w) with colour values expressed on the ASTA (American Spice Trade Association) scale, of 3,500 units maximum and 400 units maximum in decolourized types. Manufacturers have claimed replacement strengths of 1 kg of 500,000 Scoville units oleoresin for 20 kg of good quality cayenne.

(ii)   Oleoresin red pepper (*C. annuum*): Oleoresin is obtained from the longer, moderately pungent *Capsicums* used in the production of red pepper. Commercial red pepper oleoresins are usually supplied in pungency ratings ranging between 80,000 and 500,000 Scoville units (approximately 0.6–3.9% capsaicin w/w) and in a wide colour range of up to a maximum of 20,000 colour units. Manufacturers have made claims that 1 kg of 200,000 Scoville units oleoresin replaces 10 kg of good quality red pepper.

(iii)   Oleoresin paprika (*C. annuum*): Oleoresin is prepared from varieties of *C. annuum* from which paprika is produced. It has a high colour value, but little or no pungency. Commercial paprika oleoresins are available in colour strengths ranging from 12,000 to 100,000 units, the 40,000–80,000 range being the most popular. Manufacturers have claimed that 1 kg of paprika oleoresin replaces 12–15 kg of paprika powder.

   All of the three types of oleoresin are supplied by manufacturers in the free-flowing form or delivered on suitable carriers as required.

Crucial factors affecting the quality of oleoresin

As can be seen above, the colour value and pungency as well as the yield of non-volatile extract are the most important properties of these oleoresins. Provided that the fruits have been harvested at an optimum maturity and well dried, the crucial factors in extraction of the oleoresins are the state of subdivision (grinding) of dried fruits and the nature of solvent used. Mathew

Table 11.8 Effect of nature of solvent on yield and quality of oleoresins

Solvent	Oleoresin (%)	Colour as carotene (%)	Capsaicin as Scoville heat value
Hexane	8.6	0.114	99,000
Ethanol	8.2	0.052	360,000
Dichloroethane	7.5	0.106	360,000

Source: Mathew *et al.* (1971a).

Table 11.9 Analysis of important commercial chilli varieties grown in important chilli-growing countries of the world for oleoresin recovery

Variety	Country of origin	Yield of oleoresin (%)	Capsaicin content (%)	Scoville heat value (calculated from that of oleoresin)	Total colour as carotene (mg/g)
'Sannam'	India	16.5	0.33	49,500	1,450
'Mundu'	India	16.0	0.23	34,000	1,300
'Oosimulagu' (Bird chilli)	India	8.7	0.36	42,000	113
'Mombassa'	Africa	13.1	0.42	78,600	349
'Bahamian'	Bahamas	12.5	0.51	75,000	1,220
'Santaka'	Japan	11.5	0.30	55,000	1,392
'Hontaka'*	Japan	—	0.33	50,000	—
'Mombassa'*	Africa	—	0.80	120,000	—
'Uganda'*	Africa	—	0.85	127,000	—
'Abyssinian'*	Africa	—	0.075	11,000	—

Note
* Mathew *et al.* (1971a,b).

et al. (1971a) demonstrated the comparative efficiency of three solvents in the extraction of oleoresin of the 'Sannam' variety of *C. annuum*, which is grown in Madras and Bihar (Table 11.8).

Hexane appears to be a poor solvent for capsaicin, and ethanol does not extract colour efficiently. Ethanol, especially when hot, gives a product which is a semi-solid instead of a free-flowing oil. Dichloroethane is a good solvent for 'Sannam' oleoresin, and yields reach 13–14% when finer powders are used. Oleoresin chillies contain both pungency and colour. Oleoresin is prepared by solvent extraction of the powdered spice. Besides food industries, the pharmaceutical and cosmetic industries also use oleoresin of a high pungency and low colour (Table 11.9). This oleoresin known as oleoresin *Capsicum* is used in pain balms, vaporubs and linaments.

Oleoresin obtainable from Indian chillies conforming only to oleoresin red pepper with a low pungency and high colour has only limited demand in food industries. Since the capsaicin is concentrated in pericarp, high pungent oleoresin can be obtained by extraction of the pericarp alone. Know-how of a process has been developed to separate pericarp oleoresin into a high pungent, low-coloured fraction and a colour fraction free from pungency. The high pungent fraction obtained in the process is about four times more pungent than pericarp oleoresin. The separated seeds give 20% fixed oil, which constitutes a by-product of the process.

Many countries (including India) are applying more and more restrictions on the use of artificial colours. The colour fraction obtained in the process, which is a natural colour, can very well be used for colouring food products. The main product, highly pungent oleoresin, can be used for pharmaceutical purposes utilizing the medicinal properties of capsaicin. The by-product

Table 11.10 Fixed-oil content (in percentage) of chillies, *Capsicums* and paprika

Type	Whole fruit	Pericarp	Seeds	Stalk
Paprika and similar large forms of *C. annuum*	9–16	4.6–6.8	19–27	1–2.5
Capsicums	12–22	—	—	—
Chillies	15–22	—	—	—

Source: Winton and Winton (1939).

'chilli seed oil' can be used for edible purposes such as pickle-making. The solvent-extracted residue from chilli seed (seed-cake), having a protein content of 27–29%, can be used as a fertilizer or as an animal feed. Seeds are also required for cultivation.

The present world demand for highly pungent chilli oleoresin is not known. As chilli colour is a new product, its future demand is also unknown. It is expected that as the coal-tar based artificial colours are being banned and replaced, chilli colour may have a good demand as a substitute in the food industry. This fraction can also be sold as an alternative to paprika extract.

Fixed chilli seed oil

The distribution of the fixed (fatty) oil in the fruit is uneven, being mainly found in seed (Table 11.10). As with some other constituents of the fruit, the fixed-oil content gradually increases during maturation from green to the ripe red stage. Relatively few analyses of the fixed-oil composition have been reported in literature and these have been concerned exclusively with paprika and similar large forms of *C. annuum*. The fixed oil has been found to comprise triglycerides (about 60%) in which linoleic and other unsaturated fatty acids predominate. Although only a small number of samples have been analysed and some differences between cultivars are apparent, it seems that the seed-fat and pericarp fat of paprika types are distinguishable, the former possessing a somewhat higher unsaturated acid content. The fat content and composition of paprika powder and its propensity to autoxidation and perhaps also to discolouration, are, therefore, dependent upon whether seeds are removed from the pods before grinding (Portale and Nozinibia, 1967).

Salzer (1975) reported an analysis of fatty acids from the fat of a sample of chillies and suggested that chillies may be distinguished from paprika according to their relative abundance of certain fatty acids in the seed oil. However, greater sampling is necessary to test the validity of this hypothesis.

Volatile oil of Capsicum

The fruits of the *Capsicum* species have a relatively low volatile-oil content, which has been reported to range from about 0.1% to 2.6% in paprika and similar large forms of *C. annuum* (Winton and Winton, 1939). The initial volatile-oil content of freshly picked fruits is dependent largely upon the species the cultivar and the stage of maturity at harvest. The eventual volatile-oil content of the dried product, however, may be lower and is dependent upon the drying procedure, the duration and the condition (whole or ground) of its storage. Paprika powder, for example, usually contains less than 0.5% of volatile oil.

Chilli puree and chilli paste

Nowadays, many people have difficulty in finding time to purchase, wash and grind fresh chillies for domestic consumption. Additionally, hotels and restaurants also need a continuous supply of chilli puree. Consumers are more concerned about the quality standards of their foods. They wish to be able to purchase quality-guaranteed chilli powder or chilli paste puree.

The major quality factors such as pungency, colour and fineness of the paste are important. The diversifying food industry also requires chilli for a number of purposes. Considering all these facts, the demand for fresh and red chillies is high and is expected to grow. The consumption of processed and semi-processed forms of chilli (ground chilli, chilli paste and sauces) is expected to increase as households and institutional users are on the rise. To exploit these opportunities effectively, research focus was given at the Food Technology Centre, MARDI, Malaysia, for standardizing the product, developing suitable packaging materials and promotion to expand the existing markets.

Puree packed in OPP/AL/PE retained the colour, pH, TSS and vitamin C content up to 21 days when compared to PP packaging. Chilli puree stored at 10°C in OPP/AL/PE packaging showed reduced enzyme inactivation and increased nutrient retention for up to four weeks.

Processing of chilli puree

The process flow chart of chilli puree/paste is shown in Figure 11.11.

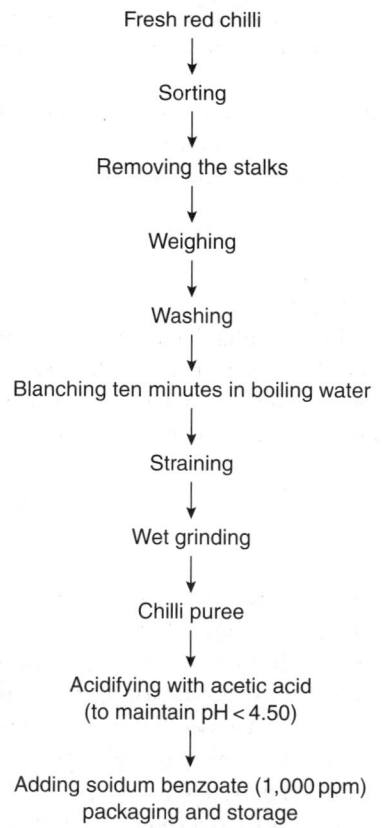

Fresh red chilli
↓
Sorting
↓
Removing the stalks
↓
Weighing
↓
Washing
↓
Blanching ten minutes in boiling water
↓
Straining
↓
Wet grinding
↓
Chilli puree
↓
Acidifying with acetic acid
(to maintain pH < 4.50)
↓
Adding soidum benzoate (1,000 ppm)
packaging and storage

Figure 11.11 The process flow diagram for chilli puree/paste (Parvathi and Aminuddin, 2000).

The developed puree can be used as a convenient product for cooking curried dishes. Processing of chilli puree and manufacturing of chilli-based processed products also gives access to distant markets and thus helps in their export promotion.

Canned or bottled pimiento or *Capsicum*

There is a demand for canned or bottled pimiento or *Capsicum* (*C. annuum*). Pimientos (*Capsicums* or chilli peppers) are processed in brine, in plain tin cans, or glass jars. The factors influencing the processing of canned pimientos, the rates of heat penetration, and suggested thermal processes have been reported and reviewed by Hight (1954) and by Powers (1961). But, in the commercial pimiento canning industry, some spoilage is still occasionally reported. Hight (1954) conducted studies to determine the thermal-death time characteristics of the micro-organisms causing spoilage and to establish processing times on the basis of these studies. The 212°F and 'Z' values were found to be 36.0 and 31.0, respectively, at pH 5.0. At pH 4.7, the corresponding values were only 7.0 and 4.2, respectively (Sane, 1950).

Acidification and calcium firming of canned pimientos were recorded. Acidification increased the drained weight of the product significantly. The use of both an acidulant and calcium chloride resulted in a highly significant increase in drained weight and in greater firmness of the product. Sane (1950) established the range of variation in pH and total titratable acidity in canned pimientos (pH 4.6–5.3, average pH 4.95, average acidity in red pimientos 0.28%, in green pimientos 0.6%). Unless the pods are fully coloured red, the pH may be as high as 5.5. As the colour becomes a deeper red, the pH decreases to 4.7. Pimientos of a good canning colour have a pH between 4.7 and 5.3. Siddappa and Bhatia (1955) reported on the satisfactory canning of green chillies.

Powers (1961) studied the effect of acid level, added calcium salts and monosodium glutamate to acid-treated pimientos and found that the organoleptic acceptability was increased. The addition of sugar improved the palatability of the product. The above conclusions are based on the exhaustive studies conducted by these authors during five packing seasons.

Chilli pickles and sauces

A number of chilli pickle are made, traded and exported from India due to their popularity. Pruthi (2001) gives full details of the different kinds of pickles such as those in oil, brine or citrus juice. Their method of manufacture, packaging, storage, storage problems and their solutions as well as a number of recipes of different pickles produced in different regions of India are also discussed.

Commercial utilization of *Capsicum* waste

Tandon *et al.* (1964) found that spent chilli meal left after oleoresin extraction contains about 28% protein, 36% carbohydrate and 29% fibre. It could go into animal feed compositions. According to Portales and Nozinibia (1967), red pepper (chilli) may be extracted with little degradation in an installation characterized by continuous rotary extractors that operate at room temperature and are fitted with indirect heating so that they can be used for drying the meal. Degumming is accomplished with H_3PO_4, citric acid and sodium tripolyphosphate. The miscella are evaporated under a vacuum, leaving an oil with 3 ppm of solvent.

Bough (1973) studied the characterization of waste effluents from a commercial pimiento (*Capsicum*) canning operation. The first stage of processing, in which the roasted peel was removed by washing, accounted for the most concentrated effluent. More concentrated effluent resulted from the citric acid dip just before the packing and closing area. Two effluents from the grading area were also obtained. The total waste produced from the canning plant contained 3.2 pounds of suspended solid, 60.2 pounds of chemical oxygen demand, and 35.4 pounds of biological oxygen demand per tonne of raw pimientos. Bough and Badenhos (1974) made a comparison of roasting and lye peeling of pimientos for the generation of waste and the quality of canned products. In general, lye peeling gives better results than flame peeling.

Economic utilization of *Capsicums*

As a food flavourant

Dry chilli is used extensively as a spice in all types of curried dishes in India and abroad. Grinding roasted dry chilli with other condiments such as coriander, cumin, turmeric and farinaceous matter makes curry powder. It is also used in seasoning egg, fish and meat preparations, sauces, chutneys, pickles, frankfurters and sausages. Bird chilli is used in making hot sauces, such as pepper sauce and tabasco sauce. 'Mandram' is a West Indies stomachic preparation made by adding cucumber, shallot, lime juice and Madeira wine to the mashed fruits of bird chilli.

As a colourant and flavourant

Paprika and red pepper, which are mild in pungency, are used to colour, flavour and garnish dishes. The use of attractive red paprika in tomato ketchup and sauces are also encouraged for improving colour. It is expected that as coal-tar based artificial or synthetic food colours are banned, the chilli/*Capsicum*/paprika colour may have a good demand as a 'natural plant colourant', as a substitute for synthetic colours in the food industry. NRDC/RRL, Trivandrum (1979) has developed a process for the fractionation of chilli-pericarp oleoresin into a highly pungent low-colour fraction for the pharmaceutical industry and a colour-rich fraction free from pungency for the colouring of foods. This colour fraction can also serve as an alternative to paprika colour extract used in food industries.

In medicine

Capsicum preparations are used as counter-irritants in lumbago, neuralgia and rheumatic disorders. Taken internally, *Capsicum* has a tonic and carminative action and is especially useful in atonic dyspepsia. It is, however, contra-indicated in gastric catarrh. Taken inordinately, it may cause gastro-enteritis. It is sometimes added to rose gargles for pharyngitis and it relaxes a sore throat. It can be administered in the form of powder, tincture, linament, plaster, ointment or medicated wool. In some of the preparations, 'Oleoresins Capsici B.P.C.' syn. 'Capsaicin', the alcohol-soluble fraction or ether extract of *Capsicum* is the active ingredient. Pharmacopoeial requirements are chiefly met by the highly pungent varieties of *Capsicum* (*C. frutescens*) grown in Sierra Leone, Nyasaland and Zanzibar. Indian *Capsicum*, known in trade as 'Bombay *Capsicum*' is used as a substitute (Wealth of India, 1985).

In pharmaceutical and cosmetic industries

Besides the use in food processing industries, pharmaceutical and cosmetic industries use chilli oleoresin of high pungency and low colour. This oleoresin known as oleoresin *Capsicum* is used in pain-balms, vaporubs and linaments since the pungent principle 'capsaicin' serves as an effective 'counter-irritant' (Wealth of India, 1985).

Seed oil (fixed or non-volatile)

Seed oil is a byproduct and can be used for edible purposes.

Seed cake

The solvent-extracted residue from chilli seed, having a protein content of 27–29% with negligible pungency, can be used as a fertilizer or as an animal feed.

Seeds

Seeds of good viability are also required for growing in nurseries or for cultivation purposes.

Dehydrated green chillies

Dehydrated green pepper is a good source of vitamin C. The pre-treatments and standard conditions for producing best quality dehydrated green chillies have been determined (Luhadia and Kulkarni, 1978). Dehydrated green chillies can find use in food industries as well.

References

Aiyer, A. S. (1944) Drying yield of chillies from rainfed crop, Special Report, Agriculture Department, TN.

American Spice Trade Association (ASTA) USA (1964) *Official Methods of Analysis of Spices*, ASTA, New York.

Anon. (2000) *Spices Market Weekly*, 14(6), 1–3, Feb 9, 2000 – 'Spice Export Review'.

Anu, A. and Peter, K. V. (2000) The chemistry of paprika. *Indian Spices*, 27(2), 15–18.

Berry, H. (1935) Examination of the extractives of capsicum, *Pharm. Pharmacol.*, 8, 470. Central Warehousing Corporation, New Delhi (1975) Manual on Warehousing of Spices.

Blazovich, M. and Spanyar, F. (1969) A method for the determination of the capsaicin in oleoresin and some other preparations with high capsicum content. *Elemisizirigalate Kozl*, 15, 358.

Bough, W. A. (1973) Characterization of some waste effluents from commercial pimento canning operations. *J. Milk Food Technol.*, 36, 371.

Bough, W. A. and Badenhos, A. F. (1974) A comparison of roasted v. lye peeling of pimentos for generation of waste and quality of canned products. *J. Food Sci.*, 39, 1105.

CFTRI, Mysore (1961) Dehydration of chillies. *J. Food Sci.*, 10, 218.

CFTRI, Mysore (1979) Improved technology for sun-drying of chillies (Tech. Note). *Indian Cocoa, Arecanut and Spices J.*, 3(1), 79.

Chen, S. L. and Gujmania, S. F. (1968) Auto oxidation of the extractable colour pigment in chilli pepper with special reference to ethoxyquin treatment. *J. Food Sci.*, 33, 274.

Daoud, H. N. and Luh, B. S. (1967) Packaging of foods in laminates and film combination pouches. IV. Freeze dried red bell peppers. *J. Food Technol.*, 22, 21.

David, I. (1982) Up-to-date methods for red-pepper processing. *Konzerv-es paprikaipar*, 1, 34–39.

De la Mar, R. and Francis, F. G. (1969) Carotenoid degradation in bleached paprika. *J. Food Sci.*, 34, 287.

Desai, S. R. P., Sharka, A. and Amonkar, S. V. (1987) Disinfestation of whole and ground spices by gamma irradiation. *J. Food Sci. Technol.*, 24(6), 321–322.

Dte. of Marketing and Inspection (DMI), Govt. of India (1982).

Ferns, R. S. (1961a) Extraction of oleoresin from pimento (Capsicum), *Glemica Acta*, 13, 391.

Ferns, R. S. (1961b) Solvents for the extraction of oleoresin from pimento, *Glemica Acta*, 13, 391.

Goldman, A. (1949) How spice oleoresins are made. *Perfum. Essential Oil Res.*, 53, 320.

Govindarajan, V. S. and Ananthakrishna, S. M. (1970) Obsevation on the separation of the capsicum from capsicums and its oleoresin. *J. Food Sci. Technol.*, 7, 212.

Grimme, C. (1954a) The binding capacity of amulin for volatile oils. *Z. Lebensm-unters-forsch*, **98**, 440.

Grimme, C. (1954b) Preservation of seasoning strength of spices during fine grinding. *Brot. Gebaeck*, **8**, 118.

Harels, S., Palevitch, D. and Ben-Gera, I. (1977) Colour retention in sweet red paprika (*Capsicum annuum* L.) powder as affected by moisture contents and ripening stage. *J. Food Technol.*, 14(1), 59–64.

Heath, H. B. (1972) Herbs and Spices for Food Manufacture. *Proc. International Conference on Spices*, London, p. 39.

Heide, L. and Bogal, W. (1987) Identification of irradiated spices with thermo and chemiluminescence measurements. *J. Food Sci. Technol.*, 24(2), 93–103.

Hight, W. N. (1954) Thermal process for pimento canned in glass jars. *J. Food Technol.*, **8**, 298.

Ilyas, S. M., Sharma, H. S., Nag, D. and Samuel, D. V. K. (1987a) Development of a chilli (*Capsicum annuum* L.) punching machine. *J. Agril. Eng.*, 24(2), 223–226.

Ilyas, S. M., Sharma, H. S., Nag, D. and Samuel, D. V. K. (1987b) Effect of punching on drying rate of chillies. Abstr. No. 222. In: Poster session *Abstracts*, J. S. Pruthi (ed.), 9–10 April 1987, Nat. Symp. on Spice Industries. AFST (India), Delhi Chapter, p. 53.

Kachru, R. P. and Srivastava, P. K. (1990a) Status of chilli processing. *Spices India*, 3(1), 13–16.

Kachru, R. P. and Srivastava, P. K. (1990b) Status of chilli processing. *Spices India*, 3(2), 10–15.

Krishnamurthy, M. N. and Natarajan, C. P. (1973) Colour and its changes in chillies. *Indian Food Packer*, 27(1), 39–44.

Lantz, E. M. (1946) Effect of canning and drying on the carotene and ascorbic acid content of chillies. *New Mexico Agric. Res. Stn Bull.*, **327** (April), 21.

Laul, M. S., Bhale Rao, S. D., Rane, V. R. and Amla, B. L. (1970) Studies on the sun-drying of chillies. *Indian Food Packer*, 24(2), 22–25.

Lease, J. G. and Lease, E. J. (1956a) Factors affecting retention of red colour in peppers. *J. Food Technol.*, **10**, 368–375.

Lease, J. G. and Lease, E. J. (1956b) Effect of fat-soluble antioxidation the stability of red colour of pepper. *J. Food Technol.*, **10**, 403.

Lease, J. G. and Lease, E. J. (1962) Effect of drying conditions on initial colour retention and pungency in red peppers. *J. Food Technol.* (USA), **16**, 104–106.

Lewis, Y. S. and Natarajan, C. P. (1980) The need to promote production of chilli varieties to suit processed chilli products. Background paper. All India Workshop on Chillies, 22 March 1980, Hyderabad, Spices Export Promotion Council. Ernakulam, pp. 11–16.

Luhadia, A. P. and Kulkarni, P. R. (1978) Dehydration of green chillies. *J. Food Sci. Technol.*, 15(4), 139.

Mathew, A. G., Lewis, Y. S., Jagdishan, R., Nambudri, E. S. and Krishnamurthy, N. (1971a) Oleoresin capsicum. *Flavour Ind.*, **2**, 23–26.

Mathew, A. G., Nambudri, E. S., Ananthakrishna, S. M., Krishnamurthy, N. and Lewis, Y. S. (1971b) An improved method for estimation of capsaicin in capsicum oleoresin. *Lab. Pract.*, **20**, 856–858.

Miller, G. (1951) New Nitrogen technique assures fine grinding of spices in one pass. *Food Eng.*, **23**, 36.

Misra, B. D. (1962) Studies on the microbial technological aspects of curry powder and spice mixtures. Assoc., Thesis, CFTRI, Mysore.

Neal, M. W. and Klis, J. B. (1964) A guide to species, natural types, processed forms and uses in food manufacture. Food processing – special report.

Parry, J. W. (1945) *Handbook of Spices*. Chem. Pub. Co., New York.

Parvathi, S. and Aminuddin, Y. (2000) Chilli – an economically profitable crop of Malaysia. *Spice India*, 13(7), 18–19.

Philip, T. and Francis, F. G. (1971a) Oxidation of capsaicin. *J. Food Sci.*, **36**, 96.

Philip, T. and Francis, F. G. (1971b) Isolation and chemical properties of capsanthin and its derivatives. *J. Food Sci.*, **36**, 1823.

Portale, J. V. and Nozinibia, P. G. (1967) Extraction of pepper oil. *Spanish Patent*, **321**, 980.

Powers, J. J. (1961) Effect of acid level, calcium salts, monosodium glutamate and sugar on canned pimento. *J. Food Technol.*, **15**, 67.

Pruthi, J. S. (1969) International collaborative studies on the pungency (capsaicin) and colour (capsanthin) content of Hungarian ground red paprika. Central Agmark Lab., DMI, Nagpur (India). Tech. Communication No. 30/69.

Pruthi, J. S. (1970) Packaging of spices and condiments. *J. Packaging India*, 3(1), 11.

Pruthi, J. S. (1980) Spices and Condiments: Chemistry, Microbiology and Technology. 1st edn, pp. 1–450, Academic Press Inc, New York, USA.

Pruthi, J. S. (1985) Recent trends in packaging of spices. *Souv. Semin. Innovations in Packaging of Processed Foods*, 17–18 May 1985, Small Industries Service Institute, New Delhi, pp. 1–7.

Pruthi, J. S. (ed.) (1987a) Spice industries: present scenario, problems and prospects. *Tech. Compendium*. AFST (India).

Pruthi, J. S. (ed.) (1987b) *Bibliography on Spices (1972–1986)*, Assoc. Food Science and Tech., Delhi, India.

Pruthi, J. S. (1989a) Spice extractives (essential oils and oleoresins). Present Scansion and Prospects. Invited status paper presented at the 11th International Congress of Essential Oils, Fragrances and Flavours, New Delhi. Proceedings vol. 6, pp. 217–243.

Pruthi, J. S. (1989b) Post-harvest technology of spices and condiments: pretreatments, curing, drying, cleaning, grading and packing – a commissioned status document presented on behalf of the Commonwealth Secretarial, London, at the 2nd International Spice Group (ISG) Meeting held in Singapore, 6–11 March 1989, pp. 1–30.

Pruthi, J. S. (1991) Advances in spice processing technology. Presented at the 3rd meeting of International Spice Group, Jamaica.

Pruthi, J. S. (1992) Post-harvest technology of spices: pre-treatments, curing, cleaning, grading and packing. *J. Spices Aromatic Crops*, 1, 1–29.

Pruthi, J. S. (1993a) Innovations in post-harvest technology of spices. In: *Advances in Horticultural Research*, vol. X, Plantation Crops (ed. K.L. Chadha), pp. 1255–1282.

Pruthi, J. S. (1993b) *Major Spices of India – Crop Management and Post-Harvest Technology*, ICAR, chapter 4 on Chillies, pp. 1–94.

Pruthi, J. S. (1998) Major Spices of India, 2nd reprint, Indian Council of Agril Res. (ICAR), New Delhi, chapter 4 on Chillies/Capsicums. pp. 180–243. Represented: 1998.

Pruthi, J. S. (1999) *Quality Assurance in Spices and Spice Products: Modern Methods of Analysis*. 1st edn. pp. 1–576. Allied Publishers Ltd, New Delhi.

Pruthi, J. S. (2001) Spices and Condiments, National Book-Trust of India, New Delhi, 6th edn, pp. 48–58.

Pruthi, J. S. (2001) Advances in Super-Critical Fluid Extraction of (SCFE) technology of spices (including chillies) – a global overview future R&D needs. *J. Bev. Food World*, 28(1), 44–55. (About 44 Spices reviewed critically.)

Pruthi, J. S. and Lakshmi Shankar (1968) Rapid determination of moisture in chillies. *Spices Bull. Chillies Seminar*, 6(3), 56–58. Spices Export Promotion Council, Ernakulam.

Pruthi, J. S. and Misra, B. D. (1963) Physicochemical and micrological changes in curry powders during drying, milling and mixing operations, *Spices Bull.*, CFTRI, Mysore, 3, 8.

Pruthi, J. S., Lakshmi Shankar and Kulkarni, R. J. (1967) Selection of moisture meter for the rapid determination of moisture in agricultural commodities (chillies-whole and ground and other spices). *Agmark Tech. Communication No. 1*, Central Agmark Laboratory, DMI, Nagpur.

Purseglove, J. W., Brown, E. G., Green, C. L., Robins, S. R. J. (1981) '*Spices*' vol. I, Longman, London.

Regional Research Laboratory (RRL), Jammu (1978) Solar drying. *RRL News letter*, 5(5), 4.

Rosebrook, D. D. (1971) Collaborative study of a method for the extractable colour in paprika oleoresin. *J. Assoc. Off. Anal. Chem.*, 54, 37.

Salzer, U. J. (1975) Fatty acid composition of the lipids of some species. *Fette-Sifer Anstrichim*, 77, 446.

Sane, R. H. (1950) The pH and total acidity of raw and canned pimentos. *J. Food Technol.*, 4, 279.

Shivhare, U. S., Narain, M., Agrawal, U. S., Saxena, R. P. and Singh, B. P. N. (1987) Some physico-mechanical properties of chillies. *J. Food Sci. Technol.*, 24(2), 98–99.

Siddappa, G. S. and Bhatia, B. S. (1955) Canning of chillies. *Bulletin,* CFTRI, Mysore, 4, 9.

Singh, H. P. and Alam, A. (1982) Techno-economic study on chilli drying. *J. Agril. Eng.*, 19(1), 23–32.

Spices Board (1995–2000) Annual Spices Export Review.

Szabo, P. (1969) Manufacturing paprika (capsicum) oleoresin. *Kenserve paprikaip. Acta*, 13, 391.

Szabo, P. (1970) Production of paprika oleoresin. *Amer. Perfum. and Cosmet.*, 85, 39.

Tandon, G. L., Dravid, S. V. and Siddappa, G. S. (1964) Oleoresin of capsicum. *J. Food Sci.*, 29, 1–5.

Van Blaricum, L. O. and Martin, J. A. (1945) Retarding the loss of red colour in cayenne pepper with oil soluble antioxidants. *Food Technol.*, 5, 35.

Wealth of India – Raw Materials, Vol. C (III) 2nd edn. Directorate of Publications and Information, CSIR, New Delhi (1985). (Under *Capsicum* genus, all species are discussed including chillies, capsicums, paprika, red pepper, Bird chillies, etc.)

Winton, A. L. and Winton, K. B. (1939) *Structure and Composition of Foods*, Vol. IV, Willey Inc., New York, USA.

Wistreich, H. E. and Schafer, W. E. (1962) Freeze grinding up products to quality. *Food Eng.*, 34, 62.

12 The storage of *Capsicum*

H. S. Yogeesha and Rame Gowda

Capsicum fruits, whether they are used as green or red ripe fruit, fleshy or dried, as medicine or dye, or for seed purposes, are to be stored for varying periods before being used. Therefore, an attempt has been made in this chapter to review the research work done on storage aspects of green and red ripe fruits in processed or whole fruit form. Moreover, special emphasis has been given to the storage of seeds, as *Capsicum* is a seed propagated crop and its productivity and quality are very much influenced by the quality of seeds used for production.

Introduction

Capsicum fruits, both red ripe and green, are used for imparting pungency and flavour to food. Sweet peppers are mainly used at the green stage as salad and for cooking. In addition to providing pungency and flavour to the food, *Capsicum* is also a good source of vitamins A and B, and has several medicinal and insecticidal properties (Zibokere, 1994). Recently, the paprika types have been used extensively as a source of natural dye because of their deep red fruits. *Capsicum* fruits, whether they are used as green or red ripe fruit, fleshy or dried, as medicine or dye or for seed purposes, are to be stored for varying periods of time before they are used. Therefore, it is very important that the fruits retain their original characteristics and suffer very little deterioration during storage. Thus, the storage of *Capsicum* needs special attention.

Storage of green fruits

There is much demand for fresh, green, pungent and sweet peppers as they are a major ingredient of spicy foods, especially of Indian food items. Demand for fresh pungent peppers has greatly increased in the United States as south-western foods have become the most popular ethnic food (Lownds *et al.*, 1994). Freshly harvested bell peppers have a shelf-life of only a few days since they lose water rapidly after being harvested (Anandaswamy *et al.*, 1959; Ryall and Lipton, 1972; Showalter, 1973; Watada *et al.*, 1987). Extending post-harvest longevity to meet the demand for fresh peppers, therefore, requires special attention.

Research on extending the shelf-life of fresh peppers is rather sparse. Loss of water content, flaccidity, changes in colour and development of diseases are some of the physical changes that occur during storage and which have a direct impact on the loss of fruit quality. The rate of water loss, flaccidity and colour development increase with an increased temperature (Lownds *et al.*, 1994). Placing pepper fruits in perforated polythene pouches at 14–20°C reduced water loss rates 20 times or even more, eliminated flaccidity and reduced colour development. The development of diseases in packed fruits, however, results in a reduction of post-harvest quality (Lownds *et al.*, 1994; Miller *et al.*, 1984; Maiero and Waddell, 1991). In contrast,

Ben Yehoshua *et al.* (1998) reported that bell peppers packaged in perforated film lost less weight during storage and maintained a higher quality than fruits stored in open boxes. They also demonstrated that lower decay levels occurred in the fruits stored in perforated films compared to the non-perforated packing.

Post-harvest water loss of fruits is the major factor influencing the quality of the fresh fruits of peppers. Water loss was found to be positively correlated with the initial water content, ratio of surface area and volume and the cuticular thickness of the fruits and negatively correlated with surface area and epicuticular wax content and hence, is cultivar dependent (Lownds *et al.*, 1993).

Shelf-life of fruits with thick exocarp is more compared to fruits with thin exocarp. However, the fruits with thick exocarp are tough, difficult to digest and, if processed, the exocarp peels off from flesh, and not preferred. On the other hand, fruits with very thin exocarp which are preferred by consumers and processing industry, are tender and readily bruised, cracked and crushed during post-harvest operations (Fischer, 1992, 1993). These mechanical injuries affect pepper quality and subsequent shelf-life which ultimately reduce the market grade. To reduce the damage during post-harvest operations all locations should be cushioned where peppers impact a hard surface and drop height should be limited to 8 cm on hard surface and 20 cm on a cushioned surface (Marshall and Brook, 1999).

Storage of fruits under controlled conditions was found to enhance their longevity (Yuen and Hoffman, 1993; Lownds *et al.*, 1994; Ahmed *et al.*, 1996; Mencinicopschi and Popa, 1999). Fruits can be stored for longer periods (14–28 days) at 8°C and more than 75% RH (relative humidity) with minimum change in chemical composition and market acceptability (Lownds *et al.*, 1994; Ahmed *et al.*, 1996). Further, storage can be extended for another six to eight months at freezing temperature with or without blanching (Lisiewska *et al.*, 1994; Mencinicopschi and Popa, 1999). However, Sundstrom (1992) reports that bell peppers are subject to chilling injury, characterized by pitting of the fruits, if stored in temperatures less than 7°C. At storage temperatures above 10°C, however, further fruit ripening and the development of anthocyanin and red carotenoid pigments will occur. At these warmer temperatures, bacterial soft rot also becomes a major problem. Hence, the temperature range of 7–10°C at 90–95% RH is ideal for the storage of bell peppers for up to three weeks.

From these studies it is observed that water loss is no longer a limiting factor of fresh fruit longevity if the fruits are stored in a modified atmosphere or in perforated polythene film. However, the limiting factor under these conditions is the incidence of diseases. Bell peppers, after harvest, are graded and frequently run through a hot (53°C) water bath containing 500 ppm chlorine to control bacterial rots. Following this process, most bell pepper fruits are sprayed with a wax emulsion to reduce moisture loss prior to being packed in cardboard cartons (Sundstrom, 1992). In another study, Fallik *et al.* (1996) found that dipping naturally infected fruits in water at 50°C for three minutes completely inhibited or significantly reduced the fruit decay caused by *Botrytis cinera* and *Alternaria alternata* without any adverse effect on fruit quality.

Since these studies are independent of storage conditions and specific to only a few pathogens, further studies are required to develop suitable storage practices in order to protect fruits from storage decay.

Storage of red ripe fruits

The quality of red chilli is based on colour, pungency and their retention during storage. The development of red colour is attributed to the presence of about 20 carotenoids, of which capsanthin is the major one. Pungency is due to the mixture of seven homologous branched chains of alkyl vanillyl amides, namely, capsaicinoids (Hoffman *et al.*, 1983) and these are produced in the glands of the fruit's placenta.

Factors affecting the quality of red chilli during storage

Colour loss during storage makes red chilli unacceptable to consumers, even though no change occurs in flavour and aroma. It has been found that colour impairment in red chilli starts when the fruits have lost about one-third of their water content and become over-ripe (Biacs *et al.*, 1989; Daood *et al.*, 1989). This colour impairment is accelerated several-fold as a result of the grinding and storage of paprika (Czinkotai *et al.*, 1989) and is probably due to the development of heat at the time grinding. Positive correlations with high regression coefficient values have been obtained between colour retention by paprika powders and the initial concentrations of tocopherol and ascorbic acid (Biacs *et al.*, 1992). Further, cultivars have showed substantial variability with respect to carotenoid composition and endogenous antioxidant content (Usha Rani, 1996; Gomez *et al.*, 1998) and those with higher levels of ascorbic acid and tocopherol have shown greater colour stability during storage in powder form. Although there is an inverse relationship between the degree of pungency and the amount of deterioration during storage, no such correlation between capsanthin content and reduction in quality could be established (Usha Rani, 1996).

Fruit drying is the most important step that determines the final quality of the product. In India, chilli growers of small holdings dry the fruits using traditional methods. The methods involve sun drying by spreading fruits on either a cement floor, a mud floor, a floor smeared with cow dung, a zinc sheet, a polythene sheet, a granite floor or on the roof of red tiles. In a study involving these traditional methods, including oven drying at 65°C, it has been observed that drying the fruits on the plant itself and in a hot air oven was found to be better as these methods have resulted in a lower percentage of whitened fruits compared to the other methods (Nagaraja *et al.*, 1998a). However, drying fruits on the plants may not always be possible because of bad weather conditions. Among the traditional methods, drying under house shade or on open ground and even on cement floors were found to be better and practically feasible, especially for large scale drying of fruits, particularly in the tropical countries where sunlight is not a limiting factor. Moreover, this type of drying would reduce the percentage of whitened fruits.

The industrial drying of pepper fruits for paprika production involves mainly two methods (Minguez *et al.*, 1994) namely, oven drying and smoke drying. Each of the drying processes has its own merits and demerits that effect the stability of the final product. One such disadvantage of rapid oven drying at high temperature is the loss of colour stability during storage (Ramakrishnan and Francis, 1973). On the other hand, slow drying using wood smoke could result in degradative processes which are associated with the post-harvest physiology, as well as rotting arising from breakage, infection, etc. The possible colour stability produced by drying at low temperatures can be diminished by the presence of highly reactive chemicals formed during the smoking process, but these can provide an efficient protection against any infection (Minguez *et al.*, 1994). However, the same authors in another study comparing the two industrial drying processes noticed a greater loss of carotenoid content in oven drying than in smoke drying. In fact, they have also observed an increase in concentration of some of the pigments during smoke drying, although the final carotenoid content of the dried fruits is less than that of fresh fruits. The loss of carotenoid content during drying and storage is attributed to a drastic reduction in the concentration of antioxidants, such as tocopherol and ascorbic acid, as a result of the antioxidation process (Daood *et al.*, 1996). Further, the rate of pigment degradation is greater at a higher storage temperature and lower relative humidity (Gomez *et al.*, 1998).

These studies suggest that the loss of fruit quality in terms of vitamins and their precursors during drying is very much dependent on the temperature and the rate of drying. Further, drying processes also determine the subsequent storability of chillies.

Dried chillies, when stored, are often attacked by the drugstore beetle *Stegobium paniceum* L. and the cigarrette beetle *Lasioderma serricorne* (Fabricius). The *Arthrodeis* species feed on dried chillies, though the loss caused by them is negligible. If the quantity of chillies infested by these storage pests is small, spreading the fruits in thin layers and exposing them to sunlight will eliminate the infection. If large quantities are infested with these pests, fumigation is the only remedy. Any good fumigant like ethyl dibromide or methyl bromide at one gram per 30 cubic metre space may be used. Fumigation should be done in airtight containers for the effective control of these stored pests (Dhamo *et al.*, 1984).

Capsicum storage for seed purpose

In India, seeds are traditionally extracted from dry fruits only at the time of sowing. The reason is that the seeds maintain viability for a longer period when they are in fruits than the seeds extracted and stored (Murthy and Murthy, 1961; Radhe Sham *et al.*, 1996). Seeds maintain a higher germination value (up to six months) when retained in fruits and stored under ambient conditions (Nagaraja *et al.*, 1998b). However, the storage of fruits for seed purposes is not commercially feasible for large-scale production, as the storage process needs large areas or storage structures and other infrastructural facilities for the maintenance of the seeds. Chilli seed is a relatively poor storer and its storability is influenced by many intrinsic and extrinsic factors.

Factors influencing the storability of chilli seeds

Several pre- and post-harvest factors influence the longevity of chilli seeds during storage. Some of these are discussed here.

Cultivars

Much variability is found among chilli genotypes for their seed storability (Thakur *et al.*, 1988; Ili, 1995; Radhe Sham *et al.*, 1996; Sharma *et al.*, 1998; Nagaraja *et al.*, 1998b; Patil and Nagaraja, 1999). It has been observed that few varieties could maintain viability even up to 18 months and some could only last for a few months when stored under ambient conditions (Nagaraja *et al.*, 1998b). This sort of cultivar difference could be made use of by the breeders while selecting the parents for the breeding programme.

Stage of fruit maturity

A seed attains maximum quality when it reaches physiological maturity. Harvesting seeds before or after physiological maturity would affect the seed quality in storage. The extent of seed deterioration is much lower if the seeds are harvested after physiological maturity rather than before. In *Capsicum*, fruits are harvested at the red ripe stage and seeds are extracted either immediately from the fleshy fruits or after the fruits have been dried. Usually red chillies are extracted after drying and wet extraction is followed in bell peppers as the fleshy pericarp of bell peppers takes a longer time to dry and is hence susceptible to fruit rotting. It is suggested that bell pepper fruits should not be harvested before the red ripe stage and seeds should remain in the fruit for a short period of post-harvest ripening to achieve maximum seed germination potential (Sanchez *et al.*, 1993). *In-situ* priming may occur during post-harvest maturation. According to Dhanelappagol *et al.* (1994), full red colour fruits in red chilli produce higher quality seeds compared to half red fruits and thus suggest that for seed purposes, fruits be picked when they have fully ripened on the plant. Demir and Ellis (1992) have observed highest seed quality, in terms of viability and seedling growth, in the seeds extracted from the fruits 10–12 days after physiological maturity and stored under ambient conditions. In another study, Oladiran and Agunbiade (2000) observed an improvement in seed germination and seedling development from *Capsicum* seeds aged for six weeks. These studies suggest that *Capsicum* seeds require a short ripening period before and after the extraction of seeds.

Fruit drying

Seed moisture plays an important role in maintaining seed viability during storage. The moisture content in seeds and fruits at the red ripe stage, i.e. the stage of harvesting, is too high and they cannot be stored unless they are dried down to safe moisture levels. In red chilli, the fruits are dried using traditional methods or by using forced air. Traditional methods of drying result in better quality seeds (Nagaraja *et al.*, 1998a). Further, the fruits should not be dried on red tiles, asbestos sheets or in shade, all of which result in the reduction of seed quality due to the development of high temperature. However, under shade drying, the seeds can become infected with mycoflora due to the slow drying process.

Seed moisture and storage temperature

Capsicum seed can withstand desiccation tolerance. Some of the earliest work on storage of vegetable seeds was done by Barton (1935) who found that pepper seed germination began to decline significantly at 10.5% moisture after two years of storage at room temperature. Further, the reduction of seed moisture to 5.2% did not change this decline but increased relative seed germination percentages for a given storage time. However, when the same seeds were held at $-5°C$ germination was not appreciably affected by storage time regardless of moisture content (Barton and Garman, 1946).

Capsicum seeds can be stored for 18–30 months under ambient conditions if they are dried to 6–8% moisture levels (Thiagarajan, 1994; Sharma *et al.*, 1998). This suggests that *Capsicum* seeds can be safely stored under ambient conditions provided the initial seed moisture is less than 8%. Storage of seeds in moisture vapour proof containers requires seeds with a lower moisture content. The seed moisture content of 6% is the upper limit for sweet pepper to be stored in airtight containers (Song *et al.*, 1999), while red chilli seeds could be stored for more than two years in 700 gauge polythene bags if the seed moisture was around 6% (Sharma *et al.*, 1998).

Seed moisture, or the RH of the storage atmosphere, and temperature are the two important factors that have profound influences on seed viability, both independently or in combination. The higher the storage temperature and RH, the higher the rate of seed deterioration in storage and *vice-versa*. The effect of one factor can substantially be altered by the other factor. This has been observed by Barton and Garman (1946) in bell peppers where the seeds at 10.5% and 5.2% moisture levels behaved differently during storage under ambient conditions but there was not much difference when stored at $-5°C$. Similarly, it has also been shown that the freeze-drying of pepper (*C. annuum*) seeds to a moisture content of 2.5% is superior to 7.3% during storage at 21–25°C or 40°C.

The method of ultra drying is gaining importance as seed longevity can be increased significantly during storage. However, not all kinds of seeds are tolerant to ultra drying. Nutile (1964) working with vegetable seeds observed severe desiccation injury in the case of *C. frutescens* L. when seed moisture was reduced from 4% to 1%. In the case of *C. annuum* L., seeds could retain higher viability in long-term storage for 15 years at 6.1% than at 2.6% moisture (Nakamura, 1975).

Shen and Qi (1998), working on short- and long-term effects of ultra drying on the germination and growth of vegetable seeds, reported the detrimental effects of drying seeds to less than 2% moisture level and concluded that there was an optimum water content of seeds for safe storage. However, Wu *et al.* (1999) have reported that the ultra drying of pepper seeds greatly improves their longevity. Based on viability and dormancy studies of tabasco pepper seeds, Sundstrom (1990) has concluded that the optimum moisture level for low temperature storage is approximately 10%.

In view of these conflicting reports on the ultra drying effects on seed longevity, an in-depth study is very much required to determine the level of drying and its parameters to prolong the storage life of chilli seeds.

Zhang and Kong (1996) have reported that sweet pepper seeds can be stored successfully at higher temperatures and medium moisture levels. For long-term storage of seeds, as in the case of germplasm preservation, the usual practice is to store the seeds at a low or ultra low temperature (liquified nitrogen, −196°C). Sweet pepper seeds can be stored for very long periods at lower temperatures if they are held in airtight containers (Thakur *et al.*, 1988; Simay and Horvath, 1992; Fischer, 1994; Ili, 1995). In contrast, Belletti *et al.* (1990) noticed an appreciable reduction in the germination percentage of pepper seeds (*C. annuum* L.) at a lower moisture level stored in liquid nitrogen. However, storing seeds at low or ultra low temperature is more cumbersome and expensive and there is every possibility of incurring a change in genetic constitution.

Packaging materials

At a given storage moisture and temperature, seed longevity may depend on the storage containers. Packaging materials used for storing seeds include aluminium foil, polythene bags of less than 700 gauge, cloth and jute bags with or without polythene lining, paper covers, friction top cans and rigid plastic containers without a gasket. *Capsicum* seeds stored in moisture proof containers are found to maintain viability for longer periods than those stored in moisture pervious containers (Oladirn and Agunbiade, 2000), provided the seeds are dried to a moisture level that is safe for such storage.

Storage pests and diseases

In storage, chilli seeds are attacked by several storage fungi, such as *Aspergillus* spp., *Penicillium* spp., *Rhizopus* spp. These storage fungi cause seed discolouration and loss of viability during storage. Storage fungi are active only when the seed moisture is above 14%. This is possible only when the seeds are stored under ambient conditions. Seeds stored under airtight containers are safe from storage fungi because the initial seed moisture is not more than 6–7%. Seed treatment with thiram has no effect on the viability of chilli seeds when stored in moisture impervious polythene bags (Sharma *et al.*, 1998), whereas the seeds treated with captan maintain germination at a higher level than untreated seeds (Thiagarajan, 1994). *Colletotrichum dematium*, an important field fungi which causes anthracnose or die-back disease, is known to be seedborne in nature and seed treatment is the only effective measure for controlling this pathogen. Seed treatment with deltan controls this pathogen very effectively. The seed treatment can be done before storage, as it does not adversely affect the viability of seeds under storage conditions. These results suggest that the chilli seed must be dried to a safe moisture level and treated with suitable fungicides to prolong their storage life under ambient conditions.

Conclusion

The storage life of green chillies harvested for vegetable purposes can be extended up to four to six weeks under low temperature and high RH conditions. However, there are contradictory reports on the ideal low temperature range for storage as chilling injury to fruits has been reported at temperatures below 6°C. This aspect of storage requires further study, with a greater number of genotypes, to know if there is any genetic variability for temperature sensitivity. Under ambient conditions, the storage life of fruits can be extended by packing them in perforated polythene film.

Loss of colour during storage is the major concern in dried red chilli. It has been found that the retention of colour during storage is directly correlated with the amount of tocopherol and ascorbic acid present in the fruits and thus varies with the cultivars. Drying temperature is another factor that affects the colour retention during storage; high temperature is harmful as it results in the degradation of antioxidants. Higher storage temperature increases the rate of pigment degradation but an increase in RH level reduces the degradation (Gomez et al., 1998).

The storage of seeds in fruits is found to maintain viability for longer periods but it is not practically feasible for large-scale seed production. The fruits harvested at the red ripe stage yield good quality seeds, although some studies showed the need for a post-ripening period before and after seed extraction. There is a lot of variation among cultivars for seed longevity during storage. The work on seed drying is negligible. Seeds with a moisture content of 6–8% and stored in airtight containers could maintain seed viability for two years. For long-term storage, seeds can be stored for many years if stored at lower temperatures in airtight containers. There are conflicting reports on ultra drying effects on seed longevity and this aspect of drying requires further study.

References

Ahmed, N., Tanaki, M. I., Mir, M. and Shah, G. A. (1996) Effect of different fruit maturity stages and storage conditions on chemical composition and market acceptability of fruits in different varieties of sweet pepper. *Capsicum Eggplant Newsletter*, 15, 47–50.

Anandaswamy, B., Murphy, H. B. N. and Iyengar, N. V. R. (1959) Prepacking studies on fresh produce: *Capsicum annuum* var. *grossum* (Sent.) and *Capsicum annuum* var. *acuminatum* (Fingh.). *J. Sci. Ind. Res.*, 18, 274–278.

Barton, L. V. (1935) Storage of vegetable seeds. *Contribution of the Boyce Thomson Institute*, 7, 323–332.

Barton, L. V. and Garman, H. R. (1946) Effect of age and storage condition of seeds on the yields of certain plants. *Contribution of the Boyce Thomson Institute*, 14, 243–255.

Belletti, P., Lanteri, S., Lepori, G., Nassi, M. O. and Quagliotti, L. (1990) Factors related to the cryopreservation of pepper and eggplant seeds. *Advan. Hort. Sci.*, 4, 118–120.

Ben Yehoshua, S., Rodov, V., Fishman, S. and Peretz, J. (1998) Reducing condensation of water in bell peppers and mangoes. In: Bon Yehoshua (ed.) *Recent Developments in Modified Atmosphere Packaging of Fruits and Vegetables*, Lasa Pages Publishing, Jerusalem, Israel, pp. 495–504.

Biacs, P. A., Czinkotai, B. and Hoschke, A. (1992) Factors affecting stability of coloured substances in paprika powder. *J. Agric. Food Chem.*, 40, 364–367.

Biacs, P. A., Daood, H., Pavisa, A. and Hajdu, F. (1989) Studies on the carotenoid pigments of paprika. *J. Agric. Food Chem.*, 37, 350–353.

Czinkotai, B., Daood, H., Biacs, P. and Hajdu, F. (1989) Separation and detection of paprika pigments by HPLC. *J. Liq. Chromatogr.*, 12, 2707–2717.

Daood, H., Biacs, P. A., Kiss-Kutz, N. and Czinkotai, B. (1989) Lipid and antioxidant content of red pepper. In: P.A. Biacs, K. Gruiz and T. Kremmer (eds) *Biological Role of Plant Lipids*, Akademia Kiado, Budapest, pp. 491–494.

Daood, H. G., Vinkler, M., Markus, F., Hebshi, E. A. and Biacs, P. A. (1996) Antioxidant vitamin content of spice red pepper (Paprika) as affected by technological and varietal factors. *Food Chem.*, 55, 365–372.

Demir, I. and Ellis, R. H. (1992) Development of pepper (*Capsicum annuum*) seed quality. *Annals. Appl. Biol.*, 121, 385–399.

Dhamo, K., Butani and Jotwani, M. G. (1984) *Insects in Vegetables*. Periodical Expert Book Agency, Delhi, India, pp. 35–42.

Dhanelappagol, M. S., Shashidhara, S. D. and Kulkarni, G. N. (1994) Effect of stages of harvesting methods of drying of chilli on seed quality. *Karnataka J. Agric. Sci.*, 7, 36–37.

Fallik. E., Grinberg, S., Alkalai, S. and Lurie, S. (1996) The effectiveness of postharvest hot water dipping on the control of grey and black moulds in sweet red pepper (*Capsicum annuum*). *Plant Path.*, 45, 644–649.

Fischer, I. (1992) The role of exocarp thickness in the production, consumption and selection of paprika for consumption. *Capsicum Newsletter* (*Spl. issue*), 106–109.

Fischer, I. (1993) The role of exocarp thickness in the production, consumption and selection of eating capsicum. *Zoldsegtermesztesi Kutato Inteset Bulletinje*, 25, 37–42.

Fischer, I. (1994) Sweet pepper gene conservation in long term storage. *Zoldsegtermesztesi Kutato Inteset Bulletinje*, 26, 47–54.

Gomez Lardronde Guevara, R., ParraLopez, V., PardoGonzalez, J. E., Saus, M. L. A. and Varon Castellanos, R. (1998) Influence of storage conditions on pigment degradation in paprikas from different greenhouse pepper cultivars. *J. Sci. Food Chem.*, 78, 321–328.

Hoffman, P. G., Lego, M. C. and Galatto, W. G. (1983) Separation and quantification of red pepper major heat principles by reverse-phase high performance liquid chromatography. *J. Agric. Food Chem.*, 31, 1326–1330.

Ili, Z. (1995) Possibility of long term storage of red pepper seed. *Selekcija, Semenarstvo*, 2, 209–211.

Lisiewska, Z., Kmiecik, W. and Jaworska, G. (1994) Suitability of Polish field grown sweet pepper cultivars for freezing and pickling. *Raczniki Postwowego zakladu Higieny*, 45, 311–320.

Lownds, N. K., Banaras, M. and Bosland, P. W. (1993) Relationships between postharvest water loss and physical properties of pepper fruit (*Capsicum annuum*. L.). *Hort. Sci.*, 28, 1182–1184.

Lownds, N. K., Banaras, M. and Bosland, P. W. (1994) Postharvest water loss and storage quality of ninepepper (*Capsicum*) cultivars. *Hort. Sci.*, 29, 191–193.

Maiero, M. and Waddell, C. (1991) Postharvest diseases of packaged green chilli peppers. *Hort Sci.*, 26, 694.

Marshall, D. E. and Brook, R. C. (1999) Reducing bell pepper bruishing during postharvest handling. *Hort. Tech.*, 9, 254–258.

Mencinicopschi, G. and Popa, M (1999) Freezing stability of some snap bean (*Phaseolus vulgaris*) and pepper (*Capsicum annuum*) varieties. In: M. Hagg, R. Ahvenainen, A. M. Evers and K. Tiilikkala (eds) *Agri-Food Quality II: Quality Management of Fruits and Vegetables – from Field to Table*, Turnku, Finland, pp. 161–163.

Miller, W. R., Spalding, D. H., Risse, L. A. and Chew, C. (1984) The effect of an imazalil to control decay of bell peppers. *Proc. Fla. State Hort. Soc.*, 97, 108–110.

Minguez Mosquerq, M. I., Jaren Galan, M. and Garrido-Fernandez, J. (1994) Influence of the industrial drying processes of pepper fruits (*Capsicum annuum* cv. Bola) for paprika on the carotenoid content. *J. Agric. Food Chem.*, 42, 1190–1193.

Murthy, N. S. R. and Murthy, B. S. (1961) Chilli seed viability in relation to period of storage. *Andhra Agric. J.*, 8, 246–247.

Nagaraja, A., Basavaraja, P. K. and Yogeesha, H. S. (1998a) Effect of drying methods on fruit and seed quality of chilli. *South Indian Hort.*, 46, 262–265.

Nagaraja, A., Jagadeesha, R. C. and Yogeesha, H. S. (1998b) Storability of seeds of chilli genotypes under ambient conditions. *Karnataka J. Agric. Sci.*, 11, 807–809.

Nakamura, S. (1975) The most appropriate moisture content of seeds for their long life span. *Seed Sci. Technol.*, 3, 747–759.

Nutile, G. E. (1964) Effect of desiccation on viability of seeds. *Crop Sci.*, 4, 325–328.

Oladiran, J. A. and Agunbiade, S. A. (2000) Germination and seedling development from pepper (*Capsicum annuum* L.) seeds following storage in different packaging materials. *Seed Sci. Technol.*, 28, 413–419.

Patil, K. N. and Nagaraja, A. (1999) Effect of storage methods on quality of chilli seeds. *Karnataka J. Agric. Sci.*, 12, 74–80.

Radhe Shyam, Arora, S. K., Tomer, R. P. S. and Sham, R. (1996) Effect of seed extraction interval on seed quality of chilli (*Capsicum annuum* L.) cultivars. *Haryana Agril. Univ. J. Res.*, 26, 183–186.

Ramakrishnan, T. V. and Francis, F. J. (1973) Colour and carotenoid changes in heated paprika. *J. Food Sci.*, 38, 25–28.

Ryall, A. L. and Lipton, W. J. (1972) *Handling Transportation and Storage of Fruits and Vegetables*, Vol. 1. AVI, Westport, Conn., pp. 11–14.

Sanchez, V. M., Sundstrom, F. J., McClure, G. N. and Lang, N. S. (1993) Fruit maturity, storage and postharvest maturation treatments affect bell pepper (*Capsicum annuum* L.) seed quality. *Scientia Horticulturae*, 54, 191–201.

Sharma, S. N., Goyal, K. C. and Kakralya, B. L. (1998) Chilli seed storage in relation to variety and container. *Seed Res.*, 26, 83–86.

Shen Di and Qi Xiao Quan (1998) Short and long term effects of ultradrying on germination and growth of vegetable seeds. *Seed Sci. Res.*, 8. (Suppl. No. 11), 47–53.

Showalter, R. K. (1973) Factors affecting pepper firmness. *Proc. Fla. State Hort. Soc.*, 86, 230–232.

Simay, E. I. and Horvath, Z. (1992) Results of seed tests XIV: on germination parameters registered for seed samples of *Capsicum annuum* L. *Capsicum Newsletter*, 11, 39–40.

Song, S. H., Zheng, X. Y. and Xing, B. T. (1999) Effect of moisture content on vegetable seed vigour in air tight package. *Acta Agricultural Boreali Sinica*, 14, 129–132.

Sundstrom, F. J. (1990) Seed moisture influence on tabasco pepper seed viability, vigour and dormancy during storage. *Seed Sci. Technol.*, 18, 179–185.

Sundstrom, F. J. (1992) Peppers, In: J. M. Swaider, G. W. Ware and J. P. Mc Collum (eds) *Producing Vegetable Crops*. International Book Distributing Co., Lucknow, India, pp. 421–433.

Thakur, P. C., Joshi, S., Verma, T. S. and Kapoor, K. S. (1988) Effect of storage period on germination of sweet pepper seeds. *Capsicum Newsletter*, 7, 58–59.

Thiagarajan, C. P. (1994) Storability of chilli (*Capsicum annuum* L.) seed. *Madras Agril. J.*, 81, 442–445.

Usha Rani, P. (1996) Evaluation of chilli germplasm for capsanthin and capsaicin contents and effect of storage on ground chilli. *Madras Agric. J.*, 83, 288–291.

Watada, A. E., Kim, S. D., Kim, K. S. and Harris, T. C. (1987) Quality of green beans, bell peppers and spinach stored in polyethylene bags. *J. Food Sci.*, 52, 1637–1641.

Wu. C. J., Fan S. Y. and Luo, G. F. (1999) A study on maintaining and improving the vigour of pepper seeds. *Acta Agricultural Universities Jiangxiensis*, 21, 505–508.

Yuen, C. M. C. and Hoffman, H. (1993) New capsicum varieties: storage suitability and consumer preference. *Food Australia*, 45, 184–187.

Zhang, H. Y. and Kong, X. H. (1996) Tests on effect of vegetable seeds for many months. *Acta Agiculturae Boreali-Sinica*, 11, 118–123.

Zibokere, D. S. (1994) Insecticidal potency of red pepper (*Capsicum annuum*) on pulse beetle (*Callosobruchus maculatus*) infesting cowpea (*Vigna unguiculata*) seeds during storage. *Indian J. Agril. Sci.*, 12, 74–80.

13 Current requirements on paprika powder for food industry

H. J. Buckenhüskes

Paprika powder is one of the most important spices surpassed by only the original pepper. Seventy per cent of the spice is used for industrial purposes, and in meat products, soups, sauces and snacks. The traditional quality attributes are taste, pungency, colour intensity and stability. Due to increasing safety requirements in food processing techniques, additional attributes like microbial status and the possible occurrence of mycotoxins are gaining importance. The modern strategies in agricultural and food technological production are guided by the "field to fork", idea. Therefore, considerations about paprika quality have to be extended to include horticultural aspects.

Introduction

The genus *Capsicum* belongs to the Solanaceae and comprises five domesticated species: *C. annuum, C. frutescens, C. chinense, C. baccatum* and *C. pubescens*. However, only members of the first three species are of industrial significance. The main distinctive features of the genus are the size of the fruits, the thickness of the fruit flesh, shape, colour and the pungency.

Capsicum fruits are known by a considerable number of names, such as *Capsicums*, peppers, Spanish pepper, Cayenne pepper, chile, chillies and paprika. All names, including the term "pepper", are misleading since *Capsicum* is not related to the true pepper, which belongs to the Piperaceae family. The common name "red pepper" is especially erroneous since it stands for the fruits of *Schinus molle* and *Schinus terebinthifolius*, which belong to the Anacardiaceae family.

Depending on its primary uses, peppers can be subdivided into vegetable and spice *Capsicums*. However, both types hold more or less pungent members. Pungency is caused by capsaicinoids, which constitute a class of plant alkaloids, which are all fatty acid amides of vanillylamine. The major components are capsaicin, dihydrocapsaicin and nordihydrocapsaicin, accompanied by several minor capsaicinoids which are present at very low levels and which are not thought to contribute greatly to overall pungency. Vegetable *Capsicums* range from the sweet, mostly large sized and thick fleshed, green, yellow, red or black coloured varieties, like garden or bell peppers, to the large, thin walled and hot green fruits. Spice *Capsicums* range from the sweet non-pungent or slightly pungent varieties, designated as paprika, to the hot species, usually designated as chillies or cayenne pepper. All varieties mentioned belong to the genus *C. annuum* and they all evolve naturally or through cultivation and hybridization from the bird pepper *C. annuum* var. *aviculare*.

The fiery hot chillies, however, belong for the most part to the species *C. frutescens* and they contain almost twice as much capsaicin as the hottest fruits of the *C. annuum* species. The best known member of this species is *C. frutescens* var. *tabasco*, which is used for the production of the famous Tabasco sauce™ in Louisiana. Chile habanero, which belongs to *C. chinense*, is said to be the hottest *Capsicum* variety known in the world.

Table 13.1 Industrial uses of *Capsicum* fruits and obtained products

Treatment	Capsicums used	Produce examples
Fresh selling	all kinds	
Canned (normally acidified with citric or fumaric acid)	*C. annuum*	Peeled pimientos
Frozen	*C. annuum*	Components of convenience food
Brined or pickled (with salt and vinegar)	*C. annuum*	Pepperoncini Tomato shaped pepper Cherry peppers Stuffed peppers
Lactic acid fermented	*C. annuum*	Tomato shaped pepper
Fermented for sauces	*C. annuum, C. frutescens, C. chinense*	Sambal oelek Tabasco™
Dehydrated	*C. annuum* *C. annuum, C. frutescens* *C. chinense, C. baccatum*	Paprika powder Chilli powder
Extracted	*C. annuum, C. frutescens* *C. chinense*	Paprika and chilli oleoresins

Capsicum fruits are widely used by the food industry in order to produce an extensive range of food. It will be clear from Table 13.1 that raw materials used for different purposes have to fulfil different requirements defined by the processing industry. It is the aim of this chapter to discuss the special requirements for *Capsicums* used for the production of paprika powder.

Paprika and chilli

Dehydrated and ground *Capsicum* fruits represent one of the most outstanding spice commodities in the world. In fact, paprika powder is the most important spice but one, surpassed only by the original pepper, *Piper nigrum*. The world supply is estimated at approximately 60,000 tons per annum, with an additional 1,400 tons of paprika oleoresin. Seventy per cent of paprika powder is used for industrial purposes, whereas the rest is used domestically.

The designation of dehydrated ground *Capsicum* products is not quite clear, neither in scientific literature nor in cookbooks. In the US, as well as in international trade, any dehydrated, non-pungent, red pepper powder is termed "paprika" (Bosland and Votava, 1999). In Europe, however, paprika represents a more or less red spice with a typical fruity taste and a degree of pungency which ranges between "free from pungency" or "scarcely pungent" and "pungent" according to the ISO 7540 standard (Table 13.2). The most comprehensive paprika range is offered by Hungary, where the spice is classified into eight classes starting with "very mild" and ending with "very hot" (Buckenhüskes, 1999).

From Table 13.3 it can be seen that in some European countries the distinction between paprika and chillies is defined by the *Capsicum* species used. Whereas paprika is normally produced from *C. annuum* varieties, chilli is at first manufactured from *C. frutescens* varieties.

Since there is no botanical reason for making a distinction between paprika and chilli, it is good practice to differentiate according to consumer expectation. From this view, however, the essential parameter is the degree of pungency: while paprika is a fruity tasting, more or less intensive red coloured, sweet, slightly pungent or very pungent spice powder, chillies are known to be extremely hot. The aroma of chilli powder is also characteristic, however, it is normally not considered important by the consumers in the western world. So it is proposed to modify the different regulations in a way that paprika, as well as chilli powder, may contain all kinds of

Table 13.2 Grading, physical and chemical requirements on ground (powdered) paprika by the ISO 7540 standard

Characteristic	Requirements on grade			
	I	II	III	Method
Degree of pungency	Free from pungency or scarcely pungent	Sweet to slightly pungent	Pungent	
Degree of fineness of grinding, mm	0.50	0.60	0.60	ISO 3588
Moisture content, % (m/m), max	11.0	11.0	11.0	ISO 939
Total ash, % (m/m) on dry basis, max	6.5	7.5	10.0	ISO 928
Acid-insoluble ash, % (m/m) on dry basis, max	0.5	0.8	1.6	ISO 930
Non-volatile ether extract, % (m/m) on dry basis, max	17.0	17.0	17.0	ISO 1108
Crude fibre content, % (m/m) on dry basis, max	25.0	25.0	30.0	ISO 5498
Capsaicin content, mg/100 g on dry basis, max	0–10[a]	20[a]	30[a]	—[b]
Natural colouring matter, g/kg on dry basis, min	2.5[a]	2.0[a]	1.5[a]	—[b]

Source: Taken from ISO 7540: 1984, reproduced with the permission of the International Organisation for Standardization, ISO. This standard can be obtained from any member body or directly from the Central Secretariat, ISO, Case postale 56, 1211 Geneva 20, Switzerland. Copyright remains with ISO.

Notes
a Tentative values.
b A method will form the subject of a future International Standard.

Table 13.3 Specifications of paprika and chilli within some European national regulations

Specification	Paprika	Chilli
ISO 7540 and ISO 7543-2	*Capsicum annuum*	Normally *Capsicum frutescens*
Codex Alimentarius Austriacus (Austria)	*Capsicum annuum*	*Capsicum frutescens, C. chinense, C. baccatum, C. annuum*
Schweizerisches Lebensmittelbuch (Switzerland)	*Capsicum annuum*	*Capsicum frutescens*
Leitsätze des Deutschen Lebensmittelbuches (Germany)	*Capsicum annuum* and other *Capsicum* species	*Capsicum frutescens* and other *Capsicum* species

Source: Buckenhüskes, 1999. Reproduced with permission.

Capsicum fruits and that the distinction between these products is only a question of their degree of pungency. According to Bosland and Votava (1999), in the US dried red pepper powders are classified into five groups based on pungency level:

Non-pungent or paprika:	0–700 Scoville Heat Units
Mildly pungent:	700–3,000 Scoville Heat Units
Moderately pungent:	3,000–25,000 Scoville Heat Units
Highly pungent:	25,000–70,000 Scoville Heat Units
Very highly pungent:	> 80,000 Scoville Heat Units

Scoville Heat Units are estimated by sensorial tests; 15 million Scoville Heat Units are set as 100% capsaicin.

General quality criteria for paprika powder

Despite a long and comprehensive tradition in paprika production, many countries lack the necessary consistency in the application of quality-oriented cultivation and adequate processing to satisfy the quality demands, especially of, the US and Germany, the two major paprika importing countries in the world.

From a general point of view the quality attributes demanded of a foodstuff are, among other things, deduced from its intended purpose. Paprika powder is a spice, which is used for its flavouring and appetizing, as well as for its colouring and colour-stabilizing properties. Among its most important applications are sausages and other meat products, soups and sauces, broiled chicken and snacks.

The most important analytical parameters of paprika, as well as its official analytical methods, are listed in Table 13.2. The required sensory properties of paprika powder may be articulated as follows:

- Fresh, fruity pleasantly aromatic smell.
- Pleasantly fruity-sweetish, aromatic flavour.
- According to the European view and the desired purpose it should be free from pungency, slightly pungent or very pungent.
- Whereas a light bitterness is a typical characteristic even for sweet paprika varieties, paprika powder should not comprise a dominantly bitter taste. The origin of the bitter taste is not quite clear and further research is needed.
- Free from off-flavours and off-odours which may be caused by the use of unsuitable raw material, by technological mistakes during processing, especially during drying, by unsuitable storage conditions, or by oxidative reactions during storage leading to a more or less pronounced rancidity.
- Intensely typical red colour, which remains stable over the shelf life.

Colour of paprika powder

The paprika powder available on the market ranges in colour from a bright rich red to a brick-red or orange-red, depending upon variety and quality. For industrial purposes rich red paprika powders that retain their colour as a ground product, are required.

The colour of paprika is composed of more than 30 different yellow, orange and red carotenoids. Antheraxanthin, β-carotene, β-cryptoxanthin, lutein, violaxanthin and zeaxanthin are the most common yellow to orange pigments, whereas capsanthin, capsorubin and crypto-capsin represent the most important red pigments.

During storage the carotenoids can be destroyed by autoxidative and oxidative attack. According to the present knowledge, the degree of pigment losses depends upon paprika variety, climatic conditions during ripening, provenance, timing of harvesting, post-harvest treatment, processing and storage conditions. The most significant factors influencing the colour stability are moisture content, temperature, light and oxygen.

From practical experience it is known that the amount of pigments in the freshly harvested fruits is not a good indication of the storage stability and the retention of the red colour. Therefore, it is necessary to breed and evaluate for both initial colour and colour retention properties (Bosland and Votava, 1999). In order to minimize the risk of a colour shift during storage from red to orange, the share of red pigments in the total colourant content should be as high as possible, whereas the share of yellow pigments should be as low as possible.

Biacs *et al.* (1987, 1993) described that the colour stability of paprika is influenced by the amount of esterified pigments and by the capsanthin/capsorubin ratio. The authors stated that varieties with a high amount of fatty acid di- and monoesters of capsanthin and capsorubin exhibit a better storage stability than varieties with a high amount of free pigments. In addition, the storability should be improved by a high capsanthin/capsorubin ratio. Since capsanthin contains fewer polar groups on its structure than capsorubin, it should be more stable against oxidative degradation.

Several investigations have shown that the oxidation of carotenoids in paprika is influenced by the water activity to such an extent that the available water slows down the oxidation process (Lee *et al.*, 1992). According to Osuna-Garcia and Wall (1998), for arid regions, a 15% pre-storage moisture content level may reduce the colour loss of stored ground products by at least 50%. However, from a microbiological point of view, such high moisture contents cannot be accepted as the risk of bacteria and mould will increase under these conditions.

Carvajal *et al.* (1998) reported that an important colour stabilizing effect can be observed from naturally available or added antioxidants like ascorbic acid and/or α-tocopherol. Another useful substance is ethoxyquin (6-ethoxy-1,2-dihydro-2,2,4-trimethylquinoline), which is permitted in the US for stabilizing the colour. In addition to its antioxidative power, ethoxyquin can be used as a fungicide. For that reason in some countries the substance is classified and permitted as a plant protecting agent which is not allowed to be used as an antioxidant in food. In Germany the residues of ethoxyquin are limited by law to a maximum value of 0.01 mg/kg fresh paprika or 0.1 mg/kg paprika powder.

In order to optimize the colour retention, along the processing line, special attention has to be paid to the drying process. Tevini (1997) reported that the drying conditions within belt-dryers often are not optimized. As a result, the paprika fruits are overheated, creating the conditions for subsequent non-enzymatic browning or the Maillard reaction which leads to brownish discolourations of the ground product.

In this respect two further requirements on paprika varieties for industrial use should be mentioned: the fruits should be thin-fleshed and should have the highest possible dry matter content in order to facilitate drying and to reduce thermal strain. Moreover, as the availability of free reducing sugars is one of the preconditions for the Maillard reaction, the content of reducing sugars in the fresh paprika fruits should be as low as possible.

Hygienic safety

In general, food safety is of increasing importance all over the world and therefore it is necessary to summarize the hygienic demands on paprika powder in detail:

- Paprika powder has to be free from toxic or hazardous substances.
- Paprika powder has to be free from living insects or larvae.
- Paprika powder has to be practically free from dead insects and insect fragments.
- Paprika powder has to be practically free from rodent contaminations.
- Paprika powder has to be free from extraneous matter, which includes all vegetable matter other than fruits of *Capsicum*, as well as colouring agents, oils or other products added to improve the quality or to mask defects.
- Paprika powder should meet the microbial values summarized in Table 13.4 (concerning *Salmonella* see "Bacteriological problems").

Table 13.4 Microbial approximative and warning values for spices which should be given to the consumer or which should be used in foods without any further germ reduction treatment

Organism	Approximative value (cfu/g)	Warning value (cfu/g)
Salmonella	—	not detectable in 25 g
Staphylococcus aureus	1.0×10^2	1.0×10^3
Bacillus cereus	1.0×10^4	1.0×10^5
Escherichia coli	1.0×10^4	—
Sulphite reducing Clostridia	1.0×10^4	1.0×10^5
Moulds	1.0×10^5	1.0×10^6

Source: DGHM, 1988. Reproduced with permission.

- Paprika powder should be free from mycotoxins (for details see "Mycotoxins").
- Paprika powder should not exceed the maximum values set up for residues of plant protecting substances prescribed in the legal regulations of different countries.

It should be mentioned that in Germany the amount of ethylene oxide and 2-chloro-ethanol in paprika powder is limited to 0.1 mg/kg. Formerly, fumigation with ethylene oxide was used as a method for the degermination of spices. However, the application has been prohibited due to the formation of toxic ethylene chlorohydrin (2-chloro-ethanol). However, sometimes small amounts of 2-chloro-ethanol is found in paprika powder that has not been fumigated, raising the question whether ethylene oxide is produced as a natural metabolite of the paprika fruits.

Bacteriological problems

There are two particular concerns of hygenic safety: the microbial contamination of spices, in particular the occurrence of salmonella, and the frequently observed interior mouldiness that poses an inherent risk of mycotoxin formation. Paprika, as well as other spices, are natural products, which contain microbial flora. Since a significant volume of spice is sold without any germ reduction, the microbial status of the product is of outstanding importance.

Among the pathogenous microorganisms which might be available in paprika powder, special attention has to be paid to the Enterobacteriaceae. Due to a salmonella outbreak in Germany in 1993 caused by paprika-seasoned potato chips (Buckenhüskes, 1998), and the perception that salmonella strains exist that are capable of changing their virulence due to the occurrence of plasmids or prophages, it must be assumed that under certain conditions one single salmonella may be infective and that any detection of salmonella must be rated as a serious health risk. Based on this knowledge it was decided by the authorities that salmonella should not be detectable in 25 g of paprika or chilli powder.

From a qualitative point of view, irradiation with ionizing beams constitutes the only really acceptable means for the degermination of paprika. Since 1999 the irradiation of spices and aromatic herbs has been permitted within the European Community, however, consumers in several countries still do not accept this procedure. What consequences these facts may have on the paprika market need not be discussed here in greater detail.

The data, which can be used as a guideline for the microbiological assessment of spices, are summarized in Table 13.4. Again, it can be seen that salmonella should not be detectable in 25 g of the spice.

Mycotoxins

Another important problem with paprika and chilli powder is the subject of interior mould growth. The topic is doubly problematic as, in many countries mouldy foodstuffs are considered unpalatable, regardless of the sanitary risk they present, and second, the presence of moulds always implies the risk of mycotoxin formation. El-Dessouki (1992) analysed 15 samples of paprika and 24 samples of chilli powder for aflatoxins. Seven of the paprika samples contained aflatoxin B_1 and B_2 and up to 15.3 µg total aflatoxins per kg sample. Thirteen of the chilli samples contained aflatoxin B_1, B_2 and G_1 contained up to 218.4 µg total aflatoxins per kg sample. According to the German legislation, the so-called Aflatoxin-Verordnung, the aflatoxin content is restricted to 2 µg aflatoxin B_1 per kg and 4 µg total aflatoxin per kg foodstuff. These values have currently been raised by the European Community for several spices including paprika and chilli to 5 µg aflatoxin B_1 and 10 µg total aflatoxin per kg foodstuff (EC Regulations No 472/2002). Bassen and Brunn (1999) analysed ten random samples of paprika for ochratoxin A residues. Only one sample was negative, the others possessed ochratoxin A in quantities between 0.5 and 8.6 µg/kg.

Against this background it is more and more important to install GAP- and HACCP-systems in the cultivation and processing lines in order to control the microbial risks for all kinds of *Capsicum* products. One of the requirements for paprika fruits is that the mechanical handling should be sufficient to ensure that the fruits are not damaged during harvesting or storage so as to prevent infection, e.g. with moulds.

Cultivation properties

Modern strategies in agricultural and food production technologies ensure a complete line from the production of the paprika fruits to the dried and ground paprika powder on the table of the consumers. In order to exploit the modern technological possibilities in an optimal way it is necessary to find a common level of communication between breeders, farmers and food technologists. For this reason the discussion of actual requirements on paprika powder for food processing should be completed with a summary of properties, which should be fulfilled by an optimal paprika variety.

- The growth period of the variety should be as short as possible so as to ensure that the fruits will thoroughly ripen on the plant before the cold season starts (e.g. in Hungary).
- In order to ensure an even maturation, the individual fruits should develop as uniformly as possible. The decision whether a fruit may be picked is to be made by the harvester in accordance with the degree of colouration. From a hygienic point of view mechanical harvesting is preferred.
- The fruits should not be prone to interior mould growth so as to prevent the occurrence of mycotoxins. It will be an interesting field of research to look at whether the comprehensive gene pool of *Capsicum* comprises natural activities against internal mould growing or not.
- Plants should be robust and have a solid stem to prevent fruits from coming into contact with the ground. The direct or indirect contact (splash water during rainfall or irrigation) of the fruits with the ground or with soil particles is one of the routes of microbial contamination.
- Maximization of hectare yields with regard to ASTA (colour) values.
- Resistance to insects and plant diseases.
- Resistance to plant-protective agents.

- Growing with equally good results in geographically differing regions. Recently it was discussed whether organizing paprika cultivation in climatically differing regions was possible so as to provide the market with new paprika every four to six months. Presupposing a defined and constant quality standard, this provision could be the basis on which to minimize the necessary storage periods between harvesting and processing.

Closing remarks

It should be clear that the partnership between breeders, farmers, food technologists, salesmen and customers has to be established on a higher level than it has been in the past. The requirements can only be met by operations and procedures which are accepted by the consumers in different countries. However, concerning the discussed context, irradiation and genetic engineering are two important methods which are not currently accepted by consumers in many countries.

References

Bassen, B. and Brunn, W. (1999) Ochratoxin A in Paprikapulver. *Deutsche Lebensmittel-Rundscha*, 95, 142–144.

Biacs, P. A., Bodnàr, J. and Hoschke, A. (1987) Separation and identification of carotenoid esters in red pepper during ripening. In: Stumpf, P. K., Mudd, J. B. and Nes, W. D. (eds) *The Metabolism, Structure and Function of Plant Lipids*. Plenum Press, New York, pp. 135–239.

Biacs, P. A., Czinkotai, B. and Hoschke, B. (1992) Factors affecting stability of colored substances in paprika powders. *J. Agric. Food Chem.*, 40, 363–367.

Biacs, P. A., Daood, H. G., Huszka, T. T. and Biacs, P. K. (1993) Carotenoids and carotenoid esters from new cross-cultivars of paprika. *J. Agric. Food Chem.*, 41, 1864–1867.

Bosland, P. W. and Votava, E. J. (1999) *Peppers: Vegetable and Spice Capsicums*. Crop production science in horticulture No. 12. CABI Publishing, Wallingford and New York.

Buckenhüskes, H. J. (1998) Hygienische Aspekte bei der Verwendung von Gewürzen. *Zeitschrift für Arznei- & Gewürzpflanzen*, 3, 21–30.

Buckenhüskes, H. J. (1999) Aktuelle Anforderungen an Paprikapulver für die industrielle Verarbeitung. *Zeitschrift für Arznei- & Gewürzpflanzen*, 4, 111–118.

Carvajal, M., Gimenez, J. L., Riquelme, F. and Alcaraz, C. F. (1998) Antioxidant content and colour level in different varieties of red pepper (*Capsicum annuum* L.) affected by plant-leaf Ti4+ spray and processing. *Acta Alimentaria*, 27, 365–375.

DGHM – Deutsche Gesellschaft für Hygiene und Mikrobiologie (1988) Veröffentlichung der Arbeitsgruppe mikrobiologische Richt- und Warnwerte für Lebensmittel. *Deutsche Lebensmittel-Rundschau*, 84, 127–128.

El-Dessouki, S. (1992) Aflatoxine in Cayenne-Pfeffer- und Paprika-Pulver. *Deutsche Lebensmittel-Rundschau*, 88, 78.

ISO – International Organization for Standardization. ISO 7540. Ground (powdered) paprika (*Capsicum annuum* L.) Specification. Ref. No. ISO 7540: 1984 (E).

ISO – International Organization for Standardization. ISO 7543-2. Chillies and chilli oleoresins – determination of total capsaicinoid content. Ref. No. ISO 7543-2: 1993 (E).

Lee, D. S., Chung, S. K. and Yam, K. L. (1992) Carotenoid loss in dried red pepper products. *Int. J. Food Science and Technol.*, 27, 179–185.

Osuna-Garcia, J. A. and Wall, M. M. (1998) Prestorage moisture content effects color loss of ground paprika (*Capsicum annuum* L.) under storage. *J. Food Quality*, 21, 251–259.

Tevini, M. (1997) Qualitätsbestimmende Faktoren in der industriellen Paprikaverarbeitung. Forschungskreis der Ernährungsindustrie e. V. 55. Diskussionstagung in Detmold, pp. 74–99.

14 Adulterants, contaminants and pollutants in *Capsicum* products

J. Chakrabarti and B. R. Roy

Chilli used as a spice has agricultural and marketing specifications and also food standards in national regulations since chilli is a food, a food supplement or a food adjunct. Concerns on quality and safety emerge on account of occasional aberrations of adulteration, contamination and pollution. Relatively stable pollutants of air, water and soil get to this plant product engendered by all these three. Needless to say, adulteration is intentional and contamination incidental. The latter exceeding the limits of good agricultural and manufacturing practices changes to adulteration even if not intentional. This chapter deals with adulteration in the whole, in the form of powder and paste of chilli. The details include microscopic detection of adulterants, estimation of carotenoids and non-volatile ether extract of extracted chilli and that influenced by addition of edible oils. Contaminants specially in respect to irradiation and added colours are included. Pollutants include trace metals, pesticides, mycotoxins and microbes. Insect-infested and insect-damaged chilli may not be rare in tropical regions.

Introduction

Chilli is not an esoteric spice and is not very costly either. Even then, it attracts its own moderate share of adulteration, which by all means requires to be taken into consideration. Both for cheapness and a very pronounced sensory bite it is utilised extensively by common and poor people specially in the developing countries. High incidence of pungency, as revealed by high Scoville units (factor of manifold dilution, which still retains pungency), is responsible for tickling the taste buds of pungency making even a bland or insipid food gastronomically acceptable. This is the first choice from the spices among poor people who consume mostly chilli in absence of multiple dishes or varieties and hence its turnover is considerable. It is the spice of the people.

Even though adulteration, contamination or pollution in chilli is not extensive, it is duly surveilled for quality control during harvesting and post-harvest processing, commercial transaction and trade, and activities towards food and flavour regulations. The powder or the paste of chilli is more vulnerable to adulteration as foreign substances can go into it visually undetected.

Adulteration in chilli

Chilli whole

The following points of adulteration may be considered:

1   Extraneous matter including calyx pieces, loose tops, dirt, lumps of earth, stones and mould growth, etc. If the mould or the sprouted spores are identified as those producing toxins, e.g. *Aspergilli* or *Penicillia*, the apprehended mycotoxins should be looked for.

2   Extraneous colouring matter (Mitra *et al.*, 1961; Banerjee *et al.*, 1974), coating of mineral oils (Sen *et al.*, 1973) or other facing and coating liquids or powders.
3   Insect-damaged foods and related matters.
4   Harmful substances.

The items are analysed as usual. While in whole chilli insect damage is estimated by refraction, that in powder or paste is estimated by colorimetry of uric acid (Martin, 1979) and by chromatography techniques (Sengupta *et al.*, 1972).

Powdered (or paste) chilli

1   Excessive moisture, dirt, dust, mould growth, insect infestation, fermentation, extraneous matter, and added colouring and flavouring agents are the main adulterants.
2   It is claimed that a small amount of edible oil is used for post-harvest processing of whole chilli and powdering or pasting chilli. Hence, Indian standard specification allows a maximum of 2% of such oils in chilli powder with the following labelling declaration "Chillies in this package contain an admixture of not more than 2% of … (name of the edible oil)" (Food Regulations, 1955).
3   Microscopic detection of adulterants: Microscopic structure of chilli powder or paste has been worked out (Winton and Winton, 1947; Parry, 1969). Thus, aleurone, chloroplasts, cuticle, cuticle grooves, endocarp, endosperm, exocarp, giant cells, hairs, hypodermis, mesocarp, oil drops, phloem, fibre, seed coat, spiral vessels, stone cells, vascular bundles, xylem fibres, etc. have been recognised and documented. It is possible for any foreign matter to be recognised by the presence of uncharacteristic structures or by detection of well-known foreign microscopic structures like different starches. Chilli powder is not known to have starch granules, though acid hydrolysed and diastase (Mitra *et al.*, 1970) treated cell contents can provide reducing equivalents not of glucose but of pentoses from pentosans and hemi-celluloses. Powdered calyces and pedicels if present in excess in chilli powder or paste will constitute adulteration.
4   Chilli is cheap and even though at present there is no major market for capsaicin, the pungent principle of chilli, or its derivatives, it will soon be in great demand due to its preservative, antioxidant and medicinal properties. There is a possibility that chilli (whole or powdered) can be partially extracted of capsaicin. Such samples can be deemed adulterated as anything abstracted or added without labelling constitutes adulteration. This deficiency can be detected by estimating the pungency in Scoville units (International Standards Organisation, 1977) and its deficiency from usual and undisputed concentration of capsaicin or from the lowest concentration as permitted in the standard specification. It is possible that capsaicin is taken as a parameter along with others for grading chilli. The deficiency will decrease the non-volatile ether extract of chilli, which is an important analytical parameter of authenticity of a sample of chilli.

     Detection and estimation of carotenoids can also be carried out in normal and extracted chilli (Banerjee *et al.*, 1973). If the normal value is established by analysing a large number of any particular variety, then the deficiency in the extracted samples can be detected.
5   It is possible that the non-volatile ether extract of chilli can be fraudulently boosted by oils. This can be ascertained by detecting if mineral oil is added (Sen *et al.*, 1973) or by estimating iodine value of lipids extracted from chilli powder. Chilli seed oil has an iodine value of 126, which will be increased or decreased according to the extraneous oil added.

Contamination

General contaminants

The possible contaminants are agricultural and biological residues, fungi, rodent hair and excreta, radioactive and radiolytic products (as chilli powder specially the old specimens can be irradiated by ionising radiations if national legislation permits), insect ova and pupa, etc.

All these contaminants will call for specific analytical exercises. While refraction will answer for visible contaminants physically present and microscopy for ova and pupa, the invisible rodent excreta of urea is detected by TLC (Roy and Bose, 1972).

Irradiation

If chilli samples are to be preserved or protected by exposure to ionising radiation, it has to be done in conformity with the statutory provision in force (Food Regulations, 1955). The dosage for chilli is 6, 14 and 10 kGy as minimum, maximum and overall average, respectively.

Extraneous colour

Colours are not added or allowed to be added in whole chilli or its powder and paste. If colours are added, the samples can be condemned as misbranded or as adulterated. The relevant analyses (Mitra *et al.*, 1961) rely on colour change in ether and alcoholic extracts with 3 : 1 or 4 : 1 HCl and on reversed-phase chromatography of alcohol (90%) extract of chilli wherein the natural and synthetic colours are resolved. While the natural colours vanish in a day or two in the chromatogram, the added synthetics are stable. Banerjee *et al.* (1974) isolated fat-soluble colours by extraction with hexane, partition by dimethyl formamide and water and chromatography on activated alumina. The isolated colours were separated and identified by thin layer chromatography and differentiated from natural colours.

Pollutants

Trace metals

The main environmental pollutants in whole chilli are toxic trace metals from agronomic sources particularly agricultural, chemical and processing operations. As per regulation (Food Regulations, 1955) the maximum allowable concentration in ppm of some metal pollutants in chilli is given below:

Metal	Maximum allowable concentration (ppm)
Lead	10.0
Copper	30.0
Arsenic	5.0
Tin	250.0
Zinc	50.0
Cadmium	1.5
Mercury	1.0
Methyl mercury	0.25
Chromium	0.02
Nickel	1.5

Pesticides

Pesticide residue can also be a class of pollutants as in other agricultural commodities, and at times it could be under surveillance as provided in regulations. When the history of use and names of pesticides in a sample of chilli are not known, the generalised methods for the particular class of pesticides can be used followed by specific tests for the individual pesticides (Chakrabarti and Roy, 1976 and by multiresidue methods, AOAC, 1990). However, the presence or qualitative detection to be determined within a short time if standardised well will be welcome; if positive it can be followed for pinpointing and quantitation of the pesticide present. Suitable insects can be utilised for the screening for only the presence of pesticides in general (Mandal and Mandal, 1999). These authors showed a possibility of laboratory rearing of an insect for preliminary detection of the presence of any pesticide.

Mycotoxins

Mycotoxins particularly aflatoxins are sometimes looked for. Our laboratory had continuous monitoring of market samples of food for aflatoxin pollution. As no positive cases on chillies came to light, no firm conclusions could be made in this regard. The usual limit of aflatoxins in chilli is 0.03 ppm.

Microbiological status

This laboratory looked for microbial spores in chilli powder among others but no significant case with potential of microbial spoilage or toxicosis was however reported (Guha and Roy, 2001). Antimicrobial effect of chilli, though mild, could help inhibit some microflora of the foods, but, on the other hand, spices are known to (it is possible that chilli too) stimulate the growth of lactic acid bacteria. This is reportedly a result of their manganese content.

If the microbial content is even moderate and if the chilli is not sterilised the food may be contaminated and to the extent spoilage will occur. Needless to say the number of thermophilic flat sour spores and thermophilic types not producing hydrogen sulphide should be low if the spice is to be used in canned products like canned curry and meat. Such type of organisms do not have the same importance in non-canned products.

Chillies showing very high microbial counts are likely to be of low sanitary quality and indicative of none-too-well hygienic practices in the production and handling. These counts can be used as a guide for judging contamination and designing uses.

References

AOAC, Association of Official Analytical Chemists Inc. (1990) *Official Methods of Analysis*, 15th edn, pp. 274–285, Arlington, VA.

Banerjee, T. S., Saha, A., Bhattacharyya, S. and Mathew, T. V. (1973) Detection and estimation of carotenoids from foods by TLC. *J. Proc. Inst. Chem.*, 45, 198–200.

Banerjee, T. S., Guha K. C., Saha, A. and Roy, B. R. (1974) Examination of oil soluble colours from food by solvent partitioning and chromatography, *J. Food Sc. Tech.*, 11, 230–232.

Chakrabarti, J. and Roy, B. R. (1976) A generalised procedure for separation and identification of pesticides belonging to different groups. *J. Indian Acad. Forens. Sc.*, 15, 20–24.

Food Regulations. *PFA Rules*. India (1955) amended upto date.

Guha, A. K. and Roy, R. (2001) *Personal Communication*.

International Standards Organisation (1977) *ISO 3513, Spices and Condiments – Chillies – Determination of Scoville Index.*

Mandal, S. and Mandal, R. K. (1999) A note on the laboratory rearing of *Spodoptera litura* with natural food in a small scale for bioassay of insecticides and insect-resistent plants. *Science and Culture*, 15, 169–171.

Martin, P. G. (1979) *Manuals of Food Quality Control*, pp. 258–259. Rome: Food and Agricultural Organisation of the United Nations.

Mitra, S. N., Sengupta, P. N. and Roy, B. R. (1961) Further studies on the detection of oil soluble coaltar dyes in chillies (Capsicum), *J. Inst. Chem.*, 33, 69–73.

Mitra, S. N., Sengupta, P. and Sen A. R. (1970) A comparative study of starch estimation in chillies. *J. Inst. Chem.*, 42, 15–16.

Parry, J. W. (1969) *Spices*, Vol. II, p. 22, London: Food Trade Press Ltd.

Roy, B. R. and Bose, P. K. (1972) Spot test for detection of urea in food and feed. *J. AOAC*, 55, 664–666.

Sen, A. R., Sengupta, P., Ghosh Dastidar, N. and Mathew, T. V. (1973) Chromatographic detection of small amounts of mineral oil in chillies. *Res. Indus.*, 18, 97–99.

Sengupta, P., Mandal, A. and Roy, B. R. (1972) Detection of uric acid by thin layer chromatography. *J. Chromat.*, 72, 408–410.

Winton, A. L. and Winton, K. B. (1947) *The Analysis of Food*, New York: J. Wiley and Sons Inc.

15 Colour differences in peppers and paprikas

Ricardo Gómez-Ladrón de Guevara, José E. Pardo-González, Andrés Alvarruiz-Bermejo, Manuel González-Ramos and Ramón Varón-Castellanos

The influence of the growing conditions of pepper (*Capsicum annuum* L.) – open air and greenhouse – on the evolution of the fruit chlorophyll content, its final colour and the quality of the obtained paprika, is discussed in this chapter. Furthermore, the chromatic coordinates $a*$ and $b*$, measured on the ripe fruits, allow a simple, fast and reliable way to estimate the quality of peppers, expressed in ASTA units. Finally, the influence of the storage conditions – temperature and humidity – on paprika colour loss are analysed and the experimental data are fitted to a first-order kinetic model. The rate constants for the Ocal variety are calculated for different growing conditions.

Introduction

The external appearance of fruits, particularly their colour, is of prime importance when considering the different attributes which define quality and, in the case of fruits destined for fresh consumption, a visual impression that does not coincide with the established standard easily leads to refusal.

In the case of pepper varieties (*C. annuum* L.) destined for the processing of paprikas and oleoresins, which are two of the most widely used natural colourants in the food industry (Fisher and Kocis, 1987), the acquirement of lines rich in carotenoids, basically red ones, is of great importance: since their quality is closely related to their colouring power (Costa, 1979, 1980; Soriano *et al.*, 1990). Thus, the ripening of pepper fruit has been the object of interest of many researchers, not only because of the spectacular change they undergo but also because of the complexity of the mechanisms which take part in the process of biosynthesis of capsanthin, a typical colourant of this fruit (López, 1972; Alemán and Navarro, 1973; De la Torre and Farré, 1975; Candela *et al.*, 1984; Fisher and Kocis, 1987; Alcaraz, 1990; Almela *et al.*, 1990, 1991; Ortiz *et al.*, 1990; Mínguez and Hornero, 1993, 1994; Biacs and Daood, 1994).

Normally, the change in colour of the pepper surface takes places as a result of both chlorophyll degradation and a considerable increase in its carotenoid content. These factors are influenced by the temperature and illumination to which the fruit is exposed. Only in the varieties containing chlorophyll retaining genes, the total chlorophyll content suddenly increases at the onset of ripening, principally due to the sharp increase of chlorophyll *b*. From this moment onwards, the content of chlorophyll (particularly of chlorophyll *a*) decreases, but not to the extent that they completely disappear in the ripe fruit (Ferrer and Costa, 1991).

As mentioned above, paprika, produced by dehydration and subsequent grinding of the whole fruits of certain pepper varieties, is one of the most widely used spices in the food sector, especially in the meat processing industry. Paprika colour and pepper colour, is the most important quality characteristic and depends on its carotenoid content (Davies *et al.*, 1970; Nakayama and

Mata, 1973). The content and relative proportion of this type of plant pigment varies greatly according to the variety, ripeness, climatic conditions and cultivation techniques used (Davies *et al.*, 1970; Nagle *et al.*, 1979; Reeves, 1987; Costa *et al.*, 1989; Almela *et al.*, 1991; Costa, 1991). Among the methods used for measuring paprika colour we can cite absorption (Soriano *et al.*, 1990; Almela *et al.*, 1991; Mínguez and Hornero, 1993; Navarro and Costa, 1993; Mínguez and Hornero, 1994) and reflection (Conrad *et al.*, 1987; Pardo *et al.*, 1996), although the most widely used method is ASTA 20.1 (ASTA, 1986). Another way to express the colour is the tint, which estimates the relative proportion of the red carotenoid pigments with regard to the content of the yellow ones in the products.

On the other hand, the use of tristimulus colorimetry is widely used. This method is quick and easy and, moreover, it allows a better classification of the samples by their external colour than the ASTA method (Conrad *et al.*, 1987). The development of methods have been attempted by Nagle *et al.* (1979) and Reeves (1987) which allow the deduction or evaluation of the real pigment content of a fruit by its tristimulus values, obtaining unlike results. It seems as if only the ratio of the chromatic co-ordinates $a*/b*$ of the paprikas is related to their total pigment content (Drdak *et al.*, 1982). In other cases, the use of more complex chromatic indices (2000 $a*/L*b*$) can satisfactorily illustrate the external visual colour of the different fruits (Carreño *et al.*, 1995). Finally, the chlorophyll evolution in the pepper paprika varieties throughout their ripening is a subject which has also been studied by several authors because of the importance of their gradual transformation into carotenoids (Camara and Brangeon, 1981; Ziegler *et al.*, 1983; Carde *et al.*, 1988; Ferrer and Costa, 1991).

Until recently, all studies examining the final quality of different varieties of paprikas and paprika peppers have used, either colorimetric absorption methods (Soriano *et al.*, 1990; Almela *et al.*, 1991; Mínguez and Hornero, 1993; Navarro and Costa, 1993; Mínguez and Hornero, 1994) or colorimetric reflection methods (Conrad *et al.*, 1987; Gómez *et al.*, 1996; Pardo *et al.*, 1996; Navarro *et al.*, unpublished). As far as we know, both methods have not been used in the same study and there is no study which compares the colour obtained in greenhouse varieties with the same varieties cultivated in the open air.

A number of pepper varieties grown in two different conditions (open air and greenhouse) have been evaluated by reflection colorimetric methods, and paprika obtained from these fruits has been evaluated by absorption colorimetric methods (Gómez *et al.*, 1998a). It was shown that the different techniques of cultivation lead to differences in the colorimetric co-ordinates of the fruits. The chlorophyll evolution throughout the ripening process of the fruit – in the two different techniques – has been evaluated, due to the importance of the presence of this pigment in the final colour.

After analysing the relationship between the pepper fruit colour and their corresponding paprika, the influence of the storage conditions of paprika on its quality has also been studied. Paprika, like most foods, loses colour during storage. This colour loss begins with the destruction of vitamins C and E and continues with the degradation of the yellow and red carotenoids due to their oxidation by free oxygen. At the same time, this oxygen is influenced by external factors which may be of a physical (temperature, humidity, light, etc.) or chemical nature (presence of metallic ions, enzymes, peroxides, etc.) (Alemán and Navarro, 1973; Salmerón, 1973).

Colour is not usually extracted from the skin of the pepper fruit immediately after harvesting and drying, but several months later. For this reason, it is very important to establish suitable storage conditions, so that colour loss might be minimized. In turn, this means that the influence of different factors (light, temperature, humidity, etc.) must be established.

The literature contains several references to the influence of temperature on the loss of colour in paprikas (Alemán and Navarro, 1973; Soriano *et al.*, 1990). Traditional pepper varieties are being substituted progressively by other varieties, obtained by breeding techniques, with greater yields and colouring potential. A selection was carried out in the Escuela Técnica Superior de Ingenieros Agrónomos de Albacete (Spain) with the Ocal variety. Since this cultivar gave good results during the latter seasons it was decided: (1) to analyse how different temperature and humidity conditions affected paprika obtained from the mentioned variety grown in the open air or greenhouse, using the ASTA 20.1 method to determine colour; and, (2) to fit the experimental data for colour to first-order kinetics and obtain paprika colour loss rate constants for each of the storage conditions studied (Gómez et al., 1998b).

Colour differences between fruits grown in greenhouse and open air

Table 15.1 shows the mean extractable value (ASTA units) and tint of paprika obtained from six pepper varieties grown in greenhouse and open air (Gómez et al., 1998a). Note that the greenhouse varieties show higher ASTA units and higher tint values in all cases.

Table 15.2 gives the values corresponding to the CIELAB colour space co-ordinates (L^*, a^*, b^*), chromatic quotients and chroma for the colour by reflection of the fruits for both growing systems. In the red fruits the greenhouse values for brightness (L^*), red (a^*) and yellow (b^*) components and chroma (C^*) are lower, due to the fact that the plant receives less sunlight (Gómez et al., 1998a). These lower values reflect darker and duller fruit colours, as expected when taking the physical significance of these parameters into account.

Figure 15.1 shows the representative points of the red greenhouse and open air fruit varieties in the colour space CIELAB, making evident the colouring difference. It is observed that the open air varieties, with better visual colours (highest C^* value), are separated from the greenhouse varieties which are on the top right-hand side of the diagram.

The chlorophyll a (C_a) and b (C_b) content of the fruits during the growing and ripening process is shown in Table 15.3 for greenhouse and open air varieties. It may be observed how the total chlorophyll content ($C_a + C_b$) decreases progressively. While the fruit colouring is only green (up to 35 days in greenhouse and 45 days in the open air), the chlorophyll a content is

Table 15.1 Average values of extractable colour (ASTA) and tint of the paprikas of the varieties cultivated in greenhouse (GH) and in the open air (OA), obtained in each case from five different measurements

Varieties	ASTA units	Tint
Datler (GH)	360 ± 5	1.009 ± 0.002
Ocal (GH)	336 ± 5	1.007 ± 0.002
Datoca (GH)	354 ± 5	0.998 ± 0.002
Larguillo (GH)	317 ± 4	1.007 ± 0.002
Numex (GH)	194 ± 3	0.996 ± 0.003
Belrubí (GH)	359 ± 5	0.993 ± 0.002
Datler (OA)	248 ± 4	0.973 ± 0.002
Ocal (OA)	268 ± 4	0.979 ± 0.002
Datoca (OA)	335 ± 4	0.968 ± 0.002
Larguillo (OA)	295 ± 4	0.979 ± 0.002
Numex (OA)	181 ± 3	0.955 ± 0.003
Belrubí (OA)	287 ± 4	0.965 ± 0.002

Table 15.2 Values of the chromatic co-ordinates CIELAB (*L**, *a**, *b**), chromatic quotients (*a*/b**, 2000*a*/L*b**) and chroma (*C**) of the red fruit varieties cultivated in GH and in OA

Varieties	L*		a*		b*		a*/b*		2000 a*/L*b*		C*	
	GH	OA	GH	OA	GH	OA	GH	OA	GH	OA	GH	OA
Datler	32.21	37.41	29.10	36.44	9.97	14.99	2.91	2.43	58.74	49.44	30.76	39.40
Ocal	34.68	35.69	32.37	35.93	11.02	13.36	2.93	2.69	54.58	52.51	34.20	38.34
Datoca	33.11	34.33	30.97	35.70	10.64	13.62	2.91	2.62	57.13	54.43	32.74	38.21
Larguillo	35.46	35.56	33.58	36.50	11.76	14.18	2.85	2.57	53.23	52.42	35.58	39.16
Numex	34.31	38.57	34.86	38.74	13.04	19.59	2.67	1.97	54.59	46.27	37.22	43.41
Belrubí	33.39	37.68	30.25	41.41	10.23	18.21	2.95	2.27	56.74	48.58	31.93	45.24

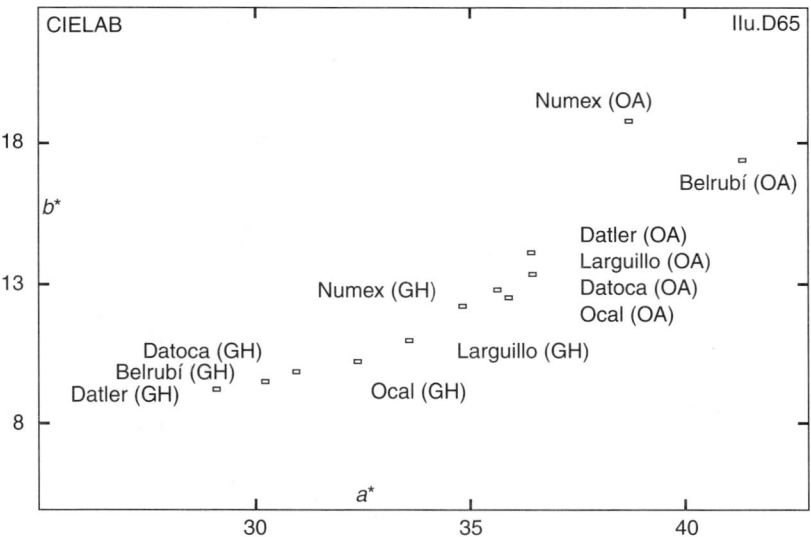

Figure 15.1 Points in the CIELAB colour space corresponding to the red fruits of the different assayed varieties in GH and in OA.

higher than the chlorophyll *b*. The appearance of the first red tones coincides with the moment in which the chlorophyll *b* content surpasses the chlorophyll *a* (45 days in greenhouse and 55 days in the open air). During the chromatic evolution of the fruit to uniform red colour, the total chlorophyll quantity decreases, but in all cases the content of chlorophyll *b* is higher than that of chlorophyll *a*. The final chlorophyll content in red fruits is higher in the greenhouse varieties, which leads to lower values of the *a** component and, therefore, to their being placed more to the left in the CIELAB colour space (Figure 15.1).

Table 15.4 illustrates the data above, showing the quotient between chlorophyll *a*/chlorophyll *b* (C_a/C_b) during the growing and ripening process. It may be noted that while the fruit presents only green tones (up to 35 days in greenhouse and 45 days in the open air), the quotient varies from 2.48 to 1.05, whereas as soon as the red tones appear (45 days in greenhouse and 55 days in the open air), these values vary between 0.66 and 0.43 due to the strong decrease of chlorophyll *a*. The completely red fruits, both greenhouse and open air fruits, show values lower than 0.55.

Table 15.3 Evolution of the chlorophyll a (C_a) and b (C_b) contents, expressed in mg/100 g weight, during the growing and ripening process of the pepper paprika fruits in GH and in OA

Varieties	15 days		25 days		35 days		45 days		55 days		65 days	
	C_a	C_b	C_a	C_b	C_a	C_b	C_a	C_b	C_a	C_b	C_a	C_b
Datler (GH)	152.91	87.21	100.58	60.26	71.04	28.55	8.15	14.59	4.50	9.86	—	—
Ocal (GH)	125.07	66.17	81.96	46.66	59.42	38.90	5.23	10.38	2.81	6.39	—	—
Datoca (GH)	68.97	41.30	54.48	31.41	32.26	17.00	4.38	9.75	3.73	7.36	—	—
Larguillo (GH)	99.84	63.15	101.07	79.97	35.40	19.75	4.78	10.93	4.00	9.91	—	—
Numex (GH)	131.43	78.05	139.46	82.45	113.52	77.63	5.68	9.52	5.32	11.06	—	—
Belrubí (GH)	138.64	80.80	136.01	80.62	14.92	14.17	11.70	22.83	6.01	12.53	—	—
Datler (OA)	91.05	67.57	62.68	45.73	49.80	40.68	16.62	10.77	4.67	7.18	2.86	5.20
Ocal (OA)	176.20	86.53	124.30	64.66	93.46	47.05	81.31	59.02	5.92	11.84	1.56	3.28
Datoca (OA)	57.55	27.45	50.58	25.00	52.56	36.47	28.13	14.57	4.25	7.84	2.36	4.83
Larguillo (OA)	92.68	56.71	71.03	52.23	42.15	33.43	18.12	13.73	3.59	6.28	1.74	3.47
Numex (OA)	93.21	57.15	78.51	43.42	66.22	38.38	45.47	27.20	2.24	3.32	1.25	2.39
Belrubí (OA)	228.12	96.56	210.19	104.79	85.87	44.56	95.15	45.62	6.88	12.98	2.25	4.58

Table 15.4 Values of the chlorophyll *a*/chlorophyll *b* quotient (C_a/C_b) during the growing and ripening process of the pepper paprika fruits in GH and in OA

Variety	Number of days					
	15	25	35	45	55	65
Datler (GH)	1.75	1.66	2.48	0.55	0.45	—
Ocal (GH)	1.89	1.75	1.52	0.50	0.43	—
Datoca (GH)	1.66	1.73	1.89	0.45	0.50	—
Larguillo (GH)	1.58	1.26	1.79	0.43	0.45	—
Numex (GH)	1.68	1.69	1.46	0.59	0.48	—
Belrubí (OA)	1.71	1.68	1.05	0.51	0.47	—
Datler (OA)	1.34	1.37	1.22	1.54	0.65	0.55
Ocal (OA)	2.03	1.92	1.98	1.37	0.50	0.47
Datoca (OA)	2.09	2.02	1.44	1.93	0.54	0.48
Larguillo (OA)	1.63	1.35	1.26	1.31	0.57	0.50
Numex (OA)	1.63	1.80	1.72	1.67	0.66	0.52
Belrubí (OA)	2.36	2.00	1.92	2.08	0.53	0.49

Table 15.5 Coefficients of correlation between the extractable colour (ASTA) of the GH and OA pepper paprikas with the chromatic co-ordinates (*L**, *a**, *b**), quotients (*a*/b**, 2000*a*/L*b**) and chroma (*C**) of the red fruits

ASTA	*L**	*a**	*b**	*a*/b**	2000*a*/L*b**	*C**
Greenhouse	−0.4202	−0.7805	−0.9436	0.9750	0.5213	−0.8602
Open air	−0.8434	−0.2764	−0.6726	0.7672	0.8472	−0.4306

Relationship between the visual colour of fruit and its pigment content

Table 15.5 gives the coefficients of the correlation between the total content of paprika pigments (ASTA units) and the chromatic co-ordinates, chroma and the quotients (*a*/b**, 2000*a*/L*b**) of the fruits. It may be observed that in the greenhouse crop the correlation between ASTA and *a*/b** is good enough, due to the better correlations of both co-ordinates considered separately with extractable colour, but the correlation is not good in the open air cultivated plants. In addition the ASTA values of the greenhouse paprikas show good correlation with chroma (*C**) since this attribute is directly related with coordinates *a** and *b**. The open air paprikas show a good correlation with brightness (*L**) and with the chromatic quotient (2000*a*/L*b**) where this co-ordinate intervenes, probably due to the favourable lighting conditions. In Figure 15.2 the points corresponding to the pairs (ASTA units, *a*/b**) in both cultivations are plotted, as well as the corresponding regression lines. The equations of these lines are:

Greenhouse	ASTA units $= -1390.24 + 594.78 \, a^*/b^*$	(1)
Open air	ASTA units $= -95.54 + 150.15 \, a^*/b^*$	(2)

being the corresponding coefficients of correlation for equations (1) and (2) those indicated in Table 15.5, i.e. 0.9750 and 0.7672, respectively.

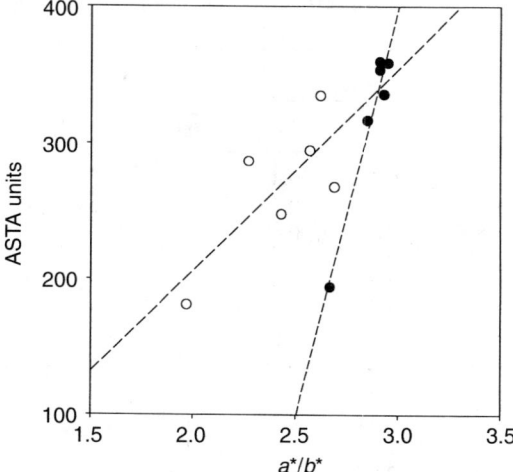

Figure 15.2 Plot of the points corresponding to the pairs (ASTA units, $a*/b*$) in both cultivation conditions, OA (○) and GH (●), as well as the corresponding regression lines.

Finally, the experimental values of the ASTA units and the chromatic co-ordinates $a*$ and $b*$, using the least square method, have been fitted for both cultivation conditions to the following equation:

$$\text{ASTA units} = c_0 + c_1 a* + c_2 b* \tag{3}$$

The equations obtained are:

Greenhouse   ASTA units $= 518.53 + 34.16a* - 115.81b*$ $\tag{4}$
Open air   ASTA units $= -122.23 + 22.89a* - 29.77b*$ $\tag{5}$

being the coefficients of the multiple correlation between ASTA units and $a*$ and $b*$ for equations (4) and (5) 0.9875 and 0.8579, respectively.

The experimental ASTA unit values are compared in Table 15.6 with the ones obtained using equations (1), (2), (4) and (5). It may be noted that both cultivation conditions show a good concordance with the estimates made from the linear multiple correlation, the greenhouse one being much better for cultivation. The obtained correlations allow a very quick and easy estimate for the ASTA units from the values of the chromatic co-ordinates $a*$ and $b*$, which are easy to obtain experimentally. These results could be applicable e.g. to decide the most appropriate moment for harvesting pepper paprika.

Influence of storage conditions on paprika colour loss

In this section a comparative study has been made of the colour and its kinetic loss in paprika made from a pepper variety, grown in the open air and greenhouse, during storage in darkness in different humidity and temperature conditions. The experimental data were fitted to a first-order kinetic. For each of the storage conditions the constant rates of colour loss and the half-lives were obtained.

Table 15.6 Comparison between the experimental value in ASTA units of the
paprikas and the values estimated by linear (*) and linear multiple (**)
correlation from equations (1) and (4), in GH, and equations (2) and (5),
in OA. The values of the correlation coefficients corresponding to each
correlation are indicated in the text

Varieties	Experimental value	Estimated value*	Estimated value**
Datler (GH)	360 ± 5	346	358
Ocal (GH)	336 ± 5	357	348
Datoca (GH)	354 ± 5	341	344
Larguillo (GH)	317 ± 4	308	304
Numex (GH)	194 ± 3	200	199
Belrubí (GH)	359 ± 5	368	367
Datler (OA)	248 ± 4	269	266
Ocal (OA)	268 ± 4	308	302
Datoca (OA)	335 ± 4	298	289
Larguillo (OA)	295 ± 4	291	291
Numex (OA)	181 ± 3	201	181
Belrubí (OA)	287 ± 4	246	284

Table 15.7 Temperature and humidity
conditions in the dryers used
in the experiment

Condition	Temperature	Relative humidity (%)
1	25	0
2	25	75
3	5	0
4	−25	0

Plant material and experimental design

The pepper variety Ocal was used for this experiment and paprika was obtained from fruits grown in thermal polyethylene and in open air.

The initial colour of the paprika was measured in ASTA units and then four 5 g samples of each type were placed in open petri dishes of 86 mm diameter. The samples were placed in dryers of the temperature and relative humidity (RH) conditions shown in Table 15.7 (henceforth designated 1–4). The RH of 75% was obtained by a saturated solution of sodium chloride placed in the dryers, while the RH of 0% (approximately) was obtained by introducing dry silica gel, which was replaced when it began to lose its bright blue colour. The temperatures of 25°C, 5°C and −25°C were obtained by placing the dryers in the appropriate stove or freezer in the absence of light. Subsamples of the paprikas were extracted at various times for colour analysis (Figure 15.3).

Rate constants of the colour loss process, supposed first order

The experimental data referring to storage time and corresponding ASTA units were fitted by non-linear regression to equation (6), corresponding to first-order kinetics (Varón *et al.*, 2000):

$$A = A_0 e^{-\lambda t} \tag{6}$$

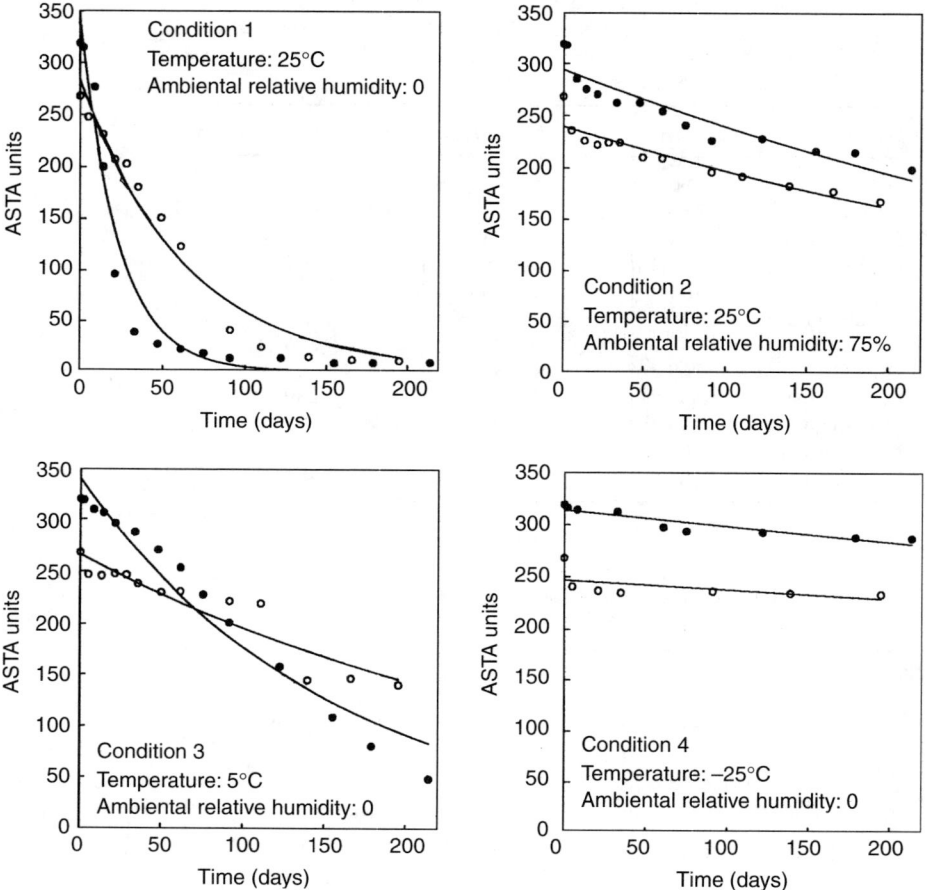

Figure 15.3 Evolution of colour in paprika obtained from the Ocal variety grown in both cultivation conditions and stored in the four conditions considered. The symbols (○) and (●) refer to OA and GH conditions, respectively.

where A is the instantaneous colour of the paprika expressed in ASTA units, A_0 is the colour at time $t = 0$ and λ is the colour loss constant of the first order. The values of the parameters A_0 and λ were obtained by non-linear regression of the experimental data of A and t to equation (6). The initial estimates of A_0 and λ, which are necessary for the fitting by non-linear regression to equation (6), were obtained from the slope and ordinate on the origin of the straight line resulting from the fit by linear regression of the experimental data of A and t to equation $\ln A = \ln A_0 - \lambda t$.

Half-lives

This parameter gives the time it takes to reduce colour by half and is obtained using equation (7):

$$t_{1/2} = \ln 2/\lambda \qquad (7)$$

The experimental half-life time, $t_{1/2,exp}$, is the time taken for A to be half of the initial experimental value of A (which generally does not coincide with the value of A_0 obtained by the fit), which we shall denominate $A_{0,exp}$. This was determined by linear interpolation between the times corresponding to the experimental values of the absorbance before and after $A_{0,exp}/2$.

Colour loss kinetics

Experimental data of Figure 15.3 were fitted to equation (6) by non-linear regression and the values of A_0 and the rate constant (λ) were obtained, as well as the half-life ($t_{1/2}$) obtained from the corresponding equation (7) and, when obtainable from the available experimental values, the experimental half-life ($t_{1/2,exp}$). For given storage conditions, the A_0 and λ values accurately characterized the process of colour loss in paprika, since they allowed us to determine the colour from equation (6), at any instant. The value of A_0 obtained from the fit was generally not equal to the value of the experimentally obtained initial colour ($A_{0,exp}$), although the difference was slight. For this reason, A_0 represents the initial colour of the paprika in the storage conditions considered. The half-life of a paprika's colour ($t_{1/2}$), will be given by equation (7). Since, according to these equations, $t_{1/2}$ decreases as λ increases, this parameter can be considered as a measurement of the speed of colour loss, the greater the value of λ the less stable the colour of the corresponding paprika. We shall now analyse the results obtained from fitting the experimental data to equation (6).

It can be seen from Table 15.8 that the colour stability of the paprika increases as follows: condition 1 < condition 3 < condition 2 < condition 4.

For conditions 1 and 2, which only differ in relative humidity, it can be observed that: (i) colour stability and, therefore, half-life ($t_{1/2}$) is much greater in condition 2; and, (ii) A_0 value is greater in condition 1.

For conditions 1, 3 and 4, which correspond to the same relative humidity but different temperatures, the following can be observed: (i) as was expected, colour stability and, therefore, half-life (both $t_{1/2}$ and $t_{1/2,exp}$, when this last value was determined), decreased as temperature rose; (ii) the value of A_0 decreased with decreasing temperatures; and, (iii) the experimental values $t_{1/2,exp}$ were higher than the corresponding theoretical ones, $t_{1/2}$.

Fitting to equation (6) showed that in every condition assayed with a higher λ value was matched by a lower $t_{1/2}$, as was seen from equation (7). Thus, in given conditions, the order of paprikas from lower to higher values of λ is the opposite of that of paprikas from lower to higher values of $t_{1/2}$.

Table 15.8 Results of fitting, by non-linear regression, the experimental data reflected in Figure 15.3 with equation (6). The values of $t_{1/2,exp}$ not obtainable experimentally are indicated by – and in all cases are in excess of 214 days

Condition	Cultivation	λ *(days^{-1})*	A_0 *(ASTA)*	$t_{1/2}$ *(days)*	$t_{1/2,exp}$ *(days)*
1	Greenhouse	$(4.34 \pm 0.54) \cdot 10^{-2}$	346 ± 17	16 ± 2	17 ± 4
1	Open air	$(1.57 \pm 0.14) \cdot 10^{-2}$	284 ± 11	44 ± 4	56 ± 27
2	Greenhouse	$(2.06 \pm 0.25) \cdot 10^{-3}$	294 ± 6	336 ± 40	—
2	Open air	$(1.98 \pm 0.23) \cdot 10^{-3}$	240 ± 4	350 ± 41	—
3	Greenhouse	$(6.45 \pm 0.52) \cdot 10^{-3}$	338 ± 10	107 ± 9	122 ± 37
3	Open air	$(3.03 \pm 0.38) \cdot 10^{-3}$	266 ± 7	229 ± 29	—
4	Greenhouse	$(5.26 \pm 0.84) \cdot 10^{-4}$	314 ± 3	1318 ± 209	—
4	Open air	$(3.89 \pm 2.60) \cdot 10^{-4}$	247 ± 6	1782 ± 691	—

A simple model of paprika carotenoid oxidation by oxygen in the air (Varón *et al.*, 1997, 2000) has been proposed for the theoretical interpretation of equation (6). This model also confirms that the first-order kinetic of colour loss is really a pseudo-first-order kinetic since, at a given temperature, (λ) is only constant if the concentration of the oxygen in contact with the paprika can be considered constant, which is reasonable to suppose in ordinary storage conditions.

The results obtained in this study show that the paprika obtained from the Ocal variety plants cultivated in greenhouse have greater initial colour (A_o). It can be seen that in conditions 1 and 3 (Table 15.7), the curves of the graphs obtained from the experimental data both from the field and greenhouse cross, while in conditions 2 and 4 (Table 15.7) they do not, since the time needed for this to happen is longer than the experimental period (214 days). This means that while the greenhouse-cultivated plants produce a paprika of greater initial colour, its colour loss is also greater. In addition, although the greenhouse paprika suffers a greater loss of colour, in conditions 2 and 4 it continues to have more colour than that obtained from field cultivated peppers throughout the experimental period. Thus, the paprikas obtained from greenhouse-grown peppers and kept for less than 200 days in conditions 2 and 4 maintain higher levels of extractable colour during storage (Figure 15.3). In conditions 1 and 3, extractable colour in one or other of the paprikas is greater according to the number of days of storage. For example, after storage for about 25 days in condition 1 and 70 days in condition 3, the extractable colour of paprikas obtained from fruit cultivated in the open air is greater than that for greenhouse-cultivated peppers, regardless of the initial value of the former.

References

Alcaraz, M. D. (1990) Carotenoides de variedades de *Capsicum annuum* L. Tesis de Licenciatura. Dep. Quím. Agríc. y Edafolog. Facultad de Químicas. Universidad de Murcia.

Alemán, J. and Navarro, F. (1973) *Estudio químico-físico de las variedades de pimientos Capsicum annuum cultivadas en Murcia*. Murcia: Centro de Edafología y Biología Aplicada del Segura, 100 pp.

Almela, L., López-Roca, J. M., Candela, M. E. and Alcazar, M. D. (1990) Separation and determination of individual carotenoids in a Capsicum cultivar by normal-phase high-performance liquid chromatography. *J. Chromatogr.*, 502, 95–106.

Almela, L., López-Roca, J. M., Candela, M. E. and Alcazar, M. D. (1991) Carotenoid composition of new cultivars of red pepper for paprika. *J. Agric. Food Chem.*, 39, 1606–1609.

ASTA (1986) *Official Analytical Method of the American Spice Trade Association*, 2nd edn, Englewood Cliffs, New Jersey.

Biacs, P. A. and Daood, J. (1994) High-performance liquid chromatography with photodiode array detection of carotenoid ester in fruit and vegetables. *J. Plant Physiol.*, 143, 520–525.

Camara, B. and Brangeon, J. (1981) Carotenoid metabolism during chloroplast to chromoplast transformation in *Capsicum annuum* fruit. *Planta*, 151, 359–364.

Candela, M. E., López, M. and Sabater, F. (1984) Carotenoids from *Capsicum annuum* fruits: changes during ripening and storage. *Biol. Plan.*, 26, 410–414.

Carde, J. P., Camara, B. and Cheniclet, C. (1988) Absence of ribosomes in *Capsicum* chromoplasts. *Planta*, 173, 1–11.

Carreño, J., Martínez, A., Almela, L. and Fernández-López, J. A. (1995) Proposal of an index for the objective evaluation of the colour of red table grapes. *Food Res. Int.*, 28(4), 373–377.

Conrad, R. S., Sundtrom, F. J. and Wilson, P. W. (1987) Evaluation of two methods of pepper fruit color determination. *Hortsci.*, 22(4), 608–609.

Costa, J. C. (1979) *Pimiento pimentonero. Selección y mejora*. H. Técnica 27. Madrid: Instituto Nacional de Investigaciones Agrarias.

Costa, J. C. (1980) Desarrollo de nuevas variedades de pimiento para pimentón con alto contenido en pigmentos y susceptibles de recolección mecánica. III Jornadas de Mejora de Tomate y Pimiento. Tenerife.

Costa, J. C. (1991) Pimiento pimentonero. Estado actual y perspectivas de futuro en la región de Murcia. *Agric. Vergel*, 3, 232–235.

Costa, J. C., Soriano, C., Nuez, F. and Navarro, F. (1989) Characterization of new red pepper cultivars for grinding. In S. Palaka (ed.) *VIIth Eucarpia Meeting in Genetics and Breeding of Capsicum and Eggplant*, Kragujevac, Yugoslavia, pp. 27–30.

Davies, B. H., Matthews, S., and Kirk, J. T. O. (1970) The nature and biosynthesis of the carotenoids of different color varieties of *Capsicum annuum*. *Phytochemistry*, 9, 797–805.

De la Torre, M. C. and Farré, R. (1975) Carotenoides del pimentón. *Anales de Bromatología*, 28(2), 149–195.

Drdak, M., Pribela, A., Zemkova, M. and Schaller, A. (1982) Connection between pigment concentration and colour of ground seasoning paprika, with particular reference to the hue and redness. *Confructa*, 26, 18–24.

Ferrer, C. and Costa, J. C. (1991) Contenido en clorofilas durante el desarrollo y maduración de frutos de *Capsicum annuum* L. *Acta Hort.*, 8, 269–274.

Fisher, C. and Kocis, J. A. (1987) Separation of paprika pigment by HPLC. *J. Agric. Food Chem.*, 35, 55–57.

Gómez, R., Pardo, J. E., Varón, R. and Navarro, F. (1996) Study of the color evolution during maturation in selected varieties of paprika pepper (*Capsicum annuum* L.). *The Second International Symposium on Natural Colorants*. Acapulco, México.

Gómez, R., Pardo, J. E., Navarro, F. and Varón, R. (1998a) Colour differences in paprika pepper varieties (*Capsicum annuum* L.) cultivated in a greenhouse and in the open air. *J. Sci. Food Agr.*, 77, 268–272.

Gómez, R., Pardo, J. E. and Varón, R. (1998b) The colour in peppers and paprikas. In S. G. Pandalai (ed.) *Recent Research Developments in Agricultural and Food Chemistry*, Vol. 2, Part II, Research Signpost, Trivandrum, India, pp. 445–477.

López, M. (1972) Carotenoides en la maduración del pimiento (*Capsicum annuum*). PhD thesis. Facultad de Ciencias. Universidad de Murcia.

Mínguez, M. I. and Hornero, D. (1993) Separation and quantification of the carotenoid pigment in red peppers (*Capsicum annuum* L.), paprika, and oleoresin by reversed-phase HPLC. *J. Agr. Food Chem.*, 41(10), 1616–1620.

Mínguez, M. I. and Hornero, D. (1994) Changes in carotenoid esterification during the ripening of *Capsicum annuum* cv. Bola. *J. Agr. Food Chem.*, 42(1), 1–5.

Nagle, B. L., Villalón, B. and Burns, E. E. (1979) Color evaluation of selected capsicum. *J. Food Sci.*, 44, 416–418.

Nakayama, R. M. and Mata, F. B. (1973) Extractable red color of chili peppers as influenced by fruit maturity and alar, giberellic acid and ethephon treatments. *Hortsci.*, 8, 16–17.

Navarro, F. and Costa, J. (1993) Evaluación del color de algunas variedades de pimiento para pimentón por espectrofotometría de triestímulos. *Rev. Esp. Ciencia y Tecnol. Aliment.*, 33(4), 427–434.

Ortiz, R., Martín, J. and Valero, T. (1990) Separación y cuantificación por HPLC de los pigmentos carotenoides del pimentón. *Química e Industria*, 36(2), 129–133.

Pardo, J. E., Navarro, F., Varón, R., Costa, J. and Gómez, R. (1996) Color characteristics in selected paprika pepper varieties (*Capsicum annuum* L.). *Bollettino Chimico Farmaceutico*, 135(3), 184–188.

Reeves, M. J. (1987) Re-evaluation of capsicum color data. *J. Food Sci.*, 52, 1047–1049.

Salmerón, P. (1973) Almacenado del pimentón. In P. Salmerón (ed.) *El Color en los Procesos de Elaboración del Pimentón*, Centro de Edafología y Biología Aplicada del Segura, Murcia, Spain, pp. 229–232.

Soriano, M. C., Navarro, F. and Costa, J. (1990) Caracterización de nuevos cultivares de pimiento para pimentón. *Agrícola Vergel.*, August, 630–632.

Varón, R., Amo, M., Gómez, R., Pardo, J. E. and Navarro, F. (1997) Loss of paprika colour. I. Kinetic analysis of a simple model. *Acta Hort.*, 16(2), 381–387.

Varón, R., Díaz, F., Pardo, J. E. and Gómez, R. (2000) A mathematical model for color loss in paprikas containing differing proportions of seed. *J. Sci. Food Agric.*, 80, 739–744.

Ziegler, H., Shafer, E. and Schneider, M. (1983) Some metabolic changes during chloroplast–chromoplast transition in *Capsicum annuum*. *Physiol. Veg.*, 21(3), 485–494.

16 Future perspectives of capsaicin research

J. Szolcsányi

Capsaicin, the pungent principle in chilli peppers has become a promising molecule for the development of a new generation of analgesic–anti-inflammatory agents with a target molecule on nociceptive primary afferent neurons. The present state of knowledge in this field is summarized in this chapter after a short historical introduction. Furthermore, the horizons of new trends in the peptidergic neuroregulatory functions of these capsaicin-sensitive afferents are outlined.

Recently the rat and human capsaicin (vanilloid) VR1 receptors have been cloned and identified in one group of sensory neurons. These nociceptive membrane proteins of the TRP superfamily are cation channels gated by noxious heat, protons, capsaicin, resiniferatoxin and endogenous ligands as anandamide and lipoxygenase products. TRPV channel subtypes with similar structure were cloned from neurons (VRL-1, VR5'sv, VR-OAc) and from non-neural tissues, but none of them were activated by the vanilloid agonists capsaicin or resiniferatoxin. The response characteristics of cell lines transfected with point mutant VR1 derivatives as well as nociceptive behaviour and sensory neuron features of VR1 knock out mice are summarized. The role of TRPV1/VR1 receptors in inflammatory pain is emphasized. VR1 agonists after prolonged opening of the cation channel induce sensory desensitization of nociceptors to all types of stimuli by inducing secondary intracellular changes (dephosphorylation of VR1 receptors, inhibition of voltage-gated Ca^{++} and Na^+ channels, mitochondrial impairment). Sensory desensitization to the pain producing effect of bradykinin in the human skin and at the level of C-polymodal nociceptors is explained by intracellular biochemical changes in VR1 expressing nerve terminals. Evidence that capsaicin stimulates then desensitizes the warmth sensors of the hypothalamus is listed. It is suggested that the signal-inducing sensory function and peptidergic efferent functions are coupled at the level of the same capsaicin-sensitive sensory nerve terminals. The excitation-induced release of tachykinins and CGRP elicit local efferent responses as neurogenic inflammation, vasodilatation and various smooth muscle responses. The release of somatostatin from these nerve endings induces systemic anti-inflammatory and analgesic effects ("the sensocrine" function of capsaicin-sensitive nociceptors). The pathophysiological relevance of these "unorthodox" regulatory mechanisms form promising perspectives for drug development. The possible physiological role of sensory neuropeptides are indicated by the vasodilatory response of CGRP and the anti-inflammatory effects of somatostatin released from these capsaicin-sensitive afferents at the subnociceptive level of stimulation.

Introduction

In recent years remarkable breakthroughs have opened up for drug research by the identification of new cellular target molecules with validated pathobiological significance. Most of these target identifications were made with the aid of wonderful exogenous lead molecules often made by nature in different plants or animal poisons. Capsaicin is certainly one of these promising lead molecules. Its powerful, selective pain producing and desensitizing effects were first described by Andreas Hőgyes (1878) and Nicholas Jancsó (1960), respectively. Subsequently, postulation of a new receptor molecule the "capsaicin receptor" in cell membrane of a subgroup of primary afferent neurons was proposed (Szolcsányi and Jancsó Gábor, 1975, 1976). This hypothesis was put forward on the basis of four sets of observations made by our group in the late sixties and early seventies (Szolcsányi, 1982, 1984a; Buck and Burks, 1986; Holzer, 1991; Maggi, 1995).

1   The potent pungent effect of capsaicin turned out to be absent in several species. Nociceptive protective reflexes were not evoked in non-mammalian species such as frog, chicken or pigeon.

2   The loss of sensation evoked by high concentration of capsaicin is highly selective, as revealed by quantitative psychophysical assessments on the human tongue after topical capsaicin application. Hot or irritant pain sensation induced by capsaicin, piperine, zingerone or mustard oil was not detectable. Noxious heat threshold was enhanced and warmth discrimination was impaired after capsaicin treatment. On the other hand, taste stimuli, difference limen in the cold range, threshold concentration of menthol and tactile sensation remained unchanged.

3   Structure–activity relationship showed that small changes in the capsaicin structure have profound effect on its agonist activity. Furthermore, pharmacophores required for the excitatory and sensory blocking effects of the capsaicin congeners were partially different.

4   The capsaicin-sensitive subgroup of primary afferent neurons was identified by the ultra-structural changes induced by systemic treatment of rats with high doses of capsaicin. Capsaicin-responsive cells were the small, dark, B-type neurons of the trigeminal, nodosal and dorsal root ganglia, as well as a subgroup of small neurons in the hypothalamic medial preoptic area, where the central warmth sensors are situated in high density. Light, A-type primary afferent neurons and satellite cells in these ganglia, or cellular elements of the sympathetic ganglia were not altered.

Subsequently, the neuroselective site of action of capsaicin was supported in our further electrophysiological and *in vitro* pharmacological studies. Close arterial injections of capsaicin in rabbit, rat and cat selectively activated cutaneous C- and A-delta polymodal nociceptors which comprise about 50% of the population of the primary afferent neurons. On the other hand, all types of mechanoreceptors, including the A-delta mechanonociceptors or C-mechanoreceptors, were not activated by capsaicin (Szolcsányi, 1993, 1996). Furthermore, neuroeffectors which mediate cholinergic, adrenergic, purinergic, peptidergic autonomic and intrinsic enteric neurotransmissions in the guinea-pig gut were not influenced by capsaicin (Maggi, 1995).

With the advent of the neuropeptide era, capsaicin gained considerable interest outside Hungary when in 1978 Tom Jessell, Leslie Iversen and Cuomo Cuello, and soon after Fred Lembeck's group in Graz and Masanori Otsuka's group in Japan, clearly showed that capsaicin releases and subsequently depletes substance P from the primary afferent neurons and their terminals but not from other parts of the nervous system (Holzer, 1991; Maggi, 1995). Capsaicin in this way turned out to be the first drug which selectively depleted a neuropeptide from a population of neurons.

Afterwards, various ways of systemic or local pretreatments with capsaicin were utilized in a broad scale to reveal the presence and role of substance P, somatostatin and the newly identified neuropeptides e.g. CGRP, VIP, bombesin, etc. in sensory neurons and in their peripheral or central axonal arborizations. Most of these data were obtained in rats pretreated with capsaicin in the neonatal age since it had been suggested that this treatment induce an acute selective cell death of the chemosensory afferent population (Jancsó *et al.*, 1977). Re-evaluation of these early data was needed, however, subsequent studies revealed a lack of selectivity in the loss of C-polymodal nociceptors in these animals (Welk *et al.*, 1984; Szolcsányi, 1990; Nagy *et al.*, 1983) which might be related to long-term phenotype changes and reorganization of the pathways. Furthermore, it was recently reported that the loss of C-afferents observed in the adult age develop after a week and not within 30 minutes as it was initally emphasized (Szolcsányi *et al.*, 1998b). The data obtained in animals pretreated in the adult age indicated a selective inactivation and loss of noxious heat responsive afferents, the major group of which is the C-polymodal nociceptors (Szolcsányi, 1990, 1993). Subsequently the capsaicin receptor was cloned (Caterina *et al.*, 1997) and the molecular biological repertoire for high throughput screening has opened the means for the discovery of the first putative analgesic acting at the level of nociceptors. The aim of the present overview is to focus on the perspectives of this drug research from a standpoint of describing the present state of knowledge in this field. Furthermore, new trends in peptidergic neuroregulatory mechanisms, which suggest that this capsaicin-sensitive substantial portion of the sensory neurons subserve local efferent and systemic "sensocrine" functions (Szolcsányi, 1996; Thán *et al.*, 2000) are discussed briefly in this chapter.

Identification of TRPV1/VR1 capsaicin receptors

During the last decade the following direct approaches supplied new information which paved way to the identification of the capsaicin receptor.

1   Resiniferatoxin (RTX) the irritant principle from the latex of a cactus-like plant *Euphorbia resinifera* (Figure 16.1) was identified as an "ultrapotent" capsaicin analogue. This homovanillic ester compound shares a common "vanilloid" structure with the vanillylamide capsaicin and is suitable for studying high-affinity specific RTX binding sites, as its Kd value is as low as 18–46 pM at DRG membranes (Szállási and Blumberg, 1990, 1999).

2   Single-channel currents evoked by resiniferatoxin and capsaicin in isolated membrane patches from dorsal root ganglion neurons revealed that capsaicin directly opens a new type of cation channel without any second messenger. An influx of calcium and sodium are responsible for the neuropeptide release and for the major component of depolarization, respectively (Bevan and Szolcsányi, 1990; Bevan and Docherty, 1993; Vlachova and Vyklicky, 1993; Oh *et al.*, 1996).

3   Capsazepine, a synthetic analog of capsaicin (Bevan and Docherty, 1993) and ruthenium red a polycationic dye (Amann and Maggi, 1991) antagonize the effect of capsaicin and resiniferatoxin on the capsaicin receptor. Capsazepine is a competitive capsaicin antagonist with moderate potency and a limited range of selectivity (IC_{50} 100–700 nM in DRG neurons). Ruthenium red antagonizes the effect of capsaicin and resiniferatoxin by blocking the cationic channel part of the receptor in a concentration range of 50 nM–10 μM and in an *in vivo* dose range of 0.5–4 mg/kg in rat (Szolcsányi *et al.*, 1991, 1993).

Capsaicin TRPV1/VR1-receptor

The capsaicin receptor-encoding cDNA was identified by functional screening of the capsaicin-induced calcium influx in transfected human embryonic kidney-derived HEK 293 cells. The cDNA library was constructed from dorsal root ganglion-derived messenger RNA and its pools

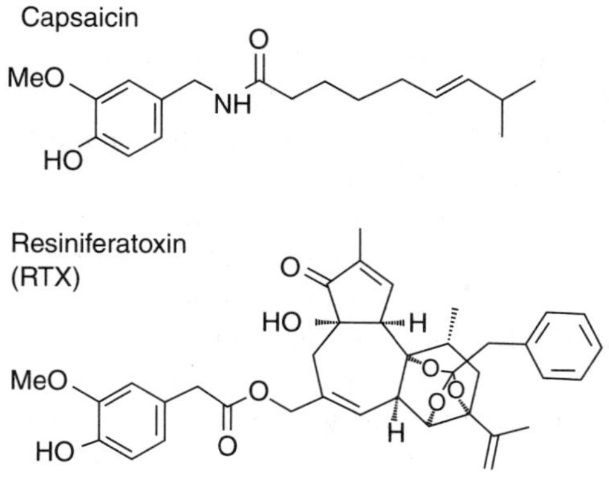

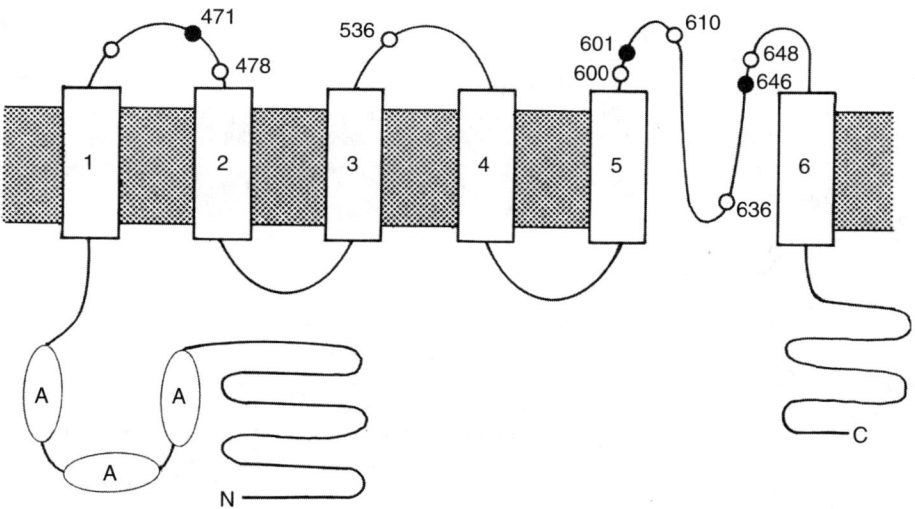

Capsaicin

Resiniferatoxin
(RTX)

Figure 16.1 Chemical structure of capsaicin and resiniferatoxin.

Figure 16.2 Domain structure of the capsaicin VR1 receptor. Positions of acidic residues in the outer loops and pore region and the N and C terminals are indicated. Open circles: glutamate, closed circles: aspartate. Open boxes with "A" delineate ankyrin repeat domains. Predicted transmembrane domains are numbered in open boxes.

of approximately 16,000 clones were transfected. Capsaicin receptor-encoding 3-kilobase newly cloned cDNA was named as VR1, for vanilloid receptor subtype 1 (Caterina *et al.*, 1997).

The term vanilloid was introduced by Szállási and Blumberg (1990) to denote two subtypes of receptors, the capsaicin (C) vanilloid receptor and the resiniferatoxin (R) vanilloid receptor. Recently, however, the inventors of the vanilloid label consider this name as "somewhat of a misnomer" since several non-vanilloid agonists were described for these receptors (Szállási and Blumberg, 1999). It has also been claimed that it is misleading to refer in the name of a cloned receptor to a chemical moiety of some exogenous lead molecules which is missing in the proposed endogenous ligand (Szolcsányi, 2000a,b). Since the capsaicin VR1 receptor belongs to the TRP superfamily (Harteneck *et al.*, 2000) the TRP Nomenclature Committee renamed it TRPV1.

The capsaicin VR receptor of rat (rVR1) is a protein of 838 amino acids with a relative mass of 95,000. It is predicted to have six transmembrane domains (β sheets) topology with an additional short hydrophobic stretch (predicted pore region) between transmembrane regions 5 and 6 (Figure 16.2). The capsaicin VR1 receptor is a cation channel with a remarkable feature of being opened by noxious heat besides capsaicin or resiniferatoxin (Table 16.1). It serves in this way as a "nociceptive molecule", a hot transducer forming the molecular basis for the selective site of action of capsaicin on noxious heat responsive nociceptors (polymodal nociceptors). Selective action on C-polymodal nociceptors has already been revealed in multi-fibre preparations (Szolcsányi, 1977) and in single unit recordings from afferent fibers (Szolcsányi, 1990, 1993, 1996). Activation of single-channel currents of VR1-expressing oocytes and that of cultured dorsal root ganglion neurons by capsaicin, resiniferatoxin and a membrane impermeable

Table 16.1 Nociceptive and non-nociceptive TRPV/VR receptor subtypes in neurons

Name	Agonist	Heat threshold (°C)	V–I relation	Antagonist	Structure	Location	Ref.
rVR1	Caps: yes RTX: yes 12-HPETE: yes Ana: yes Protons: yes PPAHV: yes	42–44	Outward rectification	Capsaze-pine: yes Ruthenium, red yes	6TM + pore, 838 aa	DRG, small cell, Brain	[1]
hVR1	Caps: yes Protons: yes PPAHV: no	42–44	Outward rectification	Capsaze-pine: yes Ruthenium, red: yes	6TM + pore, 839 aa	DRG	[2]
VRL-1	Capsaicin: no RTX: no Protons: no	52	Dual, mainly outward rectification	Capsaze-pine: no Ruthenium, red: yes	6TM + pore, r:761 aa, h:764 aa	DRG, medium, large cell, non-neural	[3]
VR.5'sv	Caps: no RTX: no Protons: no	None			6TM + pore	DRG, Brain	[4]
VR-OAC	RTX: no Ana: no Hypotonic solution: yes	None	Dual rectification		6TM + pore, r:871 aa, m:873 aa, h:871 aa, ch:852 aa	Kidney, Lung, Spleen, Trigeminal, large cell, Brain	[5]

Notes

Abbrevations:- Caps: capsaicin; RTX: resiniferatoxin; Ana: anandamide; PPAHV: phorbol 12-phenyllactate 13-acetate 20-homovanillate; 12-HPETE: 12-(S)-hydroxyeicosatetraenoic acid; r: rat; h: human; m: mouse; ch: chicken; aa: amino acid; TM: transmembrane domain; DRG: dorsal root ganglion.

References
[1] Caterina *et al.*, 1997; Jung *et al.*, 1999; McIntyre *et al.*, 2001; Savidge *et al.*, 2001.
[2] Hayes *et al.*, 2000; McIntyre *et al.*, 2001.
[3] Caterina *et al.*, 1999.
[4] Schumacher *et al.*, 2000.
[5] Liedtke *et al.*, 2000.

capsaicin analogue (DA-5018) revealed that these vanilloids activate VR1 cation channel from the intracellular domain. On the other hand, acidic solution applied to the inner surface of the patch membrane had no effect indicating that protons open the channel from outside the cell (Jung *et al.*, 1999).

More recently, the human orthologue of rat vanilloid receptor1 (hVR1) was also cloned. (Hayes *et al.*, 2000; McIntyre *et al.*, 2001). The human VR1 has a chromosomal location of 17 p 13 and is expressed in human dorsal root ganglia and to a lesser extent in some parts of the central nervous system. It has 92% homology to the rVR1 and it is also responsive to both capsaicin and noxious heat. Nevertheless, some differences in responses and pharmacology between rVR1 and hVR1 have been recently shown. Capsazepine was more effective at inhibiting the noxious heat response of hVR1 than rVR1, and it blocked only in the former case the effect of low pH. Further differences are indicated in Table 16.1. Recently, the responsiveness of native cultured dorsal root ganglion neurons and a cell line expressing VR1 receptor were compared. The native and recombinant capsaicin receptors were similar with respect to the characteristics of intracellular calcium signals and susceptibility to antagonists as ruthenium red or capsazepine. In other words, the accurate reproduction of the response characteristics of native channels with the cloned receptor support the conclusion that the VR1 gene product is responsible for the gating effects of noxious heat and capsaicin in native sensory neurons (Savidge *et al.*, 2001).

Thermal responsiveness of native trigeminal neurons, however, was more variable (Liu and Simon, 2000) and single-channel recordings from dorsal root ganglion neurons indicated that most channels respond to either heat or capsaicin, but infrequently to both. Furthermore, in contrast to the similarities observed, the cation permeability of the channel was greater for capsaicin than for heat and at the whole-cell level there was a weak correlation of amplitudes of responses to capsaicin and heat (Nagy and Rang, 1999a). For an explanation of these unexpected mismatches the following tentative list of possibilities was raised: "VR1 gene product may assume different functional states depending on other factors, including splice variants, aggregation with other membrane proteins, the presence of different multimeric species or the degree of poshophorylation or glycosylation of the channel protein" (Nagy and Rang, 1999a). These alternatives might also be the source for the varying responsiveness of the capsaicin receptor/ ion channel to different chemical agonists, particularly those which have capsaicin-like or resiniferatoxin-like chemical structures (Szállási and Blumberg, 1999; Shin *et al.*, 2001). Recently another possibility has also been raised. Considering the apolar nature of VR1 agonists including anandamide and the marked differences in binding and channel-gating characteristics of these compounds, the role of structural differences in the lipid bilayer around the VR1 protein was included in the context of a theory suggesting different sites in VR1 protein for binding resiniferatoxin or capsaicin (Szolcsányi 2000a, 2002).

These findings indicate that the polymodal nature of the VR1 capsaicin receptor with several allosteric ways of opening to polar and apolar agonists and noxious heat is a promising and challenging molecular target for analyses for the function and pharmacology of the major subset of sensory neurons of polymodal nociceptors on the one hand and on the other hand, multifunctional cell membrane receptors in general.

Vanilloid-receptor-like protein 1 (VRL-1/TRPV2)

This orthologue of VR1, the cDNA of which was identified from rat brain (rVRL-1) and in a much lower level from the CCRF-CEM human myeloid cell line (hVRL-1) share 78.4% identity with one another. It is also 49% identical to rVR1 (Caterina *et al.*, 1999). Nevertheless, the overall structure of six transmembrane domains, the putative-loop region and the cytoplasmic

N-terminal with three ankyrin-repeats are similar in both VR1 and VRL-1. In transfected Xenopus oocytes or HEK 293 cells, no response to capsaicin (100 μM), resiniferatoxin (10 μM) or to proton (pH4) was observed and the heat threshold was around 52°C. Channel opening was inhibited by ruthenium red (10 μM) but not by capsazepine (Table 16.1). VRL-1 immunoreactive sensory neurons are medium to large cells, one-third of which also contain CGRP but very few are labelled with substance P antibody or lectine IB4. It is assumed that these sensory neurons correspond to the type I A-delta mechano-heat-sensitive nociceptors (Caterina *et al.*, 1999). It is worth noting that the high-threshold, capsaicin-insensitive, mainly large-diameter dorsal root ganglion neurons were already described before the VRL-1 channel has been cloned (Nagy and Rang, 1999b).

N-terminal splice variant of the capsaicin receptor (VR.5′sv)

This splice variant of VR1 was identified (Schumacher *et al.*, 2000) from the sensory neuron cDNA library. It contains a truncated intracellular N-terminal chain with a single ankyrin domain but it maintains the six transmembrane structure of TRPV1 with the putative pore region (Table 16.1).

Conditions such as, noxious heat (53°C), capsaicin (10 μM), resiniferatoxin (0.1 μM) and pH 6.2 did not elicit inward currents in VR.5′sv transfected cell lines. The functional role of this protein is therefore not known, but it was identified in dorsal root ganglion, different brain regions and at a substantially lower level in the peripheral blood mononuclear cells.

Vanilloid receptor-related osmotically activated channel (VR-OAC/TRPV4)

This recently cloned ion channel also belongs to the TRP superfamily of ion channels (Harteneck *et al.*, 2000) like the VR1 and represents a new member of the OSM-9 family of *C. elegans* (Liedtke *et al.*, 2000). VR-OAC is a cation channel with a six transmembrane structure, a putative pore-forming region and three N-terminal ankyrin repeat domains isolated from rat, mouse, human and chicken. It is abundantly expressed in the kidney, lung, spleen, testis and fat and also at a lower level in large A-type sensory neurons of the trigeminal ganglion, in cutaneous Merkel cells and in the organ of Corti. In the brain it is mainly expressed in neurons of the circumventricular organ and in the median preoptic area of the lamina terminalis. The VR-OAC ion channel is opened by hypotonic solutions and in fact it is the first osmotically activated channel identified in vertebrates. Resiniferatoxin and anandamide, two VR1 agonists do not, however, activate this channel (Table 16.1).

Further cloned proteins with similar structure are sometimes listed among the "vanilloid receptors". They have not been shown to be expressed in sensory neurons and neither are they activated by vanilloids. Therefore, their cDNA structural resemblance to VR1 does not justify, from a functional point of view, for grouping them together with the above proteins. "Stretch-inactivated channel" (SIC) cloned from rat kidney, "epithelial calcium channel" and "growth factor-related" calcium channel (Schumacher *et al.*, 2000; Szállási and DiMarzo, 2000) belong to this group of ion channels.

Point mutations of glutamate and aspartate residues at the extracellular loops of rVR1

The responsiveness of transfected cells with VR1 point mutants were studied in order to determine protein domains important for the activation of VR1 by different stimuli as capsaicin, heat and protons. Owing to the activating and modulating effects of H^+, the first putative targets for point mutations were the extracellular negatively charged amino acids in this nociceptor-specific cation channel. In the extracellular loops of VR1 most of the charged glutamate and aspartate amino acids are in the putative pore-forming region between the fifth and sixth transmembrane domains (see Figure 16.2). In two recent papers (Jordt et al., 2000; Welch et al., 2000) point mutations of these 11 amino acids with the COOH side-chain were replaced by CONH containing glutamine or asparagine (E → Q and D → N), or in two cases (E648 and E600) by positively charged histidine (H) or lysine (K), and in one case (E648) by alanine (A).

Replacement in the first loop (E458Q, E478Q, D471N) or in the second extracellular loop (E536Q) was almost ineffective but after point mutations in the pore region various combinations of changes in the responsiveness to different stimuli occurred. Five of the mutant channels were significantly more sensitive to capsaicin than the wild type VR1. The highest sensitivity to capsaicin was at E600K, and in this case 50 nM capsaicin already caused saturation; the EC_{50} values at E600Q was 40 nM (wild type EC_{50}: 520 nM) (Jordt et al., 2000). At E636Q, D646N and E648Q mutants the EC_{50} value was 300 nM (wild type EC_{50}: 900 nM) (Welch et al., 2000). In the latter three cases, the enhanced sensitivity to capsaicin was selective and surprisingly none of these mutants differed in their activation by acidic pH or temperature. It is interesting to note that the cooperativity of capsaicin binding to VR1 (Hill coefficient ~2) was abolished only when the mutation was made near the intracellular part of the loop (E636Q Hill coefficient ~1). Furthermore, the ion selectivity of the E636Q mutant channel became seven-fold higher for Ca^{++} as compared with VR1, being the pCa^{2+}/pK ratios 17.6 and 2.6, respectively. On the other hand, the extracellular glutamate residue at E600 seems to serve as a "key regulatory site of the receptor by setting sensitivity to other noxious stimuli in response to changes in extracellular proton concentration" (Jordt et al., 2000). Furthermore, proton-independent temperature sensitivity was also markedly enhanced in E600Q and particularly in E600K, where the temperature threshold dropped to 30–32°C i.e. to a similar value which occurs when wild-type channels are heated at pH 6.3. The pH dependence of temperature gating in the E600H mutant was spanning over a pH of 9–5 while in E600D it was observed only in a limited range from 7.4 to 6.5.

The E648A mutant showed a marked and selective decrease in proton-activated currents, whereas no alterations were observed when compared to the wild-type VR1 in response to capsaicin or heat stimuli. The responses of this mutant channel also support the conclusion that H^+-induced VR1 activation and its potentiating effect on thermal responses are triggered by different conformational changes. In the case of the E648A mutant the diminished proton sensitivity was associated with normal proton potentiation. The double mutant (E600Q–E648A) combined the increased sensitivity to capsaicin and heat with a loss of proton-mediated potentiation (Jordt et al., 2000).

All these very recent interesting data are the first attempts to clarify the key molecular residues in the capsaicin VR1 receptor for operation as a nociceptor-specific cation channel. The differences in the structure and gating mechanisms among natural, point mutant or chimeric variants of TRPV receptors will be of great help in combinatorial chemistry for synthesizing promising analgesic drugs with a site of action on those nociceptors which initiate pain sensation in inflammation and in several other painful conditions. Rational molecular approaches for high throughput screening have already started. Thus, capsaicin has opened the avenue for drug research in the field of sensory pharmacology.

Immunocytochemical localization of capsaicin VR1/TRPV1 receptors

Immunoreactivity to the rVR1 receptor (Caterina et al., 1997; Tominaga et al., 1998; Guo et al., 1999; Mezey et al., 2000) and its mRNA (Helliwell et al., 1998; Michael and Priestley, 1999;

Mezey *et al.*, 2000) were in accord with the data of binding studies with [^{3}H]RTX (Winter *et al.*, 1993; Szállási and Blumberg, 1999), and revealed important new localizations as well.

Localization in sensory neurons

Immunoreactivity of rVR1 or the *in situ* hybridization of its mRNA was detected in the B-type small or medium sized neurons in colocalization with TrkA or IB$_4$ lectin markers in dorsal root ganglia and also in trigeminal, nodosal and jugular ganglia. No hybridization signal was observed in sympathetic ganglia (Helliwell *et al.*, 1998). A high proportion of colocalization of VR1 immunoreactivity was observed with substance P, CGRP, P$_2$X$_3$ purinoceptor (Guo *et al.*, 1999) and in IB$_4$ positive population with somatostatin (Michael and Priestley, 1999).

In peripheral nerves and neural plexuses, as well as in the central terminals of primary afferent neurons in the superficial dorsal horn, VR1 positive fibers were also described (Tominaga *et al.*, 1998; Guo *et al.*, 1999). VR1 immunoreactivity was associated with the plasma membrane of unmyelinated fibers having clear vesicles, as well with axonal microtubules and Golgi complex (Tominaga *et al.*, 1998; Guo *et al.*, 1999).

Localization in the brain

Although [^{3}H]RTX autoradiography failed to detect high affinity specific binding in different brain regions (Szállási *et al.*, 1995) and *in situ* hybridization of mRNA encoding VR1 did not indicate positivity in the brain, a recent elegant and more sensitive approach with reverse transcription-PCR was successful in this line (Mezey *et al.*, 2000). Whether these differences indicate a much lower level of VR1 expression in the brain than in the sensory neurons needs further research.

VR1-immunoreactivity and mRNA for VR1 positive neurons in rat and human brain were scattered in different regions, including the preoptic area of the hypothalamus. The strongest VR1 mRNA signals were detected in the following areas: all cortical areas, septum, hippocampus dentate gyrus, substantia nigra (zona compacta), cerebellum, locus coeruleus and inferior olive. In rat hypothalamus VR1 mRNA-expressing neurons and capsaicin-induced glutamate release have been reported (Sasamura *et al.*, 1998). It was striking that in rats treated with capsaicin in the neonatal age, although VR1 mRNA in DRG and nucleus of the spinal trigeminal nerve was diminished, its expression in other areas of the brain did not differ from that of the controls (Mezey *et al.*, 2000).

The functional significance of the VR1-encoding neurons in the central nervous system is a challenging new field for further research. Capsaicin-sensitive neurons in the medial preoptic area serve as warmth sensors in thermoregulatory heat loss responses, as postulated a long time ago (Jancsó-Gábor *et al.*, 1970; Szolcsányi *et al.*, 1971) and repeatedly confirmed by other groups (Szolcsányi, 1982, 1983; Hori, 1984; Pierau *et al.*, 1986). In rats after adult pretreatment with capsaicin or resiniferatoxin a similar long-lasting mitochondrial swelling was observed in the preoptic area as that in the small neurons of the dorsal root ganglia (Joó *et al.*, 1969; Szolcsányi *et al.*, 1971, 1975, 1990; Szállási *et al.*, 1989). Therefore a lower level of VR1 mRNA expression after adult pretreated rats, combined with an ultrastructural search for the characteristic cellular changes induced by capsaicin and resiniferatoxin in sensory neurons might be a possible good starting point for a morphological approach. Regarding their functional role, it should be remembered that noxious heat is not an adequate stimulus for VR1 except for the exteroceptors, while the chemical excitability feature of these neural elements might serve chemosensor functions in regulations of the internal environment.

It is worth noting that a few examples of VR1 expression have been shown also in non-neural tissues and the first data for H^3RTX binding, Ca^{++} uptake and desensitization to the effects of capsaicin and resiniferatoxin described in C_6 rat glioma cells and mast cells were in a remarkably similar concentration range as that of the DRG cells (Bíró *et al.*, 1998).

Endogenous ligands for the capsaicin VR1 receptor

Protons

Hydrogen ions activate the VR1 channel (Tominaga *et al.*, 1998) and markedly potentiate the response of both noxious heating (Tominaga *et al.*, 1998) and capsaicin (Caterina *et al.*, 1997). These two effects of protons seem to be due to different conformational changes of the VR1 molecule (McLatchie and Bevan, 2001), since point mutation of E648A markedly decreases the direct channel opening effect of low pH without affecting its potentiation on the response of capsaicin or noxious heat, as discussed earlier. Another extracellular glutamate residue (E600) has also been identified as an important modulatory site for protons (McLatchie and Bevan, 2001). It is interesting to note that pH modulates the response of VR1 to capsaicin but not to anandamide (Smart *et al.*, 2000). On cultured neurons of dorsal root ganglia the probability of capsaicin-gated channels being in an open state is greatly increased in a moderately acidic (pH 6.6) environment (Baumann *et al.*, 2000).

Anandamide

The endocannabinoid anandamide induces inward current (pEC_{50} 5.31) in VR1 expressing Xenopus oocytes and HEK293 cells (Zygmunt *et al.*, 1999) and produces an increase in Ca_i^{2+} in rat cultured trigeminal neurons (Szőke *et al.*, 2000) and in cell lines (HEK293, CHO) transfected with rVR1 or hVR1 (Smart *et al.*, 2000) with a pEC_{50} of 5.94 for the hVR1. The responses were antagonized by capsazepine but not by cannabinoid antagonist, thus the VR1 agonist effect of anandamide was clearly shown. Nevertheless, owing to its low potency, which lags behind its anti-nociceptive effect mediated by cannabinoid receptors it is questionable that anandamide might serve as an endogenous ligand for VR1 under physiological or pathophysiological conditions (Szolcsányi, 2000a,b). Certainly in this emerging field further experiments are needed. It is particularly important to obtain evidence for an effect of anandamide released from the tissues or within the sensory neurons and their terminals. Nevertheless, an endogenous anandamide-like lipid (Szolcsányi, 2000a) or synthetic structural analogues of anandamide (Szállási and DiMarzo, 2000) have already opened new perspectives in this field.

Products of lipoxygenases

Recently, an interesting scope of the intracellular activation machinery of VR1 ion channel was shown by Uhtaek Oh and colleagues (Hwang *et al.*, 2000). Single channel currents were recorded from inside-out membrane patches of HEK293 cells transfected with rVR1 capsaicin receptor and also from patches of cultured DRG neurons. It turned out that identical outward rectifying currents were evoked in these preparations by 100 nM capsaicin and 10 μM 12-(S)-hydroxyperoxyeicosatetraenoic acid (12-(S)-HPETE). It was striking that although the chemical structure of these molecules differs considerably, minimum-energy conformations as determined by molecular modelling indicated a remarkable overlap of the polar key regions of the two molecules

(Hwang *et al.*, 2000; Szolcsányi, 2002). Other lipoxygenase products investigated were less effective and no response was obtained to LTC_4, arachidic acid, PGI_2, PGH_2, PGE_2 and PGD_2. On the other hand, evidence was described that endogenous release of lipoxygenase products do activate VR1. Arachidonic acid applied to inside-out patches of cultured dorsal root ganglion neurons induced single channel activity in 50% of capsaicin-responsive patches. This response was markedly inhibited by blocking agents of 12-lipoxygenase (baicalein), 5-lipoxygenase (REV 5901) or by the non-selective lipoxygenase inhibitor of nordihydroguaiaretic acid. These results suggest that arachidonic acid metabolites activate capsaicin receptors preferentially through the lipoxygenase pathway.

Bradykinin an indirectly acting VR1/TRPV1 agonist

Bradykinin is one of the most potent endogenous pain producing substance which elicits pain when applied to the blister base of the human forearm (Figure 16.3). At low concentration (10^{-6} M), it selectively excites practically all cutaneous C-polymodal nociceptors (Szolcsányi, 1987a) when injected into the main artery of the rabbit ear (Figure 16.4) and causes discharges at a threshold concentration of 10^{-8} M when applied topically to the canine testicular polymodal receptors (Mizumura and Kumazawa, 1996). Further, a recently described capsaicin-sensitive nociceptor group for this action is the population of mechano-insensitive afferents in the human skin (Schmelz *et al.*, 2000). The sensitizing effect of prostaglandin E_2 on the bradykinin-induced discharges, as well as the dominant role of B_2 bradykinin receptors in its nociceptive effect, have been described (Bevan, 1996; Kress and Reeh, 1996; Mizumura and Kumazawa, 1996). Recently, however, evidence was obtained that bradykinin also activates the capsaicin receptors through intracellular second messengers (Figure 16.5). Bradykinin at the B_2 receptors via G-proteins activates phospholipase C (PLC), resulting in the production of inositol 1,4,5-trisphosphate (IP_3) and diacylglycerol (DAG) second messengers (Figure 16.5). DAG translocates one isoform of protein kinase C (PKC_ε) to the surface membrane within 5–30 s (Cesare *et al.*, 1999). Phosphorylation of VR1 receptors by PKC_ε induces sensitization of the nociceptors to painful heat or inflammatory chemical stimuli (Cesare *et al.*, 1999; Khasar *et al.*, 1999; Premkumar and Ahern, 2000). RTX, but not capsaicin, could produce a lasting opening of the VR1 cation channel, which also induces translocation of PKC to the surface membrane (Harvey *et al.*, 1995). The other pathway through which bradykinin could open the VR1 receptor-cation

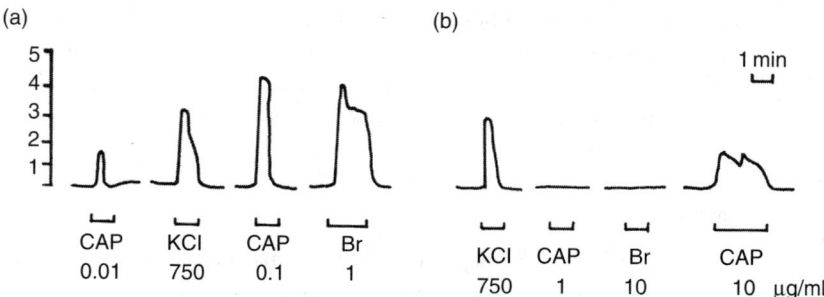

Figure 16.3 Pain responses to the application of different solutions to the blister base of the forearm before (a) and 40 min after (b) the area was bathed in a solution of capsaicin (10 mg/ml) for 5 min. Ordinate arbitrary scale which represents threshold (1), slight (2–3) or moderate (4–5) pain intensities. CAP: capsaicin; Br: bradykinin concentrations (μg/ml); and contact time (horizontal bars) are indicated. Between every two records in 3–6 min period with thorough rinse was taken (modified figure from Szolcsányi, 1977).

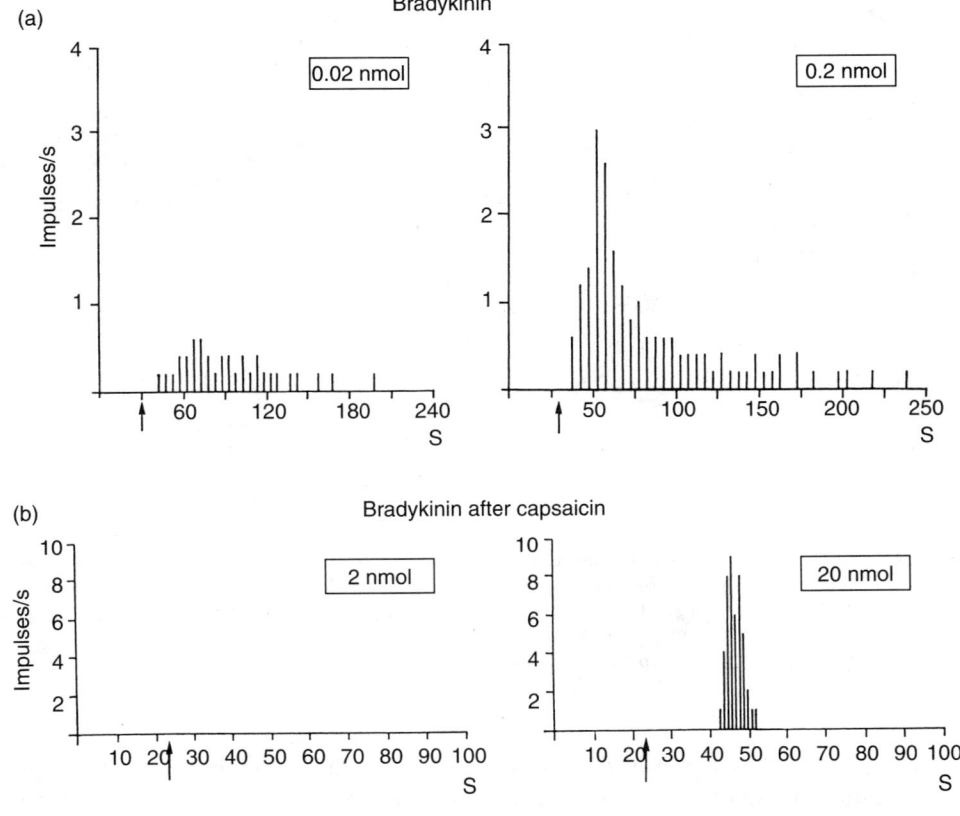

Figure 16.4 Responses of a C-polymodal nociceptor of the rabbit ear to repeated close arterial injections of bradykinin (arrows) before (a) and after (b) 5 × 200 μg capsaicin was injected into the main artery of the ear. Ordinate: number of action potentials per second, abscissa time in seconds. The doses of bradykinin are indicated at each record (unpublished details from Szolcsányi, 1987a).

channel is by the release of lipoxygenase products (12-HPETE) from arachidonic acid through activation of the phospholipase A_2 (Hwang *et al.*, 2000; Piomelli, 2001). Evidence for a release of anandamide inside the nociceptor in suitable concentration for VR1 activation has not yet been demonstrated, therefore its role as an endogenous ligand for VR1 has not been substantiated (Szolcsányi, 2000a,b). Intracellular calcium transients evoked by VR1 activation, by the second messenger of IP_3 or by opening the voltage-gated calcium channels, could also activate PKC_ε forming in this way a positive feedback loop for nociceptor sensitization (Figure 16.5). After capsaicin (Szolcsányi, 1977) or bradykinin application (Liang *et al.*, 2001), sustained sensitization of nociceptors to heat stimuli might be attributed to this VR1-mediated mechanism (Vyklicky *et al.*, 1999). The figure also shows that Ca^{2+}-dependent release of sensory neuropeptides as substance P, CGRP and somatostatin takes place without axon reflexes (Szolcsányi, 1996). Mainly Na_i^+ is responsible for depolarization, spike generation and the transmission of nociceptive signals to the central nervous system.

Pain induced by capsaicin or bradykinin were both markedly inhibited by topical pretreatment of the blister base of the human forearm with a high concentration of capsaicin. Pain

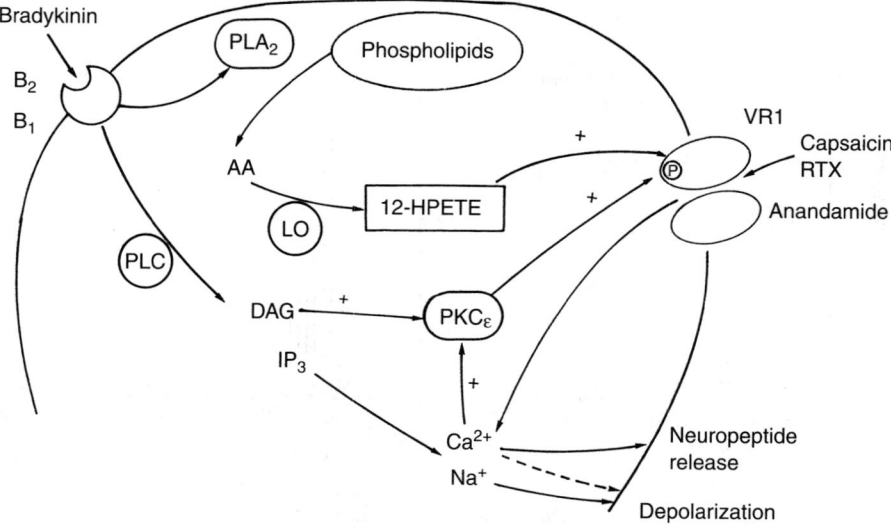

Figure 16.5 Schematic representation of intracellular pathways activated by bradykinin and capsaicin in a capsaicin-sensitive sensory neuron, B_2, B_1 bradykinin receptors; VR1: capsaicin receptor; AA: arachidonic acid; 12-HPETE: 12-(S)-hydroxyeicosatetraenoic acid; RTX: resiniferatoxin; LO: 12-lipoxygenase; PLA_2: phospholipase A_2; PLC: phospholipase C; PKC_ε: Proteinkinase C_ε; DAG: diacylglycerol; IP_3: inositol trisphosphate.

induced by KCl, however, remained practically unchanged, indicating that the capsaicin pretreatment did not induce a blockade of axonal conduction (Figure 16.3). Figure 16.4 shows a similarly strong inhibition induced by acute capsaicin pretreatment on the action potentials evoked by bradykinin in a single C-polymodal nociceptor afferent fiber of the rabbit ear (Szolcsányi, 1987a). It is worth noting that in both cases the sensory desensitization was not restricted to capsaicin and bradykinin. Other nociceptive chemical stimuli, and in the case of C-polymodal nociceptors, responses to heat and mechanical stimuli were inhibited. The fact that the desensitized units were still responsive with an elevated threshold (Figure 16.4) to one or more types of stimulation suggests an impaired function without neurotoxic degeneration (Szolcsányi, 1987a, 1993). On the other hand, further experiments are needed to decide what part is played by VR1 activation in the nociceptive effect induced by bradykinin.

Desensitizing of VR1 expressing sensory neurons and elimination of VR1 receptors

Capsaicin was introduced for the first time (before the nociceptors were discovered) as an agent which desensitizes "pain terminals" to all kinds of chemical stimuli, while their responsiveness to mechanical, thermal or electrical stimuli remained unchanged (Jancsó and Jancsó-Gábor, 1959; Jancsó, 1960; Jancsó et al., 1961). Subsequently, it turned out that this selective loss of nociception was due to the sensory desensitization of C-polymodal nociceptors to all kinds of stimuli (noxious heat, chemical, mechanical) (Szolcsányi, 1987a). Quantitative behavioural measurements also revealed a threshold elevation in nocifensive reaction to noxious heat (Szolcsányi, 1987b, 1993). Furthermore, the desensitization of peripheral and hypothalamic warmth sensors was also shown (Szolcsányi, 1982; Hori, 1984; Pierau *et al.*, 1986). The sensory neuron blocking effect of capsaicin has been exploited in a large number of experimental conditions. The following versatile ways of capsaicin applications made this substance particularly suitable for the inactivation of the capsaicin-sensitive neural elements:

1   Topical application of capsaicin to the eye or other mucosal areas of animals and to the human tongue, blister base or intact skin;
2   Close arterial injection to different organs;
3   Systemic application (usually s.c.) to rats, mice and guinea pigs;
4   Systemic injection to neonatal rats or mice;
5   Perineural application around the nerve trunk or nervous plexuses;
6   Intrathecal, epidural or intracerebroventricular injections;
7   Microinjection into the brain; and
8   *In vitro* exposure to isolated organs and tissues, to cultured cells or membrane patches.

With all these methods, a long-term blocking action of capsaicin and other stimuli were achieved, the details of which have been summarized in earlier reviews (Buck and Burks, 1986; Szolcsányi, 1990, 1993, 1996, 2002; Holzer, 1991; Maggi, 1995; Szállási and Blumberg, 1999).

The VR1 knockout mice has opened a new horizon in the field of pharmacology of capsaicin-sensitive neurons as target cells for drug research (Caterina *et al.*, 2000; Davis *et al.*, 2000). The encoding gene of VR1 seems to express all variants of the capsaicin (vanilloid) receptor which could be activated by this compound. In other words, all types of effects evoked by capsaicin or resiniferatoxin tested so far were absent in these VR1 knockout mice. Therefore, VR1 is the only subtype hitherto characterized which is responsive to vanilloid compounds (Szolcsányi, 2002). It was, however, surprising that in VR1-null mutant mice reflex latency to noxious heat temperature in the lower range of 46–48°C did not differ from that of the wild type controls, and only in the 50–58°C range exhibited an enhanced reflex latency. Considering that the lower gating temperatures for VR1 receptors are lower than that for the capsaicin-insensitive VRL-1 (see Table 16.1) this finding is intriguing and is an interesting field for further experimental approaches.

The mechanism of sensory desensitization induced by capsaicin

Although ultrastructural alterations in sensory neurons could accompany the functional desensitization induced by capsaicin, it has been suggested that at low desensitizing doses the impaired function is due to a dynamic, time dependent decrement of sensory responsiveness (Szolcsányi *et al.*, 1975; Szolcsányi, 1990, 1993) and not due to the acute cell death as induced by a neurotoxin (Jancsó and Kiraly, 1981; Jancsó *et al.*, 1985, 1987). The mechanism of long-lasting mitochondrial alterations evoked by systemic capsaicin pretreatment, both in neonatal and adult rats, is an interesting issue for further research, as is the finding that in the neonatal rat the cell death develops after a week owing to the inhibited NGF uptake through the impaired nerve terminals (Szolcsányi *et al.*, 1998b).

In vivo single unit recordings from C-polymodal nociceptors revealed that the number of action potentials evoked by noxious heat, capsaicin, bradykinin or mechanical stimuli were decreased or abolished after the close arterial injection of capsaicin to the rabbit ear over a nine-fold dose range, although each sensor still responded to at least one type of stimulation (Szolcsányi, 1987a). The threshold elevation to capsaicin or mechanical stimuli was also observed in the rat a few days after systemic capsaicin treatment (Szolcsányi *et al.*, 1988).

Details about the mechanism of acute capsaicin desensitization were discovered in studies of membrane currents of cultured dorsal root ganglion neurons (Bevan and Szolcsányi, 1990; Bevan and Docherty, 1993; Vlachova and Vyklicky, 1993; Piper et al., 1999; Szállási and Blumberg, 1999). It seems that phosphorylation of the capsaicin receptor/ion channel is necessary for a maintained capsaicin sensitivity (Koplas et al., 1997; Piper et al., 1999). Calcium entry through the activated ion channel, but not from internal stores, may promote dephosphorylation of the VR1 receptor and in this state the channel shows outward rectification which is linked to desensitization (Piper et al., 1999). The calcium-induced dephosphorylation of the capsaicin VR1 receptor is probably mediated by the activation of calcineurin (Docherty et al., 1996). As Figure 16.5 shows that the desensitized state of the VR1 receptor might result in a pronounced inhibition of the activation of the sensory nerve endings by bradykinin or lipoxygenase products.

Beyond the desensitization of the capsaicin VR1 receptor, the other scope is the functional desensitization of the sensory nerve terminal to stimuli which do not activate the VR1 receptors. One of these mechanisms is the lasting inhibition of voltage-gated calcium channels due to the markedly enhanced intracellular calcium ion concentration which takes place after VR1 activation by capsaicin or resiniferatoxin (Bevan and Szolcsányi, 1990; Bevan and Docherty, 1993). The second indirect mechanism is the inhibition of voltage-gated Na^+ channels by capsaicin. Capsaicin in a low concentration range ($K_{1/2} = 0.45$ μM) inhibits the action potential generation only in the capsaicin-sensitive subpopulation of neurons (Liu et al., 2001, unpublished observations). The third putative mechanism for sensory desensitization is related to the mitochondrial impairment which is unusually long-lasting after systemic treatment similar to the functional desensitization (Szolcsányi et al., 1975; Szolcsányi, 1993). Recently, it has been shown that after acute exposure of capsaicin to cultured dorsal root ganglion neurons, mitochondrial depolarization develops (EC_{50}: 6.9 μM) which might induce the mitochondrial swelling in vivo (Dedov et al., 2001). The role of these putative mechanisms in the capsaicin-induced sensory desensitization needs further research. In this context it is worth noting that most of these results were obtained from dorsal root ganglion neurons which were cultured in a NGF supplemented medium. Acute NGF application enhances the inward current induced by capsaicin and counteracts the development of desensitization (Shu and Mendel, 1999). Therefore, for technical reasons, the role of ion channels and mitochondria in capsaicin-induced sensory desensitization has not been tested on sensory nerve terminals under natural environmental conditions.

From the point of view of drug development, a slow channel opener VR1 agonist, which could produce desensitization without spike generation at the nociceptors might be a rational approach. Several new compounds were synthesized 25 years ago (Szolcsányi and Jancsó Gábor, 1975) to achieve this goal and recently a novel non-pungent capsaicin analogue with a clear therapeutic window has been introduced (Urban et al., 2000).

The role of VR1 receptors in the brain

The presence of the noxious heat-gated VR1 cation channel seems to have a well-defined functional relevance in nociceptors in contrast to VR1 positive neurons in different areas of the brain (Mezey et al., 2000). A high incidence of warm sensors in the preoptic area and the scattered high Q_{10} units in the midbrain and brain stem have a lower temperature threshold than the cloned VR1 receptors. Threshold temperature of warm-sensitive neurons in rat hypothalamic slices at single patches was $37.2 \pm 1.8°C$ (Hori et al., 1999). The following list of evidence suggests that capsaicin stimulates and then desensitizes the medial preoptic warm sensors in the hypothalamus (Szolcsányi, 1982, 1983; Hori, 1984; Pierau et al., 1986).

1   Microinjection of capsaicin into the preoptic area of rat and rabbit (a) elicits an immediate, coordinated heat loss response with vasodilatation and inhibition of shivering; and (b) excites the majority of warm units and decreases the firing rate of the cold ones.

2   After microinjection of capsaicin into the preoptic area (a) the impairment of regulatory responses to heat load (rat, cat) ensues, while the behavioural heat loss response is slightly affected; and (b) fall in body temperature to s.c. injection of capsaicin is reduced.

3   Systemic application of capsaicin excites the warm units and inhibits the responsiveness of cold units in the preoptic area.

4   After sensory desensitization induced by systemically applied capsaicin

   (a)   heat loss response with inhibition of shivering evoked by local heating the preoptic area is reduced;

   (b)   heat loss response to microinjection of capsaicin into the preoptic area is abolished;

   (c)   the proportion of warm and cold units in the preoptic area is reduced by about 50%;

   (d)   long-lasting similar ultrastructural changes are present in small cell types of neurons in the preoptic area and dorsal root ganglia.

5   After preoptic lesions, the fall in body temperature elicited by subcutaneous capsaicin injection is diminished but not abolished.

6   Birds are highly resistant to the effects of capsaicin, both on nociception and temperature regulation.

With the aid of immunohistochemical detection of VR1 receptor, mRNA of VR1, or by using VR1-null mutant mice, further novel approaches have been discovered to reveal the role of VR1 in the effect of capsaicin on warm sensors. The above series of experimental approaches might also serve as a guideline for versatile approaches which can be utilized in further research to form a thorough functional basis for the putative roles of VR1 positive neurons in the brain.

Local sensory-efferent and systemic sensocrine functions of capsaicin-sensitive nerve endings

Capsaicin is the lead molecule which – owing to its selective site of action – opened up new dimensions for the role of a substantial part of sensors characterized by their capsaicin sensitivity (Szolcsányi, 1984b, 1996, 2002; Maggi, 1995). Calcium entry induced by the opening of the capsaicin-gated cation channels releases sensory neuropeptides from them. The release of substance P and neurokinin A induce neurogenic inflammation, while CGRP evokes vasodilatation and enhances microcirculation (antidromic vasodilatation). These local effects take place around the activated nerve terminals and in an area which is innervated by axonal arborizations of the same nerve fibre. It has been suggested that this axon reflex arrangement terminates at endings which subserve dual signal-inducing sensory and mediator-releasing efferent functions (Szolcsányi, 1984a,b, 1996; Maggi, 1995).

Tachykinins and CGRP released from capsaicin-sensitive nerve endings are responsible for several neural effects described in a variety of smooth muscle preparations and non-vascular tissues (Holzer, 1991; Maggi, 1995; Szolcsányi, 1996). This field is still a hot topic and the role of capsaicin-sensitive afferents in diseases as in various forms of inflammation, asthma, gastroduodenal ulcer, irritable bowel, bladder hypersensitivity, and psoriasis is currently being investigated (Szállási and Blumberg, 1999).

Somatostatin, another sensory neuropeptide released from the activated capsaicin-sensitive sensors, reaches into the circulation and elicits systemic anti-inflammatory and analgesic effects

(Szolcsányi *et al.*, 1998a; Helyes *et al.*, 2000; Thán *et al.*, 2000). This "unorthodox" (Szolcsányi, 1996) endocrine-like or "sensocrine" (Thán *et al.*, 2000) function might also have relevance to pathological conditions and might be the target for drug development of the capsaicin-sensitive sensors (Helyes *et al.*, 2000).

A major representative of capsaicin-sensitive afferents are the C-polymodal nociceptors. This neural population comprises about 50% of the primary afferent neurons (Szolcsányi, 1993, 1996). The question arises why nociceptive afferent neural population of that size is needed, comparable to that of the entire peripheral adrenergic sympathetic nervous system, to signal only noxious events and pain towards the central nervous system. In intact skin, the minimum frequency of firing of C-polymodal nociceptors which elicits pain or leads to conscious perception is around 0.5 Hz (Szolcsányi, 1996). On the other hand, the optimum stimulation frequency of capsaicin-sensitive fibers which elicits antidromic vasodilatation in the skin or induce somatostatin-mediated sensocrine anti-inflammatory effect is below 0.1 Hz (Szolcsányi, 1996; Szolcsányi *et al.*, 1998a). Thus it has been suggested, that C-polymodal nociceptors at near threshold stimulation operate as effector nerve endings or endocrine cells without inducing pain sensations or evoking nociceptive reflexes. This dual role of capsaicin-sensitive nerve endings in the regulation of homeostasis is certainly a challenging issue for further research, particularly if involvement of novel sensory neuropeptides such as endothelin (Szolcsányi *et al.*, 2001), and endomorphin-2 (Martin-Schild *et al.*, 1998) as well as their role in pathological conditions such as diabetic neuropathy (Németh *et al.*, 1999) are concerned.

References

Amann, R. and Maggi, C. A. (1991) Ruthenium red as a capsaicin antagonist. *Life Sci.* 49, 849–856.

Baumann, T. K. and Martenson, M. E. (2000) Extracellular protons both increase the activity and reduce the conductance of capsaicin-gated channels. *J. Neurosci.* 20, RC80, 1–5.

Bevan, S. (1996) Intracellular messengers and signal transduction in nociceptors. In: C. Belmonte and F. Cervero (eds) *Neurobiology of Nociceptors*, Oxford University Press, New York, pp. 298–324.

Bevan, S. and Szolcsányi, J. (1990) Sensory neuron-specific actions of capsaicin: mechanisms and applications. *Trends Pharmacol. Sci.* 11, 330–333.

Bevan, S. J. and Docherty, R. J. (1993) Cellular mechanisms of the action of capsaicin. In: J. N. Wood (ed.), *Capsaicin in the Study of Pain*, Academic Press, London, pp. 27–44.

Bíró, T., Brodie, C., Moderres, Lewin, N. E., Ács, P. and Blumberg, P. M. (1998) Specific vanilloid responses in C6 rat glioma cells. *Molecular Brain Res.* 56, 89–98.

Buck, S. H. and Burks, T. F. (1986) The neuropharmacology of capsaicin: a review of some recent observations. *Pharmacol. Rev.* 38, 179–226.

Caterina, M. J., Schumacher, M. A., Tominaga, M., Rosen, T. A., Levine, J. D. and Julius, D. (1997) The capsaicin receptor: a heat-activated ion channel in the pain pathway. *Nature* 389, 816–824.

Caterina, M. J., Rosen, T. A., Tominaga, M., Brake, A. J. and Julius, D. (1999) A capsaicin-receptor homologue with a high threshold for noxious heat. *Nature* 398, 436–441.

Caterina, M. J., Leffler, A., Malmerg, A. B., Martin, W. J., Trafton, J., Petersen-Zeitz, K. R., Koltzenburg, M., Basbaum, A. J. and Julius, D. (2000) Impaired nociception and pain sensation in mice lacking the capsaicin receptor. *Science* 288, 306–313.

Cesare, P., Lodewijk, V. P., Sardini, A., Parker, P. J. and McNaughton, P. A. (1999) Specific involvement of PKC-ε in sensitization of the neuronal response to painful heat. *Neuron* 23, 617–624.

Davis, J. B., Gray, J., Gunthorpe, M. J., Hatcher, J. P., Davey P. T., Overend P., Harris M. H., Latcham J., Clapham, C., Atkinson, K., Hughes S. A., Rance, K., Grau, E., Harper, A. J., Pugh, P. L., Rogers, D. C., Bingham, S., Randall, A. and Sheardown, S. A. (2000) Vanilloid receptor-1 is essential for inflammatory thermal hyperalgesia. *Nature* 405, 183–187.

Dedov, V. N., Mandadi, S., Armati, P. J. and Verkhratsky, A. (2001) Capsaicin-induced depolarization of mitochondria in dorsal root ganglion neurons is enhanced by vanilloid receptors. *Neurosci.* 103, 219–226.

Docherty, R. J., Yeats, J. C., Bevan, S. and Boddeke, H. W. (1996) Inhibition of calcineurin inhibits the desensitization of capsaicin-evoked currents in cultured dorsal root ganglion neurones from adult rats. *Pflügers. Arch.* 431, 823–837.

Guo, A., Vulchanova, L., W. J., Li, X. and Elde, R. (1999) Immunocytochemical localization of the vanilloid receptor 1 (VR1): relationship to neuropeptides, the $P2X_3$ purinoceptor and IB4 binding sites. *Europ. J. Neurosci.* 11, 946–968.

Harteneck, C., Plant, T. P. and Schultz, G. (2000) From worm to man: three subfamilies of TRP channels. *Trends in Neurosci.* 23, 159–166.

Harvey, J. S., Davis, C., James, I. F. and Burgess, G. M. (1995) Activation of protein kinase C by the capsaicin analogue resiniferatoxin in sensory neurones. *J. Neurochem.* 65, 1309–1317.

Hayes, P., Meadows, H. J., Gunthorpe, M. J., Harries, M. H., Duckworth, D. M., Cairns, W., Harrison, D. C., Clarke, C. E., Ellington, K., Prinjha, R. K., Barthon, A. J. L., Medhurst, A. D., Smith, G. D., Topp, S., Murdock, P., Sanger, G. J., Terret, J., Jenkins, O., Benham, C. D., Randall, A. D., Gloger, I. S. and Davis, J. B. (2000) Cloning and functional expression of a human orthologue of rat vanilloid receptor-1. *Pain* 88, 205–215.

Helliwell, R. J. A., McLatchie, L. M., Clarke, M., Winter, J., Bevan, S. and McIntyre, P. (1998) Capsaicin sensitivity is associated with the expression of the vanilloid (capsaicin) receptor (VR1) mRNA in adult rat sensory ganglia. *Neurosci. Lett.* 250, 177–180.

Helyes, Zs., Thán, M., Oroszi, G., Pintér, E., Németh, J., Kéri, Gy. and Szolcsányi, J. (2000) Antinociceptive effect induced by somatostatin released from sensory nerve terminals and by somatostatin analogues in the rat. *Neurosci. Lett.* 278, 185–188.

Holzer, P. (1991) Capsaicin: cellular targets, mechanism of action, and selectivity for thin sensory neurons. *Pharmacol. Rev.* 43, 143–201.

Hori, T. (1984) Capsaicin and central control of thermoregulation. *Pharmac. Ther.* 26, 389–416.

Hori, A., Minato, K. and Kobayashi, S. (1999) Warming-activated channels of warm-sensitive neurons in hypothalamic slices. *Neurosci Lett.* 275, 93–96.

Hőgyes, A. (1878) Beitrage zur physiologischen Wirkung der Bestandteile des *Capsicum annuum. Arch. Exp. Pathol. Pharmakol.* 9, 117–130.

Hwang, S. W., Cho, H., Kwak, J., Lee, S-Y., Kang, C-J., Jung, J., Cho, S., Min, K. H., Suh, Y-G., Kim, D. and Oh, U. (2000) Direct activation of capsaicin receptors by products of lipoxygenases: endogenous capsaicin-like substances. *Proc. Natl. Acad. Sci. USA* 97, 6155–6160.

Jancsó, N. (1960) Role of nerve terminals in the mechanism of inflammatory reactions. *Bull. Millard Fillmore Hosp. Buffalo, NY* 7, 53–77.

Jancsó, N. and Jancsó-Gábor, A. (1959) Dauerausschaltung der chemischen Schmerzempfindlichkeit durch Capsaicin. *Naunyn-Schmiedeberg's Arch. Exp. Path. Pharmak.* 236, 142–145.

Jancsó, G. and Király, E. (1981) Sensory neurotoxins: chemically induced selective destruction of primary sensory neurons. *Brain Res.* 210, 83–89.

Jancsó, N., Jancsó-Gábor, A. and Takáts, I. (1961) Pain and inflammation induced by nicotine, acetylcholine and structurally-related compounds and their prevention by desensitizing agents. *Acta Physiol. Hung.* 19, 113–132.

Jancsó, G., Király, E. and Jancsó-Gábor, A. (1977) Pharmacologically induced selective degeneration of chemosensitive primary sensory neurons. *Nature* 270, 741–743.

Jancsó, G., Király, E., Joó, F., Such, G. and Nagy, A. (1985) Selective degeneration by capsaicin of a subpopulation of primary sensory neurons in the adult rat. *Neurosci. Lett.* 59, 209–214.

Jancsó, G., Király, E., Such, G., Joó, F. and Nagy, A. (1987) Neurotoxic effects of capsaicin in mammals. *Acta Physiol. Hung.* 69, 295–313.

Jancsó-Gábor, A., Szolcsányi, J. and Jancsó, N. (1970) Stimulation and desensitization of the hypothalamic heat-sensitive structures by capsaicin in rats. *J. Physiol.* 208, 449–459.

Joó, F., Szolcsányi, J. and Jancsó-Gábor, A. (1969) Mitochondrial alterations in the spinal ganglion cells of the rat accompanying the long-lasting sensory disturbance induced by capsaicin. *Life Sci.* 8, 621–626.

Jordt, S-E., Tominaga, M. and Julius, D. (2000) Acid potentiation of the capsaicin receptor determined by a key extracellular site. *Proc. Natl. Acad. Sci. USA* 97, 8134–8139.

Jung, J., Hwang, S. W., Kwak, J., Lee, S.-Y., Kang, C.-J., Kim, W. B., Kim, D. and Oh, U. (1999) Capsaicin binds to the intracellular domain of the capsaicin-activated ion channel. *J. Neurosci.* 19, 529–538.

Khasar, S. G., Lin, Y-H., Martin, A., Dadgar, J., McMahon, T., Wang, D., Hundle, B., Aley, K. O., Isenberg, W., McCarter, G., Green, P. G., Hodge, C. W., Levine, J. D. and Messing, R. O. (1999) A novel nociceptor signaling pathway revealed in protein kinase Cε mutant mice. *Neuron* 24, 253–260.

Koplas, P. A., Rosenberg, R. L. and Oxford, G. (1997) The role of calcium in the desensitization of capsaicin responses in rat dorsal root ganglion neuron. *J. Neurosci.* 17, 3525–3537.

Kress, M. and Reeh, P. (1996) Chemical excitation and sensitization in nociceptors. In: C. Belmonte and F. Cervero (eds) *Neurobiology of Nociceptors*, Oxford University Press, New York, pp. 258–297.

Liang, Y-F., Haake, B. and Reeh, P. W. (2001) Sustained sensitization and recruitment of rat cutaneous nociceptors by bradykinin and a novel theory of its excitation. *J. Physiol.* 532, 229–239.

Liedtke, W., Choe, W., Marti-Renom, M. A., Bell, A. M., Denis, Ch. S., Sali, A., Hudspeth, A. J., Friedman, J. M. and Heller, S. (2000) Vanilloid receptor-related osmotically activated channel (VR-OAC), a candidate vertebrate osmoreceptor. *Cell* 103, 525–535.

Liu, L. and Simon, S. A. (2000) Capsaicin, acid and heat-evoked currents in rat trigeminal ganglion neurons: relationship to functional VR1 receptors. *Physiol. Behav.* 69, 363–378.

Liu, L., Oortgiesen, M., Li, L. and Simon, S. (2001) Capsaicin inhibits activation of voltage-gated sodium currents in capsaicin-sensitive trigeminal ganglion neurons. *J. Neurophysiol.* 85, 745–758.

McIntyre, P., McLatchie, L. M., Chambers, A., Phillips, E., Clarke, M., Savidge, J., Toms, C., Placock, M., Shah, K., Winter, J., Weerasakera, N., Webb, M., Rang, H. P., Bevan, S. and James, J. F. (2001) Pharmacological differences between the human and rat vanilloid receptor 1 (VR1). *Br. J. Pharmacol.* 132, 1084–1094.

McLatchie, L. M. and Bevan, S. (2001) The effects of pH on the interaction between capsaicin and the vanilloid receptor in rat dorsal root ganglia neurons. *Br. J. Pharmacol.* 132, 899–908.

Maggi, C. A. (1995) Tachykinins and calcitonin gene-related peptide (CGRP) as cotransmitters released from peripheral endings of sensory nerves. *Prog. Neurobiol.* 45, 1–98.

Martin-Schild, S., Gerall, A. A., Kastin, A. J. and Zadina, J. E. (1998) Endomorphin-2 is an endogenous opioid in primary sensory afferent fibers. *Peptides* 19, 1783–1789.

Mezey, É., Tóth, Zs. E., Cortright, D. N., Arzubi, M. K., Krause, J. E., Elde, R., Guo, A., Blumberg, P. M. and Szállási, Á. (2000) Distribution of mRNA for vanilloid receptor subtype 1 (VR1), and VR1-like immunoreactivity, in the central nervous system of the rat and human. *Proc. Natl. Acad. Sci. USA* 97, 3655–3660.

Michael, G. J. and Priestley, J. V. (1999) Differential expression of the mRNA for the vanilloid receptor subtype 1 in cells of the adult rat dorsal root and nodose ganglia and its down regulation by axotomy. *J. Neurosci.* 19, 1844–1854.

Mizumura, K. and Kumazawa, T. (1996) Modulations of nociceptor responses by inflammatory mediators and second messengers implicated in their action – a study in canine testicular polymodal receptors. *Progr. Brain Res.* 113, 115–141.

Nagy, I. and Rang, H. P. (1999a) Similarities and differences between the responses of rat sensory neurons to noxious heat and capsaicin. *J. Neurosci.* 19, 10647–10655.

Nagy, I. and Rang, H. P. (1999b) Noxious heat activates all capsaicin-sensitive and also a sub-population of capsaicin-insensitive dorsal root ganglion neurons. *Neurosci.* 88, 995–997.

Nagy, J. I., Iversen, L. L., Goedert, M., Chapman, D. and Hunt, S. P. (1983) Dose-dependent effects of capsaicin on primary sensory neurons in the neonatal rat. *J. Neurosci.* 3, 399–406.

Németh, J., Szilvássy, J., Thán, M., Oroszi, G., Sári, R. and Szolcsányi, J. (1999) Decreased sensory neuropeptide release from trachea of rat with streptozotocin-induced diabetes. *Eur. J. Pharmacol.* 369, 221–224.

Oh, U., Hwang, S. W. and Kim, D. (1996) Capsaicin activates a nonselective cation channel in cultured neonatal rat dorsal root ganglion neurons. *J. Neurosci.* 16, 1659–1667.

Pierau, Fr-K., Szolcsányi, J. and Sann, H. (1986) The effect of capsaicin on afferent nerves and temperature regulation of mammals and birds. *J. Therm. Biol.* 11, 95–100.

Piomelli, D. (2001) The ligand that came from within. *Trends Pharm. Sci.* 22, 17–19.

Piper, A. S., Yeats, J. C., Bevan, S. and Docherty, R. J. (1999) A study of the voltage dependence of capsaicin-activated membrane currents in rat sensory neurons before and after acute desensitization. *J. Physiol.* 518, 721–733.

Premkumar, L. S. and Ahern, G. P. (2000) Induction of vanilloid channel activity by protein kinase C. *Nature* 408, 21–28.

Sasamura, T., Sasaki, M., Tohda, C. and Kuraishi, Y. (1998) Existence of capsaicin-sensitive glutamatergic terminals in rat hypothalamus. *NeuroReport* 9, 2045–2048.

Savidge, J. R., Ranasinghe, S. P. and Rang, H. P. (2001) Comparison of intracellular calcium signals evoked by heat and capsaicin in cultured rat dorsal root ganglion neurons and in a cell line expressing the rat vanilloid receptor, VR1. *Neuroscience* 102, 177–184.

Schmelz, M., Schmid, R., Handwerker, H. O. and Torebjörk, H. E. (2000) Encoding of burning pain from capsaicin-treated human skin in two categories of unmyelinated nerve fibres. *Brain* 123, 560–571.

Schumacher, M. A., Moff, I., Sudanagunta, P. and Levine, J. D. (2000) Molecular cloning of an N-terminal splice variant of the capsaicin receptor: loss of N-terminal domain suggests functional divergence among capsaicin receptor subtypes. *J. Biol. Chem.* 275, 2756–2762.

Shin, J. S., Wang, M-H., Hwang, S. W., Cho, H., Cho, S. Y., Kwon, M. J., Lee, S-Y. and Oh, U. (2001) Differences in sensitivity of vanilloid receptor 1 transfected to human embryonic kidney cells and capsaicin-activated channels in cultured rat dorsal root ganglion neurons to capsaicin receptor agonists. *Neurosci Lett.* 299, 135–139.

Shu, X.-Q. and Mendell, L. M. (1999) Neurotrophins and hyperalgesia. *Proc. Natl. Acad. Sci. USA* 96, 7693–7696.

Smart, D., Gunthorpe, M. J., Jerman, J. C., Nasir, S., Gray, J., Muir, A. J., Chambers, J. K., Randall, A. D. and Davis, J. B. (2000) The endogenous lipid anandamide is a full agonist at the human vanilloid receptor (hVR1). *Br. J. Pharmacol.* 129, 227–230.

Szállási, Á. and Blumberg, P. M. (1990) Specific binding of resiniferatoxin, an ultrapotent capsaicin analog, by dorsal root ganglion membranes. *Brain Res.* 524, 106–111.

Szállási, Á. and Blumberg, P. M. (1999) Vanilloid (capsaicin) receptors and mechanisms. *Pharmacol. Rev.* 51, 159–211.

Szállási, A. and DiMarzo, V. (2000) New perspectives on enigmatic vanilloid receptors. *Trends Neurosci.* 23, 491–497.

Szállási, Á., Joó, F. and Blumberg, P. M. (1989) Duration of desensitization and ultrastructural changes in dorsal root ganglia in rats treated with resiniferatoxin, an ultrapotent capsaicin analog. *Brain Res.* 503, 68–72.

Szállási, Á., Nilsson, S., Farkas-Szállási, T., Blumberg, P. M., Hökfelt, T. and Lundberg, J. J. (1995) Vanilloid (capsaicin) receptors in the rat: Distribution in the brain, regional differences in the spinal cord, axonal transport to the periphery, and depletion by systemic vanilloid treatment. *Brain Res.* 703, 175–183.

Szolcsányi, J. (1977) A pharmacological approach to elucidate the role of different nerve fibres and receptor endings in mediation of pain. *J. Physiol (Paris)* 73, 251–259.

Szolcsányi, J. (1982) Capsaicin type pungent agents producing pyrexia. In: A. S. Milton (ed.) *Handbook of Experimental Pharmacology*, Vol. 60, *Pyretics and Antipyretics*, Springer-Verlag, Berlin, pp. 437–478.

Szolcsányi, J. (1983) Disturbances of thermoregulation induced by capsaicin. *J. Therm. Biol.* 8, 207–212.

Szolcsányi, J. (1984a) Capsaicin and neurogenic inflammation: history and early findings. In: L. A. Chahl, J. Szolcsányi and F. Lembeck (eds), *Antidromic Vasodilatation and Neurogenic Inflammation*. Akadémiai Kiadó, Budapest, pp. 7–26.

Szolcsányi, J. (1984b) Capsaicin-sensitive chemoceptive neural system with dual sensory-efferent function. In: L. A. Chahl, J. Szolcsányi and F. Lembeck (eds) *Antidromic Vasodilatation and Neurogenic Inflammation*. Akadémiai Kiadó, Budapest, pp. 27–56.

Szolcsányi, J. (1987a) Selective responsiveness of polymodal nociceptors of the rabbit ear to capsaicin, bradykinin and ultra-violet irradiation. *J. Physiol.* 388, 9–23.

Szolcsányi, J. (1987b) Capsaicin and nociception. *Acta Physiol. Hung.* 69, 323–332.

Szolcsányi, J. (1990) Capsaicin, irritation and desensitization. Neurophysiological basis and future perspectives. In: B. R. Green, J. R. Mason and M. R. Kare (eds), *Chemical Senses*, Vol. 2. Irritation, Marcel Dekker, New York, pp. 141–168.

Szolcsányi, J. (1993) Actions of capsaicin on sensory receptors. In: J. N. Wood (ed.) *Capsaicin in the Study of Pain*, Academic Press, London, pp. 1–26.

Szolcsányi, J. (1996) Capsaicin-sensitive sensory nerve terminals with local and systemic efferent functions: facts and scopes of an unorthodox neuroregulatory mechanism. *Prog. Brain Res.* 113, 343–359.

Szolcsányi, J. (2000a) Are cannabinoids endogenous ligands for the VR1 capsaicin receptor? *Trends Pharmacol. Sci.* 21, 41–42.

Szolcsányi, J. (2000b) Anandamide and the question of its functional role for activation of the capsaicin receptor. *Trends Pharmacol. Sci.* 21, 203–204.

Szolcsányi, J. (2002) Capsaicin receptors as target molecules on nociceptors for development of novel analgesic agents. In: G. Keri and I. Toth (eds) *Molecular Pathomechanisms and New Trends in Drug Research*, Taylor and Francis, London, pp. 319–333.

Szolcsányi, J. and Jancsó-Gábor, A. (1975) Sensory effects of capsaicin congeners I. Relationship between chemical structure and pain-producing potency. *Arzneim. Forsch. (Drug Res.)* 25, 1877–1881.

Szolcsányi, J. and Jancsó-Gábor, A. (1976) Sensory effects of capsaicin congeners II. Importance of chemical structure and pungency in desensitizing activity of capsaicin-type compounds. *Arzneim. Forsch. (Drug Res.)* 26, 33–37.

Szolcsányi, J., Joó, F. and Jancsó-Gábor, A. (1971) Mitochondrial changes in preoptic neurones after capsaicin desensitization of the hypothalamic thermodetectors in rats. *Nature* 229, 116–117.

Szolcsányi, J., Jancsó-Gábor, A. and Joó, F. (1975) Functional and fine structural characteristics of the sensory neuron blocking effect of capsaicin. *Naunyn-Schmiedeberg's Arch. Pharmacol.* 287, 157–169.

Szolcsányi, J., Anton, F., Reeh, P. W. and Handwerker, H. O. (1988) Selective excitation by capsaicin of mechano-heat sensitive nociceptors in rat skin. *Brain Res.* 446, 262–268.

Szolcsányi, J., Szállási, Á., Joó, F. and Blumberg, P. M. (1990) Resiniferatoxn: an ultrapotent selective modulator of capsaicin-sensitive primary afferent neurons. *J. Pharmacol. Exp. Ther.* 225, 923–928.

Szolcsányi, J., Barthó, L. and Pethő, G. (1991) Capsaicin-sensitive bronchopulmonary receptors with dual sensory-efferent function: mode of action of capsaicin antagonists. *Acta Physiol. Hung.* 77, 293–304.

Szolcsányi, J., Nagy, J. and Pethő, G. (1993) Effect of CP-96,345 a non-peptide substance P antagonist, capsaicin, resiniferatoxin and ruthenium red on nociception. *Regul. Pept.* 46, 437–439.

Szolcsányi, J., Helyes, Zs., Oroszi, G., Németh, J. and Pintér, E. (1998a) Release of somatostatin and its role in the mediation of the anti-inflammatory effect induced by antidromic stimulation of sensory fibres of rat sciatic nerve. *Br. J. Pharmacol.* 123, 936–942.

Szolcsányi, J., Szőke, É. and Seress, L. (1998b) Reevaluation of the neurotoxic effect of neonatal capsaicin treatment on rat trigeminal sensory neurons. *Soc. Neurosci. Abstr.* 24, 91.12.

Szolcsányi, J., Oroszi, G., Németh, J., Szilvássy, Z., Blasig, J. E. and Tósaki, Á. (2001) Functional and biochemical evidence for capsaicin-induced neural endothelin release in isolated working rat heart. *Eur. J. Pharmacol.* 419, 215–221.

Szőke, É., Balla, Zs., Csernoch, L., Czéh, G., and Szolcsányi, J. (2000) Interacting effects of capsaicin and anandamide on intracellular calcium in sensory neurones. *NeuroReport* 11, 1949–1952.

Thán, M., Németh, J., Szilvássy, Z., Pintér, E., Helyes, Zs. and Szolcsányi, J. (2000) Systemic anti-inflammatory effect of somatostatin released from capsaicin-sensitive vagal and sciatic sensory fibres of the rat and guinea-pig. *Eur. J. Pharmacol.* 399, 251–258.

Tominaga, M., Caterina, M. J., Malmberg, A. B., Rosen, T. A., Gilbert, H., Skinner, K., Raumann, B. E., Basbaum, A. J. and Julius, D. (1998) The cloned capsaicin receptor integrates multiple pain-producing stimuli. *Neuron* 21, 531–543.

Urban, L., Campbell, E. A., Panesar, A., Panesar, M., Patel, S., Chaudhry, N., Kane, S., Buchheit, K-H., Sandells, B. and James, I. F. (2000) *In vivo* pharmacology of SDZ 249-665, a novel, non-pungent capsaicin analogue. *Pain* 89, 65–74.

Vlachová, V. and Vyklicky, L. (1993) Capsaicin-induced membrane currents in cultured sensory neurons of the rat. *Physiol. Res.* 42, 301–311.

Vyklicky, L., Vlachova, V., Vitaskova, Z., Dittert, I., Kabát, M. and Orkand, P. K. (1999) Temperature coefficient of membrane currents induced by noxious heat in sensory neurones in the rat. *J. Physiol.* **517**, 181–192.

Welk, E., Fleischer, E., Petsche, U. and Handwerker, H. O. (1984) Afferent C-fibers in rats after neonatal capsaicin treatment. *Pfügers Arch.* **400**, 66–71.

Welch, J. M., Simon, S. A. and Reinhart, P. H. (2000) The activation mechanism of rat vanilloid receptor 1 by capsaicin involves the pore domain and differs from the activation by either acid or heat. *Proc. Natl. Acad. Sci. USA* **97**, 13889–13894.

Winter, J., Walpole, C. S. J., Bevan, S. and James, I. F. (1993) Characterization of resiniferatoxin binding sites on sensory neurones: co-regulation of resiniferatoxin binding and capsaicin-sensitivity in adult rat dorsal root ganglia. *Neuroscience* **57**, 747–757.

Zygmunt, P. M., Petersson, J., Andersson, P. A., Chuang, H-H., Sorgard, M., DiMarzo, V., Julius, D. and Högestätt, E. D. (1999) Vanilloid receptors on sensory nerves mediate the vasodilator action of anandamide. *Nature* **400**, 452–457.

Index